W9-CAB-158

NUMBERS

ALSO BY ALFRED S. POSAMENTIER AND INGMAR LEHMANN

The Fabulous Fibonacci Numbers

Pi: A Biography of the World's Most Mysterious Number

Mathematical Curiosities

Magnificent Mistakes in Mathematics

The Secrets of Triangles

Mathematical Amazements and Surprises

The Glorious Golden Ratio

ALSO BY ALFRED S. POSAMENTIER

The Pythagorean Theorem

Math Charmers

NUMBERS

Their Tales, Types, and Treasures

ALFRED S. POSAMENTIER & BERND THALLER

Prometheus Books

59 John Glenn Drive
Amherst, New York 14228

Published 2015 by Prometheus Books

Numbers: Their Tales, Types, and Treasures. Copyright © 2015 by Alfred S. Posamentier and Bernd Thaller. All rights reserved. No part of this publication may be reproduced, stored in a retrieval system, or transmitted in any form or by any means, digital, electronic, mechanical, photocopying, recording, or otherwise, or conveyed via the Internet or a website without prior written permission of the publisher, except in the case of brief quotations embodied in critical articles and reviews.

Cover image © Can Stock Photo Inc./Sylverats
Cover design by Jacqueline Nasso Cooke
Unless otherwise indicated, all interior images are by the authors.

Inquiries should be addressed to
Prometheus Books
59 John Glenn Drive
Amherst, New York 14228
VOICE: 716–691–0133
FAX: 716–691–0137
WWW.PROMETHEUSBOOKS.COM

19 18 17 16 15   5 4 3 2 1

Library of Congress Cataloging-in-Publication Data

Posamentier, Alfred S.
    Numbers : their tales, types, and treasures / by Alfred S. Posamentier & Bernd
Thaller.
        pages cm
    Includes bibliographical references and index.
    ISBN 978-1-63388-030-6 (pbk.) — ISBN 978-1-63388-031-3 (e-book)
    1. Number concept. 2. Counting. 3. Arithmetic—Foundations. I. Thaller, Bernd,
1956- II. Title.

QA141.15.P67 2015
513.5—dc23

2015011662

Printed in the United States of America

We dedicate this book of mathematical enlightenments to our future generations so that they will be among the multitude that we hope will learn to love mathematics for its power and beauty!

To my children and grandchildren, whose future is unbounded,
Lisa, Daniel, David, Lauren, Max, Samuel, and Jack
                                        —Alfred S. Posamentier

To my son, Wolfgang
                                        —Bernd Thaller

CONTENTS

ACKNOWLEDGMENTS

We would like to thank Norbert Holzer, an expert in the preparation of elementary school teachers in Graz, Austria, and a specialist for dyscalculia and its diagnosis, who provided us fine insights about how children develop their ability to count. We are also grateful to Dr. Peter Schöpf, retired mathematics professor of the Karl-Franzens University in Graz, for his keen insight into the history and philosophy of mathematics. We also thank Peter Poole for his timely support with a few topics in the book.

Many thanks to Catherine Roberts-Abel for very capably managing the production of this book, and to Jade Zora Scibilia for the truly outstanding editing throughout the various phases of production, with the assistance of Sheila Stewart. Steven L. Mitchell, editor in chief, deserves praise for enabling us to approach the general readership to expose the gems that lie among the commonly known concept of numbers.

CHAPTER 1

NUMBERS AND COUNTING

1.1. A MENTAL NETWORK

We can't live without numbers. We encounter them every hour of every day. Numbers have shaped the way we think about the world. They penetrate every aspect of our life. Our whole society is organized with the help of numbers; it depends on numbers in many respects, and it has been that way since the dawn of civilization. Numbers rule our life.

We need numbers for counting, for measuring, and for doing calculations. We have numbers to describe dates and times and to tell the price of goods and services. We use numbers when we buy our meals or count our days. Numbers can be manipulated to improve statistics or to cheat in games. We are identified by Social Security numbers, license numbers, credit card numbers, and telephone numbers. Numbers describe sports records, baseball scores, and batting averages. Science, economy, and business are all about numbers, and we find numbers even in music, for example, in rhythm and harmony. To some, numbers are a never-ending source of joy and fascination, while others feel that numbers are depressing, impersonal, often incomprehensible, and without soul. Undoubtedly, people who lack fundamental skills with numbers will face diminished life chances, difficulties finding a job, and other serious impairments in everyday life, similar to people who can't read.

The immense importance of numbers should make us pause a bit and think about their nature and their origin. What are numbers? Where do they come from? Who was the first to use them? Indeed, there is more to these questions than meets the eye. In order to find answers, we will embark on a journey that visits the realms of psychology, ethnology, history, and philosophy. In the course of this journey, we will learn about ourselves, our mind, and our number sense; we will think about reality and mathematics; and we will encounter fascinating ideas and surprising facts.

Indeed, what is a number? At first, this may seem like a rather odd and unnecessary question. The symbols 1, 2, 3, 4, and so on appear so utterly familiar; their meanings seem so obvious that an explanation can only create confusion. Numbers belong to our shared knowledge about the world. We all recognize a number when we see one. It is notoriously difficult to explain something that everybody knows already, in particular if one hasn't thought of it before.

Marvin Minsky, in his book *The Society of Mind*, also muses about the nature of numbers and asks why it would be so difficult to explain meaning to others: "Because what something 'means' depends on every different person's state of mind."[1] The hope that through an explanation or precise definition, "different people could understand things in exactly the same ways" cannot be fulfilled, "because in order for two minds to agree perfectly, at every level of detail, they'd have to be identical." Nevertheless, "the closest we can come to agreeing on meanings is in mathematics, when we talk of things like 'Three' and 'Five.' But even something as impersonal as 'Five' never stands isolated in a person's mind but becomes part of a huge network."

In everyday life, there are many occasions contributing to the growth of the mental network of knowledge and meaning that is associated with a number. Numbers are often encountered in situations that have little to do with mathematics. Think for a moment of a number like four, and you will certainly come up with a lot of situations where this number plays a role (such as, the *four* wheels of a car, the *four* wisdom teeth, the *four* seasons, and so on). Even a less obvious example, like the number nine, produces a lot of associations in various contexts—there are Dante's *nine* circles of hell, Tolkien's *nine* rings of power, and the *nine* worlds of Yggdrasil in Norse mythology. Beethoven composed *nine* symphonies; a Chinese dragon has *nine* forms; Europeans like *nine*-pin bowling games; in the Caribbean Sea we find *nine*-armed sea stars; in Jewish culture, the Chanukah menorah is a *nine*-branched candelabrum; a baseball team has *nine* players on the field, and a complete game has *nine* innings. An old saying goes that a cat has *nine* lives; another, that *nine* tailors make a man; and when we are very happy, we are on cloud *nine*. Ramadan is the *ninth* month in the Islamic calendar; normal office hours start at *nine* in the morning; human pregnancy usually lasts *nine* months. Dressing nicely is often referred to as being dressed to the *nines*. Nine is a good number in Chinese mythology, but an unlucky number in Japanese culture, where its pronunciation reminds one of the word for agony or pain. And when we take the whole lot, we take the whole *nine* yards.

Figure 1.1: Various representations of the number nine

Depending on your personal background, some of these examples, and perhaps some others, will come to your mind when you think of *nine* (see figure 1.1). And similar or even larger amounts of rich associations come with many other numbers, giving them individuality and meaning. These numbers, forming parts of every individual's mental network, are not that impersonal after all. "Numbers have souls, and you can't help but get involved with them in a personal way," writes Paul Auster (1947–) in his novel *The Music of Chance*.[2] And when he emphasizes this point, the statement even gets a slightly absurd touch:

> After a while you begin to feel that each number has a personality of its own. A twelve is very different from a thirteen, for example. Twelve is upright, conscientious, intelligent, whereas thirteen is a loner, a shady character who won't think twice about breaking the law to get what he wants. Eleven is tough, an outdoorsman who likes tramping through the woods and scaling mountains; ten is rather simpleminded, a bland figure who always does what he's told; nine is deep and mystical, a Buddha of contemplation.

1.2. WHAT IS A NUMBER?

As this is probably difficult to answer, let us ask a different question: "Can you give an example of a number?" Probably, the answer will be something like 5, or *five*. But then, what about V or ||||| or 3 + 2 or *cinque*?

Clearly, the symbol 5 is not a number—it is just a symbol. It is a common mistake to take a symbolic representation for the "real thing." But this mistake is very understandable because our everyday language does not distinguish between them and calls everything a number. But as long as we talk about the "meaning of numbers," we have to be precise: A symbol, like 5, serves to designate a number, but it is not the number itself. Indeed, the number five can be represented by quite different symbols—for example, by the Roman symbol V or the Chinese 五. The number five can even exist without any written symbol at all—it was probably used by *Homo sapiens* long before the invention of writing and expressed by showing the fingers of one hand.

In the same way, the spoken word *five* (a combination of sounds) and the written word *five* (a combination of letters) are just representations of the number five. The number itself is an abstract idea, and it can be expressed in many different ways and by other words. For example, the word for five in French is *cinque*, in German it is *fünf*, and in Japanese it is *go*. In any case, all these different representations—symbol, word, sound, or even a dot pattern like ⁙—should evoke the same idea of the number five. In linguistics, a word designating a number, like *five*, or *twenty-four* (no matter whether it is spoken or written), is called a *numeral* or a *number word*.

So far, we have not really explained what a number is; rather, we have said what it is not: It is not a symbol or a number word, which are just names. We are going to distinguish between the abstract idea *number* and the words or symbols used to designate numbers. The abstract idea is unique and invariable; symbols and words are a mere matter of convention and hence quite arbitrary. Moreover, there is a difference between the idea of a number and its different (although related) applications. The number described by the symbol *5* could be used, for example, to describe the fifth place in a sequence (as an *ordinal number*) or the number of objects in a collection (as a *cardinal number*) or the length of a flagpole in yards (as a *measuring number*).

In this chapter, we want to describe the "thing behind the symbol," the genesis, true meaning, and scope of the abstract idea *number*, which belongs to the greatest inventions of humankind.

In order to approach this concept, we shall first concentrate on the most basic aspect of numbers: their ultimate and original raison d'être. A first reason for the existence of numbers is that they can be used for counting.

1.3. COUNTING

The numbers that can be used for counting are denoted by 1, 2, 3, 4, 5, and so on, and they are called *natural numbers*. Sometimes, *zero* is also included in the list of natural numbers, in order to be able to express the absence of things. On the other hand, for the Greek philosophers of antiquity, counting started with two objects, hence *one* was not regarded as a number. But no matter where we let them begin, the natural numbers are the basis for the understanding and mathematical construction of other types of numbers, such as negative numbers, rational numbers, and even real numbers—the numbers used for measuring quantities. German mathematician Leopold Kronecker (1823–1891) has best described the fundamental role of the natural numbers in his often-quoted dictum: "God made natural numbers; all else is the work of man."

We typically make our first conscious acquaintance with natural numbers when we learn to count. Whether or not one likes mathematics, the ability to count has become second nature to us. As soon as we acquire this ability, we forget about this tedious learning process. Counting, then, appears to be a simple exercise, and we are usually not aware of its inherent complexities. But, in fact, counting is a rather delicate process, and it takes some maturity in abstract reasoning to describe it in more detail.

Can you estimate the number of pebbles in figure 1.2? If you want to know exactly, you will have to count them. By observing ourselves when counting, we find that this task consists of several steps:

1. We start with an arbitrary object in the collection and say "one."
2. We mark this object as "already counted" (at least in our mind, in order to avoid counting it twice).
3. We select a new object (either by pointing with a finger or simply by looking at it).

4. We say the next "number word" (using number words always in the same strict order).
5. We go back to step 2 and repeat until there are no more uncounted objects. The last number word obtained in that way describes the number of objects.

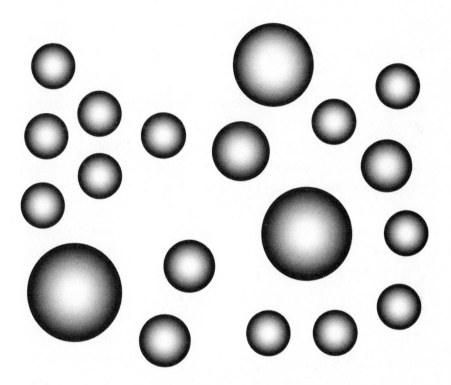

Figure 1.2: Explicit counting of a set of pebbles.

Counting is a process of associating number words with objects in a collection. One of the more difficult tasks involved here is that one has to divide the collection of objects into those that have already been counted and those that still remain to be counted. This is fairly easy if we can put the objects in a row, but it could be impossible if the objects were moving and kept changing places.

When counting nonpermanent objects or events—for example, the chimes of a clock striking the hour—we typically say a number word as the event occurs. When the events are separated by long time intervals, we normally have to create a permanent record of that event—for example, tally marks on a sheet of paper—and finally determine the number of events from the record.

1.4. THE COUNTING PRINCIPLES

The act of counting is governed by five principles. They describe the conditions and prerequisites that make counting possible. We call them the "BOCIA" principles—from the words **B**ijection, **O**rdinality, **C**ardinality, **I**nvariance, and **A**bstraction. They were proposed by Rochel Gelman (1942–) and C. R. Gallistel (1941–) within the field of cognitive psychology, where they can be used to describe and classify typical counting errors of children. Every child who learns to count masters these principles intuitively, through practice and experience, by trial and error.

In this section, we give a brief description of each of these principles. In the following sections, we elaborate on these principles in more detail and show how they are related to some fundamental mathematical observations. An awareness of the inherent complexities of the counting procedure will also help us to better understand the psychological dimension of the number concept in chapter 2, the intricacies of the historical development described in chapter 3, and the philosophical problems with the foundations of mathematics in chapter 11.

1. Bijection principle (one-to-one principle):

When we count the objects of a collection, we associate these objects with number words. We do this in a one-to-one manner—that is, we pair each object with a unique "counting tag." See figure 1.3.

Figure 1.3: Counting is a process of pairing objects with counting tags.

In practice, counting is often done by pointing a finger at each object while reciting the well-known sequence "one, two, three, . . . ," and so on. When we do so, we have to be careful about the following two points:

- We have to point to *each* object once and only once. (In that way, no element is left without a counting tag and no element receives two.)
- We must use each of the number words only once. (In that way, two different elements of the collection cannot receive the same counting tag.)

This results in a unique one-to-one correspondence, a "pairing" between the objects and a set of counting tags, as illustrated in figure 1.3. In mathematics, a one-to-one correspondence or association is called a *bijection*, hence the name of this principle.

2. Ordinal principle (stable-order principle):

When we count, we do this in some order. At least in our minds, we first arrange the objects to be counted in a certain (but arbitrary) order, before we apply the counting tags to each object in turn, as shown in figure 1.4. The set of counting tags is also ordered. Typically, the name or label for the first counted object is *one*, then follows *two, three,* and so on. The order of the counting tags must not be changed when counting is repeated or when another collection is counted.

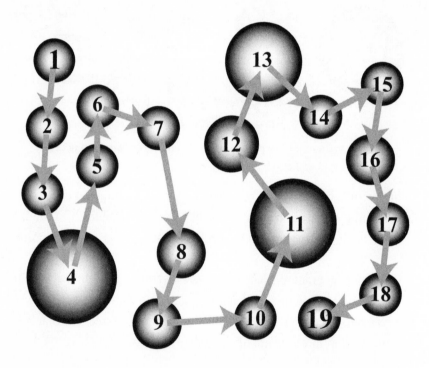

Figure 1.4: Counting by numeration (ordinal principle).

Whenever we count something, we have to use the same set of ordered counting tags. Even when the collection is apparently in dis-order, we have to decide upon the order in which the objects receive their counting tags, as symbolized by the arrows in figure 1.4. In that way, the counting tags, which always follow the same order, describe or even create the order of the objects within the collection: One of the objects will be the first—where counting starts—then each object has a unique successor, until we reach the last one—where counting ends. In mathematics, numbers used to label things in a row are called *ordinal numbers*, hence the name of that principle.

In order to apply this principle, we have to know the sequence of number words by heart. One must be able to recite the number words in their correct order. The commonly used sequence of number words is constructed in a very systematic way, with a strict built-in ordering and without limit. Once the system is understood, one can produce as many number words as needed, and one can always name the next after any given number word in the sequence. Our number words thus provide a useful reservoir of ordered counting tags that is never exhausted, no matter how large the collection we want to count.

3. Cardinal principle:

When we start counting with *one*, then the last number word reached after having counted all elements of the collection has a very special meaning: It not only is the counting tag of the last counted item but also describes a property of the collection as a whole. The last counting tag is the result of counting. In everyday life we would call this the "number of objects in the collection." The property of a collection that is described by the last number word is sometimes called its *numerosity*. Mathematicians call it *cardinality*, hence the name of that principle. The cardinality of the collection of disks in figure 1.4 is 19, or *nineteen*.

For young children, it is a difficult task, and a great achievement, to make the transition from the mechanical use of number words during the counting procedure (as expressed in the ordinal principle, where numbers just serve for tagging objects) to an understanding that a number word can also denote a numerosity. They have to learn that the last number word is not just the name of the last thing but a property of the whole collection. It is the answer to the question "How many?"

4. Invariance principle (order-irrelevance principle):

For the final result (that is, for the cardinality of a set), it is completely irrelevant where in the set we start counting and in which particular order we count the elements. It does not matter whether we start counting from left to right or from right to left. Figure 1.5 shows a procedure analogous to the one in figure 1.4, but it starts at another first element and proceeds in a different order, and it is a different element that receives the final counting tag, *nineteen*. Yet for the purpose of counting the set of pebbles, these two procedures are completely equivalent and lead to the same result. Bertrand Russell (1872–1970), in his book *Introduction to Mathematical Philosophy*, describes this as follows: "In counting, it is necessary to take the objects counted in a certain order, as first, second, third, etc., but order is not of the essence of number: it is an irrelevant addition, an unnecessary complication from the logical point of view."[3]

Indeed, the result of counting—the cardinality of a collection—is invariant under rearrangements of the objects of the collection. The cardinality of a collection does not change if we change the order of the objects. The invariance principle thus implies that cardinality is a property of the collection and not a property of the particular counting procedure.

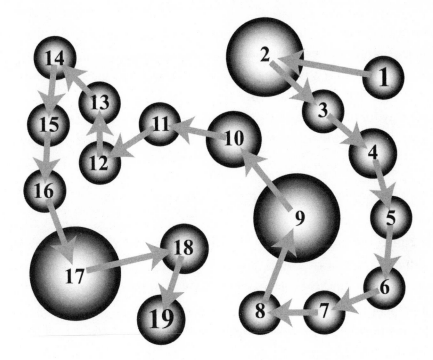

Figure 1.5: Counting in a different order (invariance principle).

5. Abstraction principle:

While the first four principles tell us *how* to count, the abstraction principle tells us *what* we can count. This principle simply states that one can count everything. Any collection of well-distinguished objects can be counted. The process of counting does not depend on the nature of the things to be counted: they may be tangible, like apples or people, or intangible, like ideas or actions. Likewise, the size of the collection to be counted is not limited (as long as it is finite). In theory, we could even count the stars in the sky or the sand of the sea, provided we have a large enough reservoir of number words so that we would not run out of counting tags.

For a child, it is again a great achievement to realize that all kinds of things can be counted, and that quite different things can be combined for counting, like toys of different shapes or immaterial things like games or actions—even numbers can be counted. And two collections that are totally different could nevertheless have the same number of elements—that is, the same cardinality. Understanding this paves the way for perceiving a number as something that has a meaning of its own, something that is independent of a concrete collection of objects.

1.5. COLLECTIONS OF OBJECTS, ELEMENTS IN A SET

At first sight, the counting principles may seem elementary and scarcely worth mentioning, but they contain subtle observations that are useful for further considerations. They show that counting, although a familiar operation, is logically rather complex. In the following, we use the counting principles as a guideline for a deeper analysis of these complexities.

Obviously, in order to be able to count, we need something that can be counted. The abstraction principle is perhaps the most fundamental of the counting principles because it describes *what* we can count—collections of objects, such as apples in a basket or people in a room. So far, we have not been very specific about what we mean by a "collection." In mathematics, a collection, or group of things, would be called a *set*. The crucial property of a set is that it is a collection of well-distinguished objects.

German mathematician Georg Cantor (1845–1918) has defined a set as an aggregation into a whole of definite well-distinguished objects of our intuition or our thinking. The members of a set are called the *elements* of the set. The elements of a set are, thus, objects of our intuition or thinking, and this includes not only material objects but also ideas, numbers, symbols, colors, people, or actions, and so on. The elements of a set could even be sets themselves—for example, figure 1.1 shows a set of elements, and among these elements is a group (which is a set in itself) of people, and also a set of black squares. In mathematical notation, sets are often indicated by putting a list of its elements between braces. So {A, B, C} would be the set containing the letters *A*, *B*, and *C* as elements. The set {1, 2, 3, 4, . . .} would be the set of natural numbers, an example of an infinite set.

A set is formed either by actually "putting apples in a basket" or just by definition—that is, by using some descriptive property. For example, we can define the "set of the blue objects on this table." In any case, it must always be clear which elements belong to a set and which do not. Moreover, every element must occur in a set only once. For this reason, Cantor emphasized that the elements of a set must be distinguishable from each other.

Whenever we count something with the procedure described earlier, we count the elements of some finite set. This is what makes the concept of a set important to us: Counting always deals with a set, even if the set is often not defined explicitly. And the abstraction principle states that any finite set can be counted—any finite collection of distinguishable objects.[4]

The invariance principle states that the result of counting a set does not depend on the order imposed on its elements during the counting process. Indeed, a mathematical set is just a collection *without any implied ordering*. A set is the collection of its elements—nothing more. For example, if you shuffle a set of playing cards, it retains its identity as the same set of cards.

People have often wondered why mathematics is able to describe many aspects of our world with high precision and accuracy. In a sense, this is not so astonishing, because from the very beginning, mathematical concepts have been formed on the basis of human experience— an experience, in turn, formed by the world surrounding us. We can see this even at the basic level, when the concept of a set is defined— perhaps one of the most important concepts of modern mathematics.

But what property of our world, what kind of human experience about the world, would be reflected by the mathematical definition of a set?

To begin with, the notion of a set would make little sense to us if we hadn't made the observation of temporal stability. Typically, objects endure long enough that it makes sense to group them together and consider the whole collection as a new "object of our thinking." Hence, for example, we can put objects into a box and know from experience that they are still there, even if we can't see them. The existence of permanent objects is helpful for the idea of grouping them together to form a new whole, a set. But the notion of a set is general enough to include nonpermanent elements. Time is not mentioned in Cantor's very general definition of a set as a collection of "objects of our thinking." Therefore, temporal permanence of objects is not necessary as a prerequisite for combining them in a set. We can also form a set of events like drumbeats or strokes of a clock, and we can define, for example, the set of days between one new moon and the next.

A very basic observation concerning a fundamental property of the world we live in is *the existence of objects that can be distinguished from each other*. For the definition of a set, it is indeed of crucial importance that things have individuality, because in order to decide whether objects belong to a particular set they must be distinguishable from objects that are not in the set. Without having made the basic experience of individuality of objects, it would be difficult to imagine or appreciate the concept of a set.

It is perhaps worthwhile to make a little thought experiment. How might life be in a radically different universe? Imagine, for example, a vast ocean populated by intelligent protoplasmic clouds, containing no solid objects at all. Let us assume that whenever these cloud-beings meet, they would mix and merge into a new cloud-being. Would those cloud-beings, however intelligent, develop any concept of numbers and counting? Even if they did, numbers would probably have little importance and would appear as a very exotic idea. Arithmetic would appear rather impossible because for our cloud-beings 1 + 1 would be just 1 in most cases. Quite certainly, mathematics would have evolved in a different direction. We learn from this that statements like "1 + 1 = 2," which appear to be so true and obvious to us, need not be true and obvious to everybody, under all circumstances. In our universe, however, one of the first observations a child makes about its surroundings is that there are well-defined objects that can be seen either alone or in pairs or in groups. We have learned to form sets, and we have learned to count, because our world contains objects with a certain permanence and individuality.

Moreover, there is still another important observation that seems to be essential for the idea to group objects into a set: This is the human ability to recognize similarities in different objects. Usually, a collection, or group, consists of objects that somehow belong together, objects that share a common property. While a mathematical set could also be a completely arbitrary collection of unrelated objects, this is usually not what we want to count. We count coins or hours or people, but we usually do not mix these categories. If you hear that 4 people watched 3 movies within 2 days, you would hardly want to know that there are 4 + 3 + 2 = 9 well-distinguished objects—this information appears to be rather useless, or even meaningless. When we count, we usually group objects by similarity, and count the number of people *or* the number of movies *or* the number of days. We automatically tend to group similar objects and perceive them as belonging together. This ability is the basis for defining a set by a common property of its elements (for example, the set of all blue objects).

1.6. THE BIJECTION PRINCIPLE AND THE COMPARISON OF SETS

A bijection is a one-to-one correspondence that describes a perfect match between two sets. Figure 1.6 shows flies sitting on the pebbles.

There is no pebble without a fly, no two flies sit on the same pebble, and no fly is without a place to sit. Hence there is a one-to-one correspondence—a pairing—between pebbles and flies. In mathematical terminology: There is a bijection between the set of pebbles and the set of flies in this image.

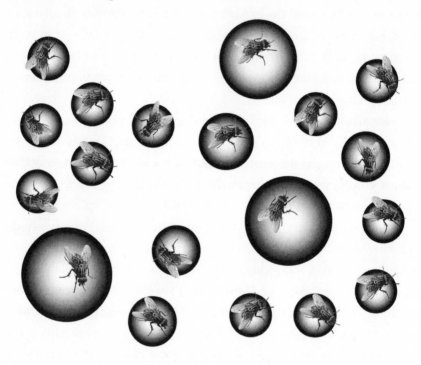

Figure 1.6: Counting via one-to-one correspondence (bijection principle).

As a consequence, we can say immediately and *without counting* that there are as many flies as there are pebbles. Whenever there is a bijection between two finite sets, we know that they contain the same number of elements.

We can also use the bijection principle to find out whether one set is bigger than another. Obviously, the set of flies is bigger than the set of pebbles if, after pairing the flies with pebbles, all the pebbles get used and some flies are still left over. Or we could also draw the opposite conclusion: When there are more flies than pebbles, then either some flies are without a pebble or there must be at least one pebble with two or more flies on it. In mathematics, this observation is called the *pigeonhole principle*. It states that when we put n objects (pigeons) into m holes, and n is bigger than m, then at least one hole will contain two or more objects (pigeons). We can use this simple observation to answer the following question: Are there two people in New York City who have exactly the same number of hairs? The answer is yes. There are more than eight million people in New York City, and even the hairiest among them will have fewer than a million hairs. Hence, the possible hair numbers range from zero to fewer than a million, and there are more people than there are possible hair numbers. By the pigeonhole principle, when we attempt to assign to each person a hair number, at least one of the hair numbers will have to be assigned to two or more persons.

1.7. UNUSUAL COUNTING TAGS

According to the bijection principle, any act of counting establishes a bijection between the objects to be counted and a set of counting tags. Typically, the counting tags are number words, but, as far as the bijection principle is concerned, they could be anything, even objects of another set. And for small collections, instead of number words one could use letters of an alphabet or the words of a counting-out rhyme, such as "eeny meeny miney mo." Japanese people, for example, while having a perfectly logical system of number words, occasionally also use a particularly poetic alternative for enumerating up to forty-seven items. They use the syllables of a famous poem, the *Iroha*, as counting words. The Iroha is an ingenious piece of poetry from the Heian period (794–1185), in which every possible syllable, and thus every sound of the Japanese language, occurs exactly once. Usually written in hiragana (a Japanese writing system), it starts as follows:

い	ろ	は	に	ほ	へ	ど		ち	り	ぬ	る	を
i	ro	ha	ni	ho	he	do		chi	ri	nu	ru	wo...

and it goes on to use every single hiragana character once, and only once, without repetitions. The Iroha is a poem about the transience of all being. In English, this line roughly means: "Although the colors (of blossoms) smell, the paint scatters away." The poem is sometimes used, even today, for teaching the Japanese syllabary. But the fixed order of syllables defined by this poem and the uniqueness of the syllables makes it also suitable for counting (see the bijection principle and the ordinal principle, described above). Indeed, it is sometimes used, for example, to number the seats in a theater:

i, ro, ha, ni, ho, he, do, . . . su
1, 2, 3, 4, 5, 6, 7, . . . 47

The bijection principle enabled people to deal with numerosities even in a time before number words were invented. In ancient times, a shepherd counted his flock by putting a pebble in a bag for each animal leaving shelter. Therefore, he established a one-to-one correspondence between pebbles in the bag and animals in the herd. Although the shepherd could not count, he would know exactly the number of eventually missing animals. When he took a pebble out of the bag for each animal returning to shelter, pebbles remaining in the bag would precisely indicate the number of lost animals.

Another prehistoric method of counting involved the marking of notches on a tally stick. This created a bijection between marked notches and things to count. This method could also be used to count events. Events are "objects" that lack the property of permanence. Once an event has occurred, it is gone and exists only in memory. In order to count events, they have to be remembered, which is difficult because of the limited capacity of our memory. Therefore, it is a good idea to create a permanent record of each event. For example, we can create a tally mark for each passing day. The oldest tally sticks probably used in that way are some thirty thousand years old. Even more recently, prisoners counted days with scratches on the wall, thereby creating a one-to-one correspondence between days in prison and tally marks.

A common way of counting in prehistoric times was by using body parts. They used not only fingers, but also wrists, elbows, shoulders, then toes, ankles, knees, and hips, all in a fixed order.[5] Probably the earliest counting tags were simply the names of the corresponding body parts spoken together with appropriate gestures. Prehistoric people developed concrete procedures with pebbles, tally marks, or body counting, or combinations of these, in order to trade goods or fix the dates of religious festivals. They did all this even before they had developed any abstract knowledge of numbers or sufficient vocabulary for counting in the sense that we understand it today.

1.8. CARDINAL AND ORDINAL NUMBERS

The main ideas of the BOCIA principles are shown in figure 1.7. As we inspect them once again, we observe that numbers seem to play a double role.

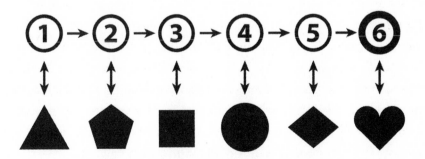

Figure 1.7: Counting principles.

- Any collection of arbitrary objects can be counted (*abstraction principle*). Figure 1.7 shows a collection of certain shapes.
- Counting is a process of associating unique "counting tags" with each of the objects. This one-to-one association between objects and counting tags is represented by the vertical double arrows (*bijection principle*).
- The counting tags must have a fixed order. In figure 1.7, the counting tags are simply the natural numbers 1, 2, 3, and so on, and their ordering is symbolized by the horizontal arrows.
- The act of counting puts the objects to be counted in a certain order, but the ordering of the objects is irrelevant (*invariance principle*). Rearranging the shapes would not change the final outcome.
- When we start counting at *one*, the cardinality (numerosity) of the set is described by the last of the number words (*cardinal principle*). The set of shapes in figure 1.7 has the cardinality six (a *cardinal number*).

As a first observation, we note that in the process of counting, number words are used in two different ways. On the one hand, number words tag the individual objects, giving them a certain order; on the other hand, one of these number words will denote the final result of counting and describe a property of the collection as a whole. The number words that serve as labels during counting represent *ordinal numbers*. Generally, ordinal numbers denote the position of an element in an ordered sequence. This is a very important concept, and on some occasions we use special number words for indicating an ordering, like *first, second, third,* . . . , *seventeenth,* and so on. Typically when we ask "which one?" or "which position?" the answer would be an ordinal number ("object number two" or "the third man").

While ordinal numbers describe the position in a sequence, *cardinal numbers* are used to describe the size of a collection. The cardinal number of a collection is given by the number word obtained when, during counting, we reach the last object of the collection. This is the final result of the counting process. It represents what we called the numerosity, or cardinality, of a set. Whenever we ask "how many?" the typical answer will be a cardinal number (such as "five").

In everyday life, you will frequently encounter a third type of numbers. These are called *nominal numbers*, and they just represent names, like telephone numbers or zip codes or item numbers in a catalog. A nominal number does not describe a size or quantity and need not imply any natural order. It is just for naming things and cannot be used in arithmetic operations. Certainly, it is of minor interest to mathematicians.

In order to appreciate the distinction between cardinality and ordinality, assume a moment that you have to relearn counting using letters of the alphabet instead of using numbers. Most probably, you will know already how to recite the sequence of twenty-six letters of the (English) alphabet in the correct order (A, B, C, . . .), hence you will certainly master the mechanical procedure of "counting" sets of up to twenty-six elements with letters instead of numbers. That means that you have mastered the ordinal principle, but probably not the cardinal principle. You can easily check your understanding of cardinality with the following questions: Do you have an intuitive feeling for the "number K"; that is, do you have an understanding of the size of a set containing K elements? Can you estimate the weight of F apples? And how large is a set that is F objects smaller than a set with K objects? Unless you have special training, it will not be easy to give the answer without resorting to counting with fingers. This is how it feels when you know how to count but still can't properly associate number words with cardinalities: You can't perform even simple computations without using your fingers.

1.9. ABSTRACTION

Considering the counting procedure and its results, we come to understand a natural number as a property of a set, describing numerosity. For example, the number *nine* will appear as the cardinal number of any collection of nine objects or any group of nine people or any set with nine elements. With experience, one learns to associate a certain "size," or "numerosity," with the cardinal number nine. Through repetition and by force of habit, numbers can gain an abstract connotation independent of the concrete collection of objects. And finally, one is able to think of *nine* without picturing any particular group of nine objects. The (cardinal) number nine has finally achieved a meaning of its own. It describes what nine apples, nine people, or nine black dots on a sheet of paper have in common. It has become an abstract concept that represents a *property* of certain sets, namely all those sets for which counting ends with the number word *nine*. And the particular nature of the set, the type of objects in the collection, is completely irrelevant, as long as there are precisely nine elements in the set. Therefore, this concept of *nine-ness* of a set has become something that lives in our imagination and does not need a concrete realization any longer. A number seems to be a property of a collection in a similar sense that a color is a property of an object.

Forming abstractions is a natural process in our language. Consider, for example, the word *table*, which is the end result of a process of abstraction that starts with concrete objects. These concrete tables will differ in shape and material. But the abstract notion of a "table" makes no distinction and refers to all concrete tables simultaneously. The word *table* lets us think of an object that typically has a flat horizontal surface and is supported by one or more "legs." Without further information, we do not know whether the word *table* refers to a dining-room table, a coffee table, a billiard table, or a workbench. The word *table*, therefore, contains less information than a reference to a particular table. Obviously, abstractions are created by reducing the amount of concrete information. Therefore, the process of forming abstractions can be seen as a process of simplification, and it is very handy, because we can now talk about tables in general without having to refer to any particular table.

The abstraction leading from special groups of objects to a number is a quite similar process. It is the process of removing information about the concrete nature of the counted objects. And as soon as we have successfully performed this abstraction, we have achieved a simplification. When we think of *four*, we do not need to think of four apples or four persons or four corners any longer. *Four* refers to *all* sets with four elements. We can work with numbers without having to think of their concrete realizations. And this is a prerequisite for successfully doing computations with numbers. We can perform computations, such as $9 + 4 = 13$, without having in mind real manipulations with concrete objects. And while concrete realizations of computational tasks are still possible with smaller numbers like 9, 4, and 13, it becomes impractical, or even impossible, for larger numbers. When you think of the number 2,734, you will probably think of "many," but it isn't always useful to imagine a particular collection of 2,734 objects. From a practical point of view, the transition from concrete realizations to an abstract number concept becomes absolutely necessary when dealing with really large numbers.

There is indeed some evidence from neuropsychology that numbers are represented in our brain in an abstract way. That means whenever we see the symbol *4* or hear the word *four* or see a collection of four dots, the same group of neurons in the same part of our brain gets activated. No matter how the number is presented to us—whether in verbal or nonverbal form—the end result of the information processing in our brain is always the same neuronal activity pattern. The fact that all the different sensual inputs invoke the same representation of "four" in our brain is the neuronal origin of our perception of the number four as an abstract "mathematical" object.

Thinking of numbers as abstract objects in the sense described above belongs to the first and most fundamental concepts of mathematics. When the abstraction described here is not learned, then a number is not perceived as a concept independent of the things that are counted. Numbers would be inseparable from concrete objects, and one could not understand that the four seasons and the four wheels of a car have something in common. The old language of Fiji Islanders, where ten boats would be *bola* and ten coconuts would be *koro*, provides examples of number words that cannot be detached from objects and did not acquire an abstract meaning. A vestige of this state of human development can still be found in modern languages—for example in Japanese, where different (although related) counting words are used for counting different types of things:

ippon, nihon, sanbon, yonhon, . . .   for counting long, cylindrical objects

ichimai, nimai, sanmai, yonmai, . . . for counting flat, thin objects

ikko, niko, sanko, yonko, . . .   for counting small, compact objects

ichidai, nidai, sandai, yonday, . . . for counting machines, vehicles, etc.

ikken, niken, sangen, yonken, . . .   for counting houses, buildings

ippiki, nihiki, sanbiki, yonhiki, . . . for counting small animals

ittou, nitou, santou, yontou, . . .   for counting large animals

are just a few examples.

1.10. COUNTING BY THE ORDINAL PRINCIPLE

The ability to do things systematically in some order is a prerequisite for counting. It is already present in primates and is probably connected with what one can do with one's hands. When manipulating several objects at a time, it becomes necessary to do this in a certain order, thereby avoiding doing something at the wrong time or unnecessarily twice. Indeed, primates can be observed to proceed in a very systematic and orderly way when harvesting fruits from the branches of a tree or when searching each other's fur in mutual grooming activities.

More directly related to counting is the observation that objects in a finite set can always be brought in a certain order. This can be done, for example, by arranging them in a row or by sorting them according to size, weight, or some other property. The ordinal principle for counting states that even in case of an apparently disordered set like the one in figure 1.2, we have to take the objects in a certain order, as first, second, third, and so on. In figures 1.4 and 1.5, this order is symbolized by the chain of arrows leading from one object to the next.

But the set of objects to be counted need not have any predefined order. Indeed, the set of pebbles in figure 1.6 is a typical set without natural predefined ordering. During counting, we associate each object in the chain with a unique counting tag. The counting tags are always used in a fixed order; that is, they must be from an ordered set whose elements are in a predefined, strict, and invariable order. Counting tags are useful only if they are always used in the same order. Then the last tag given to the last object in the chain would describe the cardinality of the set.

In mathematics, a set with this kind of strict ordering is called a *sequence*. In a sequence there is a unique "first element," and every element in a sequence has a unique successor.

The most natural counting tags are, of course, the familiar symbols (1, 2, 3, 4, 5, . . .). They have a natural, predefined order. The set of natural numbers used for counting starts with 1, and then every number has a unique successor:

$$1 < 2 < 3 < 4 < 5 < \ldots \text{ (natural order of natural numbers).}$$

The strict order of the cardinal numbers makes them suitable as counting tags, and hence the (finite) natural numbers serve as ordinal numbers and as cardinal numbers at the same time. In the process of counting, each ordinal number is at the same time the cardinal number of the set of already-counted objects.

Obviously, the bijection principle and the ordinal principle closely work together in the process of counting. The counting tags (number words or number symbols) form an ordered sequence with a unique first element, and each counting tag has a unique successor. And when we reach a certain number word during counting, we have actually recited all number words that were first in the sequence. For example, reaching "six" as a result of counting means that we have counted "one, two, three, four, five, six." The counting procedure thus establishes a one-to-one correspondence (or bijection) between a given set of six elements and the set of number words up to six, as shown, for example, in figure 1.7. We learn from this that a sentence like

"This box contains six items"

is actually a very brief account of the activity of counting. This statement actually means something like

"I have just found a one-to-one correspondence between the set of objects in the box and the following set of number words {*one, two, three, four, five, six*}."

And this just means that the set of objects in the box contains exactly as many elements as the initial section of the sequence of number words, which (in virtue of the strict ordering) is uniquely determined by the word *six*.

1.11. SYSTEMATIC ENUMERATION

The vocabulary of counting developed over many thousands of years in order to meet practical needs during the process of humankind's settling down. Expressing even large quantities by an exact number word became necessary to keep track of provisions or animals in a large herd, and for commerce. Hunters and gatherers had only a few number words, and then words like *few* and *many*. But if one only has a few counting words at one's disposal, such as names of body parts, one cannot count larger sets.

People need a set of counting words that cannot be exhausted—at least in principle, every natural number needs a unique name. Still, it should be easy to recite the sequence of number words in the correct order. In order to avoid a huge load on the human memory, the counting words have to be constructed in a systematic way based on simple logical repetition. Any such systematic method of creating number words is called a *numeral system*.

A basic idea for creating such a numeral system is the grouping of a large number of objects into manageable parts; for example, in such a way that each part can be counted with fingers. One can use this idea to count precisely and communicate the result, even at a stage of development where the language knows no number words at all, and even if the numbers involved are fairly large.

We do not know under what circumstances the first numeral system was developed. As a typical example, let us consider again the situation of the prehistoric shepherd who counted sheep by putting a pebble in a bag for every animal in his herd. With a growing number of sheep, the pile of pebbles might become unmanageable, and the method is not well suited for communication. So the shepherd follows a slightly different procedure. Let us assume he counts with his fingers, for example, by forming a fist and extending a finger for each animal coming through the gate. Once he has lifted all fingers of both hands, he puts aside a wooden stick and starts anew, counting the next group. So, for every group of 10 sheep, he would put a stick on a pile. For the final group, which probably does not amount to a full 10, he adds the corresponding number of pebbles. Eventually he will end up with 7 sticks and 9 pebbles, a handy and light-weight representation for a total of 79. He thus knows the number of sheep in his herd, probably without being able to name that number.

Once this process of grouping has started, it can be continued on a higher level. For counting larger quantities, people would take, for example, a bone for every 10 wooden sticks, and a big stone for every 10 bones. A collection of 1 big stone, 7 bones, 7 sticks, and 6 pebbles would thus symbolize the number 1776, as in figure 1.8. The progressive grouping is necessary to describe larger numbers. It is the basis for a systematic naming of numbers—a numeral system. It is not difficult to recognize our own numeral system in the shepherd's counting method: Just replace the word *stick* by the word *ten*, *bone* by *hundred*, and *stone* by *thousand*.

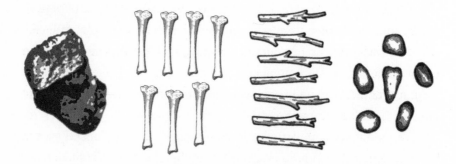

Figure 1.8: A natural representation of the number 1776.

We can also see how other base-*n* systems could have evolved. Had the shepherd used only the fingers of one hand, he probably would have counted in groups of five, and this would have been the origin of a numeral system with base-5—as it is, for example, in use in the Epi languages of oceanic island nation Vanuatu. Remnants of such a system are also visible in the system of roman numerals, where one has special symbols for five, fifty, and five hundred. Their symbols for the first numbers are I, II, III—easily recognizable as pictograms of fingers or counting sticks. The special symbol V, denoting *five*, represents a hand, and the letter X for *ten* obviously consists of two hands.

If the shepherd had used all his fingers and all his toes for basic counting, this would have been the origin of a system with base-20—as it is found in the Mayan culture and among the Celts in Iron Age Europe. In some European languages, the linguistic structure of the names of certain numbers still shows the Celtic heritage; for example, in French the word *quatre-vingt* for *eighty* means "four-twenty." The Yan-tan-tethera ("one-two-three") was a sheep-counting system in use in northern England until the Industrial Revolution. It was derived from an earlier Celtic language and used number words only for numbers up to twenty. For counting larger numbers, a shepherd would drop a pebble into his pocket every time he counted to twenty—that is, for each score. The word *score*, actually, comes from Old Norse, where it meant a notch on a tally stick. In a base-20 system, the twentieth notch was made larger, and this finally gave the meaning *twenty* to the word *score*.

If the shepherd had a system of tabbing with his thumb each of the three phalanges of the four opposing fingers, he would have created a system with base-12. A combination of phalanx counting on the right hand with finger counting on the left hand would lead to a system with base-60, as was used in ancient Mesopotamia. And in Toontown, the home of Bugs Bunny and Donald Duck, where all the cartoon characters have only four fingers on each hand, most probably a system with base-8 would have evolved.

How would the prehistoric ancestor of Donald Duck have represented the number 1776? To him, a stick would represent 8 items instead of 10. Consequently, he would replace 8 sticks by a bone (thus, a bone would represent 64 items), and a stone would represent 8 bones—or $8 \times 64 = 512$ items. It is not too difficult to figure out that he would need 3 stones, 3 bones, and 6 sticks, as in figure 1.9, because $1776 = 3\times512 + 3\times64 + 6\times8$.

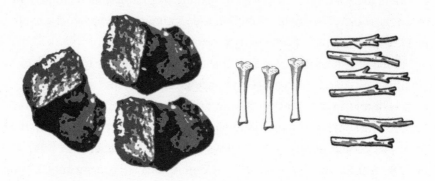

Figure 1.9: Representation of the number 1776 if we had only eight fingers.

What about other systems? A Sumerian using a base-60 system would have needed just 29 sticks and 36 pebbles ($1776 = 29 \times 60 + 36$). On the other hand, with a small base—like 5—we need 14 stones, 1 bone, 1 pebble, and no sticks. And in order to be consistent, the 14 stones have to be represented with the help of the next higher category—say, pearls—where each pearl represents 5 stones ($1776 = 2 \times 625 + 4 \times 125 + 25 + 1$).

1.12. THE NUMERAL SYSTEM IN WRITING

Once culture developed among early settlers, the society became more complex, goods were produced, and division of labor began. Yet resources were unequally distributed, and therefore trade and exchange between communities became necessary. The need arose to communicate what could be offered in what quantity or how much of a certain good was demanded in exchange. Notched tally sticks or bags with pebbles and bones to quantify number soon became impractical and number words came into use. With the development of writing, number symbols were also invented, and the spoken numeral system was translated into a written form.

We have seen that a numeral system is obtained by hierarchical grouping. In the example of the previous section, every item was represented by a pebble, every 10 pebbles by a stick, every 10 sticks by a bone. This is the foundation of a base-10 numeral system. From here it is still a long way to a symbolic representation of numbers in writing.

A systematic method to write arbitrary numbers typically uses arrangements of basic symbols that we call *digits*. We are used to the ten digits—0, 1, 2, 3, 4, 5, 6, 7, 8, 9—that also serve as symbols for the first natural numbers. In order to express larger numbers, we use combinations of the basic symbols. This can be done in quite different ways. Let us see how this problem was resolved in our own culture. (In chapter 3 we will describe some other historically interesting methods of symbolizing numbers.)

Currently, we still use two different systems of writing numbers. One is the system familiar from everyday use and the other, although rarely seen, is the Roman system. In spoken language, the Roman numerals are not fundamentally different from the numerals in English. The number forty-nine would be *quadrāgintā novem* in Latin, which would literally translate into *forty nine*. In writing, however, these two systems take completely different approaches to symbolizing numbers—just compare 49 with its equivalent XLIX in the Roman system! It is worthwhile to take some time and describe these different methods in more detail.

Consider, for example, the numeral 1776. It is written by putting the digits one-seven-seven-six in a row. It is a very ingenious notational trick that allows us to write a relatively large number in such a compact form, using only a few symbols. We all know immediately that 1776 actually means one thousand seven hundred seventy-six, or

$$1 \times \text{thousand} + 7 \times \text{hundred} + 7 \times \text{ten} + 6 \times \text{one},$$

or in the language of the shepherd from the previous section: one stone, seven bones, seven sticks, and six pebbles.

In our way of symbolizing numbers, every digit in 1776 has a meaning that is given not only by its numerical value but also by the place where it appears. The digit 7 even appears twice, but each time its meaning is quite different. Reading the number from left to right, the first 7 is seven hundreds, and the second 7 means seven tens. The actual value of every digit depends on its position. The rightmost digit always counts the *ones*, the next counts the *tens*, and so on. Each digit contributes with a certain value to the final numerical meaning of 1776. Because the value of a digit depends on the place where it is written, our numeral system is called a *place-value system*. The most important consequence of the place-value system is that we do not need special symbols for ten, hundred, thousand, and so on.

In order to write a number like

$$1 \times \text{thousand} + 7 \times \text{hundred} + 6 \times \text{one},$$

we need a special symbol that denotes the absence of a position. We cannot simply omit the place describing the tens, because 176 would be something completely different. And leaving a gap, as in 17 6, is bound to create confusion. Therefore, using the symbol 0 as a place holder, we write 1706 to indicate that there are no tens. Without that symbol, it would be very difficult to distinguish between 176, 1076, 1706, and 1760.

In our place-value system, the numerical value of a numeral is determined by two operations:

1. multiplication of every digit with its place-value (one, ten, hundred, . . .)
2. addition of the results from step 1.

It is quite different from the ancient Roman system, which uses addition (and sometimes subtraction). In the Roman system, each power of ten has a separate symbol: 10 is written as X, 100 is written as C, 1000 is written as M. With additional symbols for 5=V, 50=L, and 500=D, the base-10 numeral 1776 would be written as

$$
\begin{aligned}
\text{MDCCLXXVI} &= M + D + C + C + L + X + X + V + I = \\
&= 1000 + 500 + 100 + 100 + 50 + 10 + 10 + 5 + 1 \\
&= 1776
\end{aligned}
$$

Therefore, in order to find the numerical value of a Roman numeral, we just have to perform addition, no multiplication. There are some exceptions to this rule: As a shortcut, one writes IV ($V - I = 5 - 1 = 4$) instead of IIII, and similarly in other cases, where the symbol with a smaller value is written first to indicate subtraction instead of addition: IX = $X - I = 9$, and so on. Note that we need no placeholder symbol to write a Roman numeral; 1706 would simply be MDCCVI.

Today, we still encounter Roman numerals occasionally; for example when denoting the year of construction on the cornerstone of a building, or sometimes the year of production at the end of a movie.

1.13. BASE-10

Our place-value system is a written representation of a base-10 numeral system. The first number after *nine* plays a particular role because it is the first number that requires a symbolic representation consisting of more than one digit—namely 10, which is one times ten plus zero times one. Moreover, the other place values can be obtained as powers of 10:

hundred	$= 100 = 10 \times 10 = 10^2$
thousand	$= 1,000 = 10 \times 10 \times 10 = 10^3$
ten thousand	$= 10,000 = 10 \times 10 \times 10 \times 10 = 10^4$

and so on. With this notation we can write

$$1776 = 1\times10^3 + 7\times10^2 + 7\times10 + 6,$$

and with the common definition $10 = 10^1$ and $1 = 10^0$ we obtain the unified notation

$$1776 = 1\times10^3 + 7\times10^2 + 7\times10^1 + 6\times10^0.$$

In our place-value system, any natural number, no matter how large, can be written with the help of a finite number of digits, say,

$$d_0, d_1 \ldots d_{n-1}, d_n,$$

and each of these digits is taken from the set of ten symbols {0, 1, 2, 3, 4, 5, 6, 7, 8, 9}, except d_n, which is usually assumed to be not zero. The numeral representing the given number is then formed by writing the digits in a row:

$$d_n d_{n-1} \ldots d_2 d_1 d_0.$$

The whole expression is then just a shortcut for digit-times-place-value addition:

$$d_n \times 10^n + d_{n-1} \times 10^{n-1} + \ldots + d_2 \times 10^2 + d_1 \times 10^1 + d_0 \times 10^0.$$

1.14. MEASURING MAGNITUDES

Natural numbers are "counting numbers." They can be used to count every finite collection of arbitrary things; they measure the size of a set. But there is an even more important aspect of numbers that we have not covered yet, and that goes beyond just natural numbers. In everyday life, we use numbers not so much to measure sets but to measure and describe quantities and magnitudes. These entities cannot be counted in a strict sense. For example, the length of a line segment is a magnitude—in itself it is something quite different from a natural number, which, as we have just seen, describes the cardinality of a set. How is it that we can express lengths through numbers?

The key idea here is to measure a length in multiples of unit lengths. The choice of units is a mere matter of convention. In practice, the unit is provided by the measuring device that we use to measure the given quantity. For measuring lengths we have, for example, a measuring tape or a ruler with the units printed on it. By comparing a given length with the ruler and its units, we determine a measuring number that describes the length. For example, in figure 1.10, we find that seven unit lengths in a row add up to give us the length of a stick. We see that measuring lengths is just another form of counting.

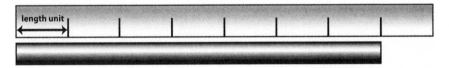

Figure 1.10: Measuring lengths is done by counting units.

What do we do when the length of the object does not fit exactly to a whole number of units, as in figure 1.11? In this case, we first count the number of whole units that measure the length of the stick. This gives two inches. Now there is a remainder that is shorter than one inch. The given length is not a multiple of the unit. In that case, we divide the unit into suitable subunits and count the remainder in these subunits. A subunit is always a fraction of the given unit, which means the subunit is given as one n^{th} of the unit. For example, one could take $\frac{1}{8}$ inch as a subunit for one inch, as in figure 1.11, where we count seven of these subunits. This means that the length of the stick is two inches plus seven times $\frac{1}{8}$ of an inch, which is usually written as $2\frac{7}{8}$".

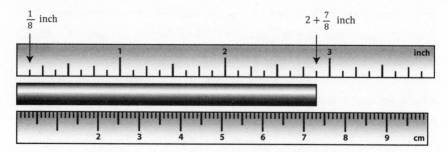

Figure 1.11: Measuring lengths with units and subunits.

Metric units are usually divided into ten parts. One tenth of a centimeter is a millimeter. The same stick would then have a length of 7 cm and 3 mm. For the division of units into ten parts, we have a convenient notation that is an extension of the place-value system described in the previous section: We write *7.3* (seven point three) when the magnitude is given by seven units and three tenths of that unit. And if we need a higher accuracy, we could divide smaller subunits even further. With more precise methods, we would find that a stick of 2 and $\frac{7}{8}$ inches would measure 7.3025 cm, which means 7 units + 3 tenths + 2 thousandths + 5 ten-thousandths.

Notice that we have the role of the zero symbol as a place holder, denoting the absence of hundredths in this case.

We call a number that measures a quantity by comparison with a given unit a *real number*. An important special case is a *rational number*. A rational number describes a quantity as a multiple of a unit plus a multiple of a subunit, as, for example, in 2 plus $\frac{7}{8}$ inches or 7 plus $\frac{3205}{10000}$ centimeters or 5 plus $\frac{1}{3}$ gallons. In all these cases, it is possible to express the quantity as an integer multiple of a suitably chosen subunit:

$$2.875 = 2 + \frac{7}{8} = \frac{23}{8}, \quad 7.3205 = 7 + \frac{3205}{10000} = \frac{73205}{10000}, \quad 5.33333\ldots = 5 + \frac{1}{3} = \frac{16}{3}.$$

It has come as a surprise that this cannot be done in all cases. Not every length can be described as an integer multiple of a unit plus an integer multiple of a subunit. Such a number would be called *irrational*—as distinguished from the rational numbers shown in the examples above. In other words, an irrational number cannot be expressed as a fraction. An example of an irrational number is the length of the diagonal of a unit square—one whose side length is one. The length d of the diagonal of the unit square is the square root of two or, written symbolically,

$d = \sqrt{2} = 1.41421356\ldots$

This means that the length of the diagonal equals the length of the side plus four tenths of the side, plus one hundredth of the side, plus four thousandths, plus two ten-thousandths, and so on. The chain of digits behind the dot will never end. This alone would not make d irrational, as the example with $5.33333\ldots = 5\frac{1}{3}$ shows. This number has also infinitely many digits in the base-10 representation, but these digits are all the same. In contrast, there is no repetitive pattern in the digits of $d = \sqrt{2}$. And one can show that it is not possible to write d as a multiple of a certain fraction of the unit, which makes it irrational.

Another famous example of an irrational number is

$\pi = 3.14159\ldots$

The measure of the circumference of a circle, using the diameter of that circle as the unit of length equals three times the diameter of the circle, plus one tenth of the diameter, plus four hundredths of the diameter, plus. .... Again, the chain of digits would never end and follows no regular pattern (no periodic repetitions). Thus the number π is irrational: it cannot be written as a multiple of a fraction of the unit. More about this amazing number π will be discussed in chapter 10.

Provided that a unit (a "gauge") is fixed, magnitudes of all kinds can be measured by the same type of real number, and in quite a similar way as we measure lengths. We measure areas by square meters and volumes by cubic meters or similar units. We measure time by counting hours, minutes, and seconds. It is quite interesting that hours and minutes have sixty subunits, which reminds us of the ancient Babylonian numeral system (see chapter 3).

Thus, we now have a better understanding of the nature and development of how numbers were represented over time to where we now have a rather sophisticated system for representing quantities.

CHAPTER 2
NUMBERS AND PSYCHOLOGY

2.1. CORE KNOWLEDGE OF THE WORLD

How would life be in a primitive foraging society of hunters and gatherers? Can we think of circumstances that would obviate the need for counting, numbers, and arithmetic? Today, it is almost impossible to find people who haven't been influenced by modern civilization, but there are still smaller groups living in remote regions of the Amazon jungle that apparently haven't invented counting. If we lived in such a society, where cultural needs would not force us to learn how to count, how would that affect our knowledge about numbers?

As we have seen, the number concept is based on some fundamental knowledge of the world. The objects in our environment have permanence and individuality and are encountered either alone or in pairs or in larger groups. Evolution has probably ingrained some of this rudimentary knowledge in the neuronal structure of our brain, to the extent that it is helpful for surviving. Thus, it seems natural to assume that some apprehension of number is already hardwired in the brain of a newborn.

The field of science that is concerned with these questions is called *mathematical cognition*. It has been established in recent years as a new domain of cognitive science. Its subject is to investigate how the human brain does mathematics. An impressive amount of research has been devoted to, among other things, the following questions: What are the neuronal structures in the brain that represent numbers? Do animals have a sense of number? Is there inherited knowledge about numbers? How do we obtain knowledge about numbers? What mathematical faculties are obtained through culture and learning, and what is innate? Is the understanding of numbers and arithmetic connected with the ability to speak a language, or does it have preverbal roots? When do we learn to count, and which steps do we master in that process?

With these questions we also enter the domain of "genetic epistemology." Epistemology (from Greek *episteme*, meaning "knowledge") investigates the nature and scope of knowledge and was long considered the exclusive domain of philosophy. Jean Piaget (1896–1980), perhaps the most influential developmental psychologist of the twentieth century, argued that epistemology should also take into account the findings of cognitive science about the psychological and sociological origins of knowledge. He developed the field of genetic epistemology to investigate how knowledge is achieved through cognitive processes taking place in every individual.

In the mid-twentieth century, Piaget held the opinion that we are born without any knowledge, the brain being an empty page, fully ignorant of its environment but endowed with some fundamental mechanisms for learning. Sensual input would trigger processes of mental organization and adaptation in our brain, creating internal representations of aspects of the external world. Further refinements and adjustments of these mental concepts are constructed by the human brain in a continuous effort to harmonize internal representations with sensual impressions.

Concerning the abilities of newborns, we have a quite different view today. There is now sufficient evidence that we are already born with inherited neuronal structures that originated in evolutionary processes. These "core-knowledge systems" represent some basic knowledge of the external world and help us to interpret sensual inputs and guide our acquisition of further abilities. Research in mathematical cognition has identified two basic mental representations of the number concept: an exact representation for small numbers up to three or four and an approximate number sense for larger quantities. Here, we describe these neuronal foundations of number very briefly. Those seeking more detail are referred to the excellent book *The Number Sense: How the Mind Creates Mathematics* by Stanislas Dehaene (1965–), one of the pioneers in the field of cognitive neuroscience.

2.2. OUR BUILT-IN OBJECT-TRACKING SYSTEM

The ability to simultaneously track objects through space and time is certainly essential for survival. When you try to cross a street, you will probably have to observe the motion of several cars at the same time. Animals would need the same mental skill when tracking the motion of several predators. Obviously, in a potentially hostile environment, individuals would have diminished chances for survival if they lacked this particular ability. Evolution would thus have selected individuals with this skill over those without it. Therefore, neuroscientists regard the "object-tracking system" as a basic functionality of the brain, an inherited cerebral mechanism. It is a mental device that keeps perceptual information of up to three or four individual objects in the working memory. Our consciousness seems to know automatically the number of objects stored in the object-tracking system. Thus, it seems that this system is responsible for the following important manifestation of our inborn "number sense":

Subitizing (from the Latin word *subitus*, meaning "sudden") is the mental ability to enumerate small sets rapidly without counting. When we see a small group of not more than three or four objects, we often know their number immediately. This perception of number appears to be automatic, effortless, and exact, and it occurs without conscious counting. For example, look at the fields in the first row of figure 2.1.

Even if there is no regularity whatsoever in the placement of stars, we see immediately if a field contains two or three stars (some people also have no difficulty with four objects). This feeling of instantaneous recognition is absent when we look at the fields in the second row, where determining the number of stars is much more difficult and requires actual counting.

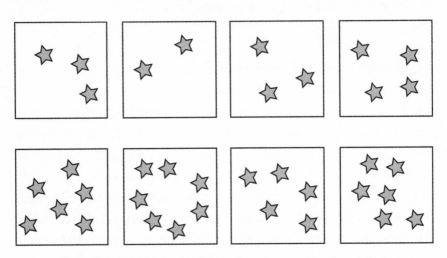

Figure 2.1: Subitizing—Which numbers do you recognize without counting?

Like the object-tracking system, subitizing is limited to three or four objects. Within that range, perception of numbers "at a glance" is not only fast but also highly accurate, and errors seldom occur. It is quite different from the counting described in chapter 1, and it does not require that we direct our attention from one object to another. Eye-tracking experiments have indeed shown that during subitizing, one doesn't look at the objects individually. Instead, a single glance at the whole group is sufficient to know the number. When the number of objects reaches four or five, eyes start moving around in order to scan the collection, either for counting the objects or to search for familiar arrangement patterns.

With training, one can learn to subitize sets with a slightly higher number of objects, like six or seven, but this remains different from the subitizing of small sets. The effect can be measured exactly through the reaction time of test persons. Consider a person who is asked to determine as fast as possible the number of dots presented on a computer screen. Within the subitizing range, it takes only a reaction time of about one-half of a second until the test person starts giving the answer. This reaction time increases only a little from one to three dots, but after that it starts to increase significantly by roughly a quarter of a second for each additional dot on the screen. At the same time, the error rate increases with the increasing number of dots. This indicates that beyond three or four, one obviously relies on other mechanisms for determining the number, such as finding familiar patterns or explicit counting.

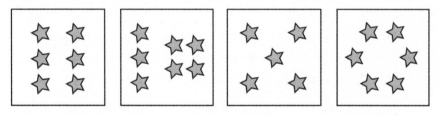

Figure 2.2: Pattern recognition may help to enumerate objects.

One might think that subitizing also has to do with pattern recognition, because two objects are always arranged in a line and three objects either form a line or a triangular shape, which is easily recognizable. You can see in figure 2.2 that pattern recognition indeed facilitates the task of determining the number. Here the stars are arranged either in familiar patterns or in quickly recognizable subgroups, so that it becomes much easier to tell the number without counting. Because there are so many possibilities for different spatial arrangements with higher numbers of objects, the probability to encounter familiar patterns decreases. Subitizing, however, does not need static patterns; it also works if the objects are moving and change places, but it seems not to work as well for sequentially presented stimuli (like drum beats). All this indicates its connection with the object-tracking system, which has the purpose to track simultaneously perceived objects through space and time.

2.3. THE APPROXIMATE-NUMBER SYSTEM

Although useful in some situations, an object-tracking system limited to the number four is certainly not sufficient. Very often one needs approximate knowledge of higher numbers. When a tribe of early humans met a rivaling group, they had to decide quickly whether they should stay and fight or run away if they were outnumbered by their enemies.

Consider the two collections of dots in figure 2.3. Can you say without counting which set is larger? When presented with this, and similar tasks, most adults will usually give the right answer. In figure 2.3, the collection of dots on the right is about 27 percent greater than the collection on the left—a difference that is well within the capacity of the approximate-number sense.

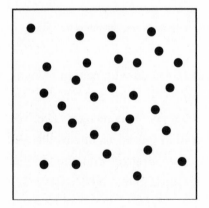 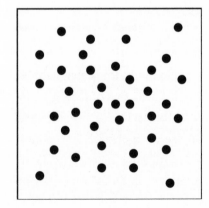

Figure 2.3: The approximate-number sense tells you which field contains more dots.

This strong intuition, the ability to estimate and compare approximate numbers, is another innate mental mechanism that contributes to our preverbal understanding of number. This second core-knowledge system is called the *approximate-number system*. Experiments have shown that it is already active in newborn babies, and it becomes more and more accurate with age and experience.

The approximate discrimination between the number of two sets can only be done when the two sets differ by a certain percentage. Babies can only distinguish two numbers when one of the sets is more than twice as large, while adults can discriminate with some certainty when one set is about 20 percent larger than the other. In all cases, the error rate decreases when the two sets differ by a higher percentage.

This shows that the approximate-number system follows the Weber-Fechner law, which applies to sensual stimuli in general. This states that the same impression of change is created when the stimulus is increased by a certain percentage. So if you are able to distinguish between the numbers 12 and 15 with an error rate of, say, 10 percent, then you will be able to distinguish between the numbers 120 and 150 with the same error rate. The Weber-Fechner law can also be stated as follows: One perceives the same difference between two collections when their numbers have the same ratio (e.g., $\frac{15}{12} = \frac{150}{120}$), or equivalently, if they differ by the same *percentage* (15 is 25 percent larger than 12, and 150 is 25 percent larger than 120).

Approximate knowledge of number is sufficient in most situations of everyday life. Sometimes, the approximate character is explicit. Whenever we say "about a dozen," we usually do not care whether it is 11 or 13. Even seemingly exact numerical information is often meant in an approximate sense. When you drive at 50 mph, it could be, in fact, 48 mph or 53 mph. In particular, large round numbers in statements like "this village has a population of 500" or "the galaxy has 400 billion stars," are automatically understood as approximate numbers. For large numbers, we have no sense of exactness. Would you expect that anyone can tell you the number of hairs on a particular dog? From our intellect, we know that the dog must have an exact number of hairs. But we don't really conceptualize that. To us, this number is a very fuzzy quantity. Moreover, the dog would continuously shed hairs and grow new ones, so the number would not remain constant even for a short time. We are perfectly happy with an approximate answer like "Could well be about 10 million," but we would have the same reaction if the answer is twice as much or only half as much. The concept of exact number is irrelevant for this question. If you had never learned to count and had to rely completely on your approximate-number sense, you would feel that way in view of much smaller numbers: Numbers beyond the subitizing range could be perceived only in a fuzzy or approximate way.

Today, there are indeed still tribes in the Amazon jungle that have never yet encountered counting. We know about these tribes from research done by psycholinguist Peter Gordon, who visited the Pirahã Indians, and Pierre Pica (1951–), who visited the Mundurukú to carry out experiments codesigned by the French neuroscientist Stanislas Dehaene (1965–). The Mundurukú have only number words up to five and, beyond that, words for "a few" and "many." They normally do not count at all and use the number words inconsistently, making occasional mistakes already with four or five items. They use their word for five, which literally means "a handful," also for six to nine items. They have never heard of addition or subtraction. This makes them ideal to test hypotheses about an innate number sense. Some simple experiments were done by presenting them with various "number games" on the screen of a solar-powered notebook computer. It turned out that they were able to compare large groups of dots as skillfully, and with about the same precision, as educated people integrated into a Western culture. For example, when presented with two groups of objects that were subsequently hidden, they could compare the sum with a third number shown to them. They obviously have an inherited capacity to understand how collections of objects behave in operations like taking away or joining. But, unlike people who have learned to count, they fail with exact arithmetic beyond the number 3, being able to give only approximate answers. Dehaene concludes that the Amazon natives share our innate number sense. This number sense already provides us with an arithmetic intuition that is sufficient to master the major concepts of arithmetic, like larger-smaller relations, addition, and subtraction. Number words are not necessary for understanding these concepts in an approximate sense.

The Pirahã studied by Peter Gordon are linguistically even more restricted than the Mundurukú. The Pirahã have number words only for *one* and *two*, which are probably also synonymous with *few* and *many*. They did not even master the bijection principle (i.e., one-to-one correspondence), because when the number of objects exceeded three, they could not compare quantities exactly by this one-to-one correspondence; they always did this by estimation. Some linguists believe that these people are forever unable to grasp the concept of a number beyond two or three because they lack the essential language tools. Without suitable number words, people cannot count.

In any case, verbal counting seems to help with the integration of the approximate number representation and the discrete number sense of the object-tracking system. At age three or four, children learning to count realize that each number word refers to a precise quantity—and, as Dehaene formulates it, this "crystallization" of discrete numbers, out of an initially approximate continuum of numerical magnitudes, seems to be exactly what the Mundurukú and Pirahã are lacking.

2.4. GOING BEYOND THE CORE SYSTEMS

It appears that the two core-knowledge systems—the object-tracking system and the approximate-number system—constitute our innate number sense. Like the uneducated tribesmen of the Amazon jungle, we would have these faculties even without culturally driven learning opportunities. Moreover, it has been shown that even babies and some animals do have these abilities.

The two core systems give us two quite different impressions of "number." The object-tracking system provides us with a precise mental representation of a small number of individual objects. This representation is discrete and tells us about the exact number, with 2 being perceived as fundamentally different from 3 or 1. This system gives us a precise mental model of what happens when we add or remove one item.

The approximate-number system, on the other hand, represents large numbers as a continuous quantity. It gives only an approximate and vague impression of number. There is no fundamental difference between 12 and 13, and the difference between 200 and 300 is rather a difference in the intensity of number perception, as it would be with other continuously varying quantities like size or density.

The inherited number sense represents a rather primitive knowledge and is a long way from the culturally refined understanding of number that children might have acquired already at the age of three or four. The core-knowledge systems, however, influence and guide later learning activities. Humans have the ability to go significantly beyond the limits of the core-knowledge systems and develop new cognitive capacities. For example, children in our culture soon learn to reconcile the two different impressions of number: They can apply the idea of discreteness provided by the object-tracking system to large numbers for which the approximate-number system only gives the vague feeling of a continuously varying quantity. Soon they realize that 12 and 13 are different in the same sense as 2 and 3. Even if they cannot count that far, they know that a large number, like 50, is changed by adding or taking away one item. Obviously, the idea of the discreteness of number can soon be applied to large collections.

One factor that might help in applying the idea of discreteness to large numbers is that the small numbers 1, 2, and 3 seem to be represented by both core-knowledge systems, so that we feel no discontinuity in our perception of numbers when they increase beyond the limit of the object-tracking system. Hence, for example, the idea of adding one item to obtain a new number can be easily transferred to higher numbers. Even monkeys trained to order small sets according to their size can generalize this ability immediately to larger sets of up to nine items. It appears, however, that the ability to think of larger numbers as discrete units is unique to humans and is not shared by animals.

But this still does not explain how children acquire these additional insights, and, consequently, this is a matter of ongoing research. It is probable that other core-knowledge systems help in this process—for example, systems related to social interaction and the ability to acquire language. In particular, language seems to be important for being successful in combining the mental representations from different core systems, like the discrete representation of small numbers and the continuous representation of large numbers. In this process, children develop a sense for the exact cardinality of arbitrary collections.

2.5. HOW WE LEARN TO COUNT

We have seen that children initially have only an approximate concept of large numbers. In order to develop the idea of exact large numbers, they have to break the limitations of their core-knowledge systems. The approximate-number system tells them that the numbers 12 and 13 are essentially indistinguishable, that they are the same. But they can track small numbers up to three or four exactly; here a difference by one creates a completely different sensual impression. Then, at the age of three or four, when they learn verbal counting, they also learn to combine these two concepts. Even if they cannot count very far, they understand that every number word designates an exact cardinality and does not apply any longer when a single object is removed from, or added to, the collection. There is no evidence for this type of human learning in animals.

Verbal counting according to a systematic numeral system is unique to humans living in a highly developed culture. Learning to count is nevertheless a complicated process with several stages, which is still a matter of ongoing research in mathematical cognition. It is of particular interest, and particularly rewarding, for parents to observe their own children in their individual approach to counting and their understanding of the number concept. Parents should help their children through that process because children who master all the hurdles early often have fewer, or even no, difficulties with mathematics in school. Based on well-known research results by the American mathematics-education professor Karen Fuson (1943–) in 1988, we will first consider the typical steps in the acquisition of verbal tools and number concepts.

When children learn to talk, roughly at the age of two, they also learn number words, which are first used without any understanding of cardinality. They learn to recite the sequence of number words "one-two-three-four-five . . ." like the words of a rhyme, as a single, whole word. Indeed, there are several nursery rhymes that are great for learning the number-word sequence:

> *One, two, three, four, five,*
> *Once I caught a fish alive,*
> *six, seven, eight, nine, ten,*
> *Then I let it go again.*

Gradually, children become more fluent in reciting the sequence of number words, but they do not yet use it to count. Next they begin to understand that the chain can be broken into individual words arranged in a particular order. They can start using the number-word sequence for counting as soon as they understand the rule "exactly one number word for exactly one object" (bijection principle).

Perhaps by the age of three or four, they can name the successor of a number—for example, say which word comes after *six*—without going back to the beginning and reciting the whole sequence starting from *one*. They begin to associate *smaller/less* with numbers that come earlier in the sequence and *larger/more* with numbers that come later in the sequence. This must also mark the beginning of the association of numbers with positions on a mental number line with a built-in direction. They are able to understand simple arithmetic and associate increase in number with "going forward" on the number line, and "taking away" with "going backward" on the number line or in the number-word sequence. So far, however, only the ordinal aspect of numbers is understood (ordinal principle).

Now comes the big step, perhaps happening at the age of four or five: As a result of increased experience in the game of counting, children begin to understand that numbers indicate not only a position in the counting sequence but also the (cardinal) number of the objects counted. So the number *four* not only is the fourth position reached during counting but also indicates that a group of four objects has been counted, and that this group also contains one, two, and three objects (cardinal principle). During this time, their skill in handling number words also improves: They can name the successor and the predecessor of a number, can recite the number sequence starting with any number, and are partially successful in counting backward.

But there are big individual differences, and there are three-year-old children who can count better than some five-year-olds. At the age of four, children of our culture can typically count to 10 and are learning to count up to 20. Beginning at about five years of age, they learn to understand the systematic and repetitive structure of the number words between 20 and 100. At this time, they do not need to memorize every single counting word and its position in the sequence, but they need to understand the rule according to which the number words are generated. Understandably, it always takes longer to learn how to count backward. With insight into the general structure of counting words comes the insight that the sequence of counting words never ends. For every counting word, one can produce a next one, just by following the general rule.

The integration of the ordinal and the cardinal aspect of numbers, the realization that the last-recited number word tells us about the numerosity of the set, is not achieved by all children without problems. This might well be a source of dyscalculia in school. When these children count a set, they still cannot answer the question "how many?" because they associate the last counting word like a name only, with the last-counted object and not with the collection of all counted objects. (As reported by Karen Fuson, they would point to the last toy car and say, "This is the five cars," instead of "This is the fifth car."[1]) When they answer the question "how many?" just by counting the objects again, this indicates that they consider number in the sense of the bijection principle. They just represent the set of toy cars by the corresponding number of counting tags. Instead of "five cars," it is "one-two-three-four-five cars." The number words are used just like tally marks on a counting stick.

When children understand the importance of the last counting word for the set as a whole, and that every number of the counting sequence describes the cardinality of the set of already-counted objects, they can start counting from every point of the number sequence. When asked to add five and three objects, they do not need to count every group separately starting with one; instead, they understand that the first result "five" denotes the numerosity of the first group, and they would continue counting the second group with "six, seven, eight" and give the answer *eight*.

Later, an understanding of differences in number is developed, and numerical relationship between the whole and its parts is understood. Starting from any number, the child now can count forward or backward without difficulty, and, accordingly, the child begins to develop an understanding for the "same distance" between subsequent numbers. And this development, normally reached during the first school year, also paves the way for more sophisticated strategies of calculation than simple counting. Failing to learn the relationship between the whole and its parts, and how a group can be decomposed into subgroups, can be another source of dyscalculia. A child needs to understand that a group of five can be decomposed (for example, into a group of two and another group of three). This is an important prerequisite for understanding computational strategies in arithmetic (for example, the observation that 8 plus 5 equals 8 plus 2 plus 3).

2.6. LOGICAL FOUNDATIONS FIRST?

According to an earlier psychological model by Piaget, certain logical faculties must have developed before it makes sense to teach numbers to children. Learning and understanding concepts (including the idea of number) develops through the active and constructive cognitive processes that our brain constantly performs. The goal of these processes is to harmonize the internal mental representations with sensual impressions.

According to Piaget, the cognitive development of children is not gradual but is marked by certain qualitative changes in cognitive abilities and logical understanding, which indicate that a new stage has been reached. At the age of six or seven, a child should reach the so-called concrete operational stage and have the logical faculties necessary for a working knowledge of numbers. According to Piaget, it makes little sense to teach numbers to children before that. The required logical insight would include an understanding of the concept of a set. One needs the ability to recognize eventual similarities between objects and to group them into sets of similar things that belong together, like a group of marbles or a group of persons. Piaget calls this process "classification." Next, one must be able to order objects, from first to last. One can order things simply by their position on the table, or from shortest to longest, or from smallest to largest, or according to some other criterion. This skill is called "seriation," in Piaget's terminology. A sense of invariance must be developed together with these faculties. For example, if the objects of a set are rearranged with larger distance between them, so that the set appears larger, the set will nevertheless contain the same number of objects. So the appropriate training for children to help them learn numbers would be to engage them in exercises involving invariance, classification, and seriation. An understanding of number would arise from their synthesis only when these concepts are fully mastered. This "logic-first" approach was clearly influenced by the strictly logical structure of mathematics. It was very influential in the second half of the twentieth century and was one of the motivating factors for the introduction of the "new math" movement in schools during the 1960s. Suddenly, children had to learn set theory instead of traditional counting. But, unfortunately, the logic of a child's development is quite different from the logic of mathematics, and this approach was later abandoned.

Later research has shown that, actually, the traditional way of learning mathematics wasn't a bad idea at all. The developmental stages are not as homogeneous and stringent as claimed by Piaget. Some abilities are actually acquired much earlier or are not really necessary in the initial use of numbers. The ordinal aspect of understanding numbers develops before the cardinal aspect, and verbal counting with ordinal numbers has a much higher importance than was admitted by Piaget.

But Piaget was right in that the child's own mental activity is the central component of learning. Numbers need not be explained; children use their own cognitive processes to develop abilities and understanding. The process of learning numbers is a very complex process of learning the series of number words, counting by associating an object in a row with a number word, and finally understanding the aspect of quantity and the whole-part relationship. It is not "logic-first" but "counting-first," through which a logical understanding is achieved with time.

2.7. THE NUMBER LINE

Experiments carried out by Stanislas Dehaene and his coworkers have revealed some interesting effects concerning the processing of numbers in the human brain.

When comparing two numbers according to size, the time it takes to decide which of two numbers is larger depends on their difference. For example, in a reaction test we all need longer to decide whether 6 is larger than 5 than it takes us to decide whether 9 is larger than 4. In the case of two-digit numbers, it takes longer to decide if 81 is larger than 79 than it takes to decide whether 85 is larger than 76 (even if in both cases one could make the decision by only looking at the tens digit). Dehaene calls this the "distance effect." It is more difficult to sort numbers according to size when they are closer together, and this fact cannot be changed even by extensive training. Another aspect of the distance effect is that 85 and 76 appear to be closer together than 15 and 6, although their difference is the same. This is an effect of the Weber-Fechner Law discussed earlier. When sorting two numbers, one has roughly the same reaction time and error rate when the two numbers differ by the same percentage (such as 9 and 12, versus 60 and 80).

An explanation seems to be that during learning numbers, the brain creates an analog representation of quantity in its neuronal network. This representation is a vague association of numbers with a spatial arrangement. Numbers appear to be arranged in space according to their size, so that 2 is "in between" 1 and 3, and likewise 12 is between 11 and 13. The idea of a spatial arrangement of number is a result of cultural learning. In Western culture, people usually arrange numbers increasing from left to right, the usual direction of Western writing, while people in cultures with another direction of writing also tend to arrange numbers in that direction. It is commonly believed that the activation of that mental number line is automatic and unconscious yet influences our perception of numbers.

The SNARC effect ("spatial numerical association of response codes"), discovered by Dehaene in 1993, provides another indication of the role of the number line in the mental representation of numbers. A testing subject has to answer by pressing a button either on his right side or on his left side, to decide as fast as possible whether a number that appears on a computer screen is even or odd. The response time shows a difference between answers given with the buttons on the right side and answers given on the left side. For small numbers, people are able to give the answer faster on the left side, and for large numbers they are faster on the right side. And this effect does not depend on whether the test takers are right- or left-handed or whether they provide the answer with their arms crossed. It is not the hand with which we give the answer, it is the side of the body where the answer buttons are located that determines the reaction time. On the left side, the side that is associated with small numbers, we are faster to discover properties of small numbers. With larger numbers, we have a shorter reaction time on the right side than on the left side because the right side corresponds better to our inner representation of numbers as being arranged on a mental number line. Dehaene believes that the mental number line is invoked unconsciously during the perception of any number, even when the number line is irrelevant for the task at hand, such as to determine whether a number is even or odd.

The effects of the internal spatial organization of numbers have often been considered. This internal spatial organization cannot be changed by training, and it is independent of the mathematical preparation of the test person. The mental number line is closely related to the ordering of numbers according to size. The placement of cardinal numbers in a spatial arrangement certainly makes it easier for an individual to harmonize the cardinal and ordinal aspects of number. Although the mental number line is deeply ingrained in our neuronal structure, and is often activated without conscious control, it is not innate but a result of cultural learning.

The mental number line as a neuropsychological effect has to be distinguished from the mathematical number line that we learn in school. Young children, if asked to draw the numbers on a line, where the end points are marked as 1 and 10, would produce something like figure 2.4, with 3 in the middle and the higher numbers placed closer together, with overlapping uncertainties.

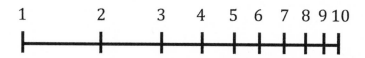

Figure 2.4: Mental number line.

Early in a child's development, the number line contains only vague information about the direction of getting larger. It will perhaps contain discrete points for one, two, and three, corresponding to the exact subitizing range of the innate number sense. For larger numbers, the number line becomes blurred, and the individual numbers appear to be the closer together as they get progressively larger, representing approximate quantities. Later, when the child learns to count, and an understanding of differences in number is developed, a conscious image of the number line is formed and gradually adjusted. When the child can count forward or backward starting from any number, the child begins to develop an understanding for the "same distance" between subsequent numbers. Older children would, therefore, tend to draw the numbers on a line with equal distances between numbers. This, of course, corresponds to the mathematical number line, for which a measuring tape would be a good model (figure 2.5).

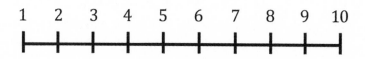

Figure 2.5: Mathematical number line.

The conscious perception of the number line is flexible and adaptive, depends on the cultural situation and on individual preferences, and changes through learning and experience, when the number concept is further developed. Not everybody prefers a linear arrangement. For some purposes, or in some intervals, numbers can be mentally arranged in a nonlinear way—for example, as on the face of a clock or sometimes even with colors as additional attributes.

2.8. EVOLUTION OF NUMERALS

Counting is a cultural invention. The beginning of systematic counting can already be seen at a very primitive level, based on the number two. It is reported that at the beginning of the twentieth century, some indigenous tribes in Australia, South America, and South Africa still had a number-word limit of two but were able to express numbers up to four using the scheme one, two, two-one, two-two. It would be easy to go beyond four with that method, such as two-two-one for five, and two-two-two for six. However, there seems to be no real need for this in a foraging society; hence, it is usually not done. It is commonly assumed that the language for counting was developed after people became sedentary. A nonterminating progression of counting words was not invented overnight. Rather, it was an intricate and long-lasting process, which took place in times of unrecorded history. Number words belong to the oldest parts of the vocabulary, and many languages still reflect some of the early encountered difficulties, which give indirect evidence of this development process. For example, the English words *eleven* and *twelve* are related to the Gothic *ain-liv* and *twa-lif,* which mean "one-left" and "two-left." This hints at an early stage in the development of a Proto-Germanic language, where ten was the upper limit of number words and people faced a situation in which, after counting to ten, one or two objects still remained.

The numbers one, two, three, and four play a special role in many languages. In social life, they correspond to the elementary ideas of "me/alone," "you/pair," "someone else," and "two pairs." Thus, they belong to the oldest words in any language, and they are the only number words that are occasionally changed according to the gender and case of the objects to which they belong. In Latin, the first four numbers (*unus, duos, tres, quattuor*) are declined like adjectives, while beginning with five (*quinque*) the Latin numerals are invariable. Even in today's German, *one* would be *ein* in the masculine form and *eine* if it refers to a feminine noun. Two and three were also inflected like real adjectives in Old and Middle-High German, but they have lost their variability in modern German. An old word for the masculine form of two in German is *zween*, which survived in English as *twain* and *twenty*. Today's German word *zwei* was originally the neutral gender, while the old feminine form *zwo* is still occasionally heard today in counting, but only for clarity of speech. Also, bear in mind that more than half of the words in the English language stem from the German language.

In English, the special role of one, two, and three is still seen in the ordinal words that are usually created by adding a *th* (like fourth, fifth, etc.). But the ordinal number words corresponding to one, two, and three are *first, second*, and *third*.

The special role of numerals up to three or four might well be related to the subitizing limit mentioned earlier. It is also visible in writing. In most writing systems, the numerals 1, 2, and 3 are derived from a symbol, which can be thought of representing one, two, or three fingers or counting sticks: The corresponding Roman numerals are I, II, and III. In Chinese, it is 一, 二, and 三. Beyond three, the Chinese have symbols of different origin—四, 五, 六, 七, 八, 九. It would be too difficult to distinguish and recognize at a glance symbols made of four, five, or six parallel lines. Therefore, other types of symbols, more useful for practical purposes, came into use for the representation of higher numerals. It is most likely that two and three parallel lines, which over time became connected in writing, also were the origin of our digits 2 and 3 (see figure 2.6).

Figure 2.6: Our digits 1, 2, and 3 probably evolved from the corresponding number of lines.

2.9. IDIOSYNCRASIES OF LANGUAGES

A useful sequence of number words has to follow some principles: For example, all words have to be unique. There should be no words used for numbers that sound the same, and refer to various numbers. Moreover, the counting sequence should be nonterminating. This can be achieved on the basis of a systematic, hierarchical method of grouping, as explained in chapter 1. The system of counting with the help of sticks, shells, pebbles, and the like, would then be reflected in the construction of number words. A general system of forming number words with base number 10 is shown in table 2.1. This system was invented in various parts of the world and is still in use in China. Although the English counting system is somewhat different, it is very similar to the system of table 2.1 in the case of higher numbers. For example, "nine-hundred-nine-ten-three" in table 2.1 is obviously the same as the English numeral "nine-hundred-ninety-three."

Table 2.1 shows a general scheme of creating number words using addition ("ten-two" means "ten plus two") as well as multiplication ("two-ten" means "two times ten"). Because the grouping is based on the number 10, this counting system needs number words for the basic digits from 1 to 9, and words for the higher units, such as ten, hundred, thousand, and so on. Higher numerals are typically formed by combining words from a higher rank (like "two thousand") with words from a lower rank (like "three hundred" and "sixty-seven") to form a new word ("two thousand three hundred sixty-seven").

one	two	three	. . .	nine	ten
ten-one	ten-two	ten-three	. . .	ten-nine	two-ten
two-ten-one	two-ten-two	two-ten-three	. . .	two-ten-nine	three-ten
. . .	. . .	. . .	. . .	. . .	. . .
nine-ten-one	nine-ten-two	nine-ten-three	. . .	nine-ten-nine	hundred
hundred-one	hundred-two	hundred-three	. . .	hundred-nine	hundred-ten
. . .	. . .	. . .	. . .	. . .	. . .
hundred-nine-ten-one	hundred-nine-ten-two	hundred-nine-ten-three	. . .	hundred-nine-ten-nine	two-hundred
. . .	. . .	. . .	. . .	. . .	. . .
nine-hundred-nine-ten-one	nine-hundred-nine-ten-two	nine-hundred-nine-ten-three	. . .	nine-hundred-nine-ten-nine	thousand

Table 2.1: A systematic number-word sequence.

Life would be easier for children learning arithmetic if the number words would always follow this strict building rule, as it is indeed the case in China, Japan, and Korea. The Chinese number system is built with an exceptional regularity and follows (almost) exactly the system of table 2.1. They have special words for the numbers 1, . . . , 10, and then for 100, 1,000, and 10,000, and they construct other numbers recursively: After counting to ten, they start again by adding the number words *one* to *nine* as a suffix to the base ten, until one reaches "two-ten," and so on, exactly as shown in table 2.1.

This scheme of table 2.1 combines multiplication and addition. Compare two-ten (meaning two times ten) and ten-two (meaning ten plus two). The multiplier is always a word of lower rank, set before a word of higher rank. A word with lower rank set behind the higher-rank word is meant additively. This is also the building principle for higher numbers, where some powers of 10 have no special name. While there is a special numeral for *ten thousand* in Chinese, there is no such word in English, and the numeral is constructed by the multiplicative principle (ten thousand = ten times one thousand). Similarly, *two hundred thousand* means two times hundred times thousand, but *two thousand hundred* means two times thousand plus hundred.

The English system of number words essentially follows this scheme, as far as larger numbers are concerned—as is the case with many other languages. But there are many exceptions for smaller numbers. In English, it is *twenty* and not "two-ten," and it is *thirteen* instead of "ten-three." It should be noted that in American English the usage of the word *and* anywhere in a numeral is discouraged ("five hundred seventy-eight," except when indicating a decimal point's placement, as in "fifty-six and three tenths"), whereas in British English it is common to say "five hundred and seventy-eight." Moreover, there are some irregularities in vernacular use, like "two-oh-seven" in American English instead of "two hundred seven" or "twenty-two fifty-one" instead of "two thousand two hundred fifty-one."

It can be assumed that more idiosyncrasies and exceptions lead to more difficulties for children to grasp the overall organization of the counting sequence, and to more difficulties in understanding the relationship of number words to their written form. And it makes arithmetic more difficult ("ten plus two is ten-two" would be easier to learn than "ten plus two is twelve"). Comparative studies have indeed shown that first graders in China, Japan, and Korea have a better understanding of the base-10 structure of the numeral system and of the place-value system in number notation than their American counterparts.

Similar deviations from the logical structure of the counting system occur in many languages, but, as in English, the exceptions are always restricted to the small numbers. Lower numbers appeared first in history, and they are often used on a daily basis. Therefore, they are often easily changed by idiomatic usage and manners of speaking.

For example, in Latin, subtraction appears in some places. The Latin word for nineteen is *un-de-viginti* (one-off-twenty) and the word for eighteen is *duo-de-viginti* (two-off-twenty). In Finnish, the words for numbers 11 to 19 are constructed by adding *toista*, meaning second. So the numeral for sixteen describes that this number has the sixth place in the second block of numbers (*kuusi-toista* = "six second"). In French, the numbers between 60 (*soixante*) and 100 (*cent*) reflect an old use of a *vingesimal* system (a numeral system with base-20). So they have no word for seventy, using *soixante-dix* (sixty-ten) instead. The word for eighty is *quatre-vingt* (meaning four-twenty), and ninety would be *quatre-vingt dix* (meaning four-twenty-ten). Italian has special words for 11 to 16 (*un-dici*, . . . *se-dici*) and then changes the building principle for 17 to 19 (*dici-assette, dici-otto, dici-annove*).

In German, there is a separate building principle for all words up to 100. The word for 23 would be *dreiundzwanzig*, meaning "three and twenty," which reverses the written order of digits. It also violates the general rule that words of lower rank set before a word of higher rank would imply a multiplication. Therefore, the addition is spelled out explicitly (the *und* in between *drei* and *zwanzig*).

As we have seen in chapter 1, a base-10 system, although predominant in our time, is not the only possible numeral system. Moreover, from the spoken numerals, it is still a long way from being a useful system of notation. In the next chapter, we will look into the history of various systems of writing numbers.

CHAPTER 3

NUMBERS IN HISTORY

3.1. NUMBERS IN BABYLON—THE FIRST PLACE-VALUE SYSTEM IN HISTORY

The first highly developed culture was in the land of Sumer, in southern Mesopotamia (today's Iraq). The first towns and cities there were built by the Sumerians more than five thousand years ago, and it was there that the earliest known writing in history appeared. It was probably developed to organize economic processes and record them in durable form. Also, writing was needed to facilitate administration that had become very complex, so that human memory alone became insufficient. The cuneiform script was developed from the first versions of writing using pictographic symbols. In the often-flooded areas between the Euphrates and Tigris rivers, there was plenty of clay available, and the Sumerians formed clay tablets, engraving symbols with a wedge-shaped tool in the still-soft clay. The name *cuneiform* means, in fact, "wedge-shaped" (Latin *cuneus*, or "wedge"). Gradually, the number of different symbols was reduced, and the cuneiform writing became less pictographic and more and more phonological, with symbols describing the sounds and syllables of the spoken language. When the Akkadians conquered Sumer in about 2300 BCE, the two cultures merged, and the Sumerian writing system came into use everywhere in Mesopotamia, where the Babylonian empire soon emerged. From the innumerable clay tablets that were created by Sumerians, Akkadians, and Babylonians, at least half a million tablets have survived until today. Among them, about four hundred have mathematical content or deal with mathematical problems. It is from these clay tablets that we know about Babylonian science.

It is most interesting that the Sumerians and the later cultures in Mesopotamia used a sexagesimal numeral system, which means that the base of the numeral system was 60, rather than the base-10 that we are all quite accustomed to in our daily lives. It is the only culture of the world where such a large base number was chosen.

It has been speculated that the reason for the use of a base-60 system was due to astronomy and astrology. Babylonian priests observed the precise locations of the planets, the moon, and the sun on the celestial sphere. The location of the sun with respect to the stars can be determined during the short period before sunrise, when the brightest stars are still visible. They found that the sun moves in the firmament along a great circle, the ecliptic, in roughly 360 days. This would be the origin of dividing a complete circle into 360 degrees. A circle is easily divided into six parts, each of which is 60 degrees, by marking off the radius length along the circumference (as in the construction of the hexagon). Sixty days roughly corresponds to two lunar periods and has some other nice properties as a unit—for example, it can easily be divided into 2, 3, 4, 5, and 6 parts. Therefore, it was chosen as the base of their number system.

This explanation might be appealing, but it has the disadvantage of being most probably wrong. As Georges Ifrah, in his book *The Universal History of Numbers*, has noted, it presupposes that a highly developed science like astronomy was available before the development of a number system.[1] However, a number system fulfills some very basic needs of an emerging culture, and all historic evidence indicates that a number system would be much older than any systematic observation of the sky. Also, a numeral system is not chosen according to sophisticated and abstract mathematical considerations, such as the number of divisors. People get accustomed to a grouping scheme by systematic use and cultural habit, in a phase of development that precedes any advanced knowledge of numbers. Ifrah hypothesizes that the Sumerian sexagesimal system had developed when two prehistoric groups of people merged, one accustomed to a quinary system (base-5), the other with a duodecimal system (base-12). As discussed in the first chapter, such a combination of base-12 and base-5 can be easily realized when counting with fingers. Pointing with the thumb of the right hand to the phalanges of the four opposing fingers lets you count to 12. Counting the groups of 12 with the five fingers on the left hand would give 60 as a natural unit that can be counted with the fingers of both hands and can be understood by people accustomed to base-5 as well as by people accustomed to base-12. Moreover, as Ifrah demonstrates, remnants of a base-5 system of counting seems to appear in the number words for 1 to 10 of the spoken Sumerian language.

A base-60 numeral system has the disadvantage that is poses a huge load on the human memory because one would have to memorize different names for all of the numbers from 1 to 60. But the Sumerians overcame this difficulty with the trick of using 10 as an intermediary base of their numeral system. So they named the numbers between 1 and 60 essentially by using the same principle as we do, combining special names for each multiple of 10 below 60 with the names of the numbers from 1 to 9. This amounts to using a decimal system for numbers up to 60. This also influenced the way that numerals were later written using cuneiform symbols.

The Akkadian-speaking people, who dominated the later Babylonian empire in Mesopotamia, were of Semitic origin, and they were used to a decimal numeral system. They continued to use it, but in writing they first adopted the Sumerian base-60 system and only slowly transformed it for everyday use to a notation that was better adapted to oral number names with a decimal structure. This process was facilitated by the Sumerian use of base-10 as an intermediary system for the numbers up to 60. But the base-60 system continued to be used by Babylonian scholars throughout the second and first millennium BCE.

On the basis of the Sumerian numeral system, the Babylonian astronomers and mathematicians developed, probably around 1900 BCE, a very advanced method of writing numerals in cuneiform script. This written numeral system was the first place-value system in history. They used only two different cuneiform symbols, a vertical wedge 𒁹 meaning 1 and a chevron 𒌋 meaning 10. From these basic symbols they created composite symbols, which played the role of "digits" with values from 1 to 60. These digits were created by using as many chevrons and wedges as needed. For example, 56 would be written as in figure 3.1:

Figure 3.1: The "digit" representing the number 56 in cuneiform script.

In this manner, a symbolic notation was created for all the basic numerals from 1 to 59. In a place-value system, larger numbers are written by positioning the basic numerals next to each other in a row. This is the way we currently combine our digits. The Babylonian "digits" thus had a place value. The single wedge 𒁹 could mean 1 or 60, or even 60 × 60 = 3600, depending on its position in a numeral. For example, the chain of cuneiform symbols in figure 3.2 combines the "digits" for 12, 35, and 21 in a single numeral.

Figure 3.2: The number 45321 in cuneiform script.

The numerical value of the combined numeral is obtained by adding the place-values of each "digit." Thus the number represented in figure 3.2 is $12 \times 60^2 + 35 \times 60 + 21 = 45321$.

At first sight, this might seem like a complicated way to represent a number. But, actually, it is quite familiar to us. Whenever we measure times in hours, minutes, and seconds, we follow exactly the same technique. Assume that the number in base-60 shown in figure 3.2 represents a time measured in seconds, and its meaning becomes quite clear. The next sexagesimal unit would be 60 seconds (a minute), and 60 minutes gives an hour. Hence the number in figure 3.2 simply means 12 hours, 35 minutes, and 21 seconds. For us, this is much easier to interpret than in the decimal system as 45,321 seconds. But we are not very comfortable in using sexagesimal units. The next higher unit, 60 hours, would be a unit of two and one-half days, for which we have no name, so this is where our system for time measurement deviates from the sexagesimal system.

For a long time the Babylonians had no "zero," that is, no symbol to indicate the absence of a unit in a number, which certainly posed some problems. If we had no symbol for zero, we could not distinguish 1 from 10 or 100. So whether a wedge Ⅰ meant 1 or 60 or 3600 had to be inferred from the context. If a unit of a given order of magnitude was missing in the middle of a numeral, they sometimes left a blank space at the corresponding position. It was probably in the third century BCE that a symbol for zero emerged, invented by Babylonian scholars. In figure 3.3, the symbol ⟍⟍ denotes the absence of the position for $60^2 = 3600$.

Figure 3.3: The number 1 × 60³ + 0 × 60² + 54 × 60 + 23 = 219,263.

It has to be mentioned that the Babylonian system was also used to express sexagesimal fractions in the place-value system. In our decimal system, a decimal fraction would be written, for example, as 1.11, which means $1 + \frac{1}{10} + \frac{1}{100}$. Whenever the digit 1 appears behind the decimal point, it could mean $\frac{1}{10}$ or $\frac{1}{100}$, and so on, depending on its position. This is completely analogous to the Babylonian base-60 system. The only problem was that the Babylonians did not invent something like a decimal point. What the writer had in mind had to be guessed, which was not always easy. The chain of symbols in figure 3.2, which we interpreted as 12×3600 + 35×60 + 21, could also mean 12×60 + 35 + $\frac{21}{60}$, or $12 + \frac{35}{60} + \frac{21}{3600}$, and so on. It had to be decided from the context which variant was intended by the scribe. This certainly required an increased attention and logical thinking on the part of the reader, and it was also a source of errors.

The abstract Babylonian way of writing numerals had a profound influence on the scholars of antiquity. Greek astronomers, although used to a decimal system, translated the cuneiform script into their own "alphabetical" way of writing digits. However, they adopted the Babylonian system, in particular, for expressing negative powers of 60. It would have been too much work to convert thousands of astronomical tables into a decimal system. It is for that reason that we still measure units of time as well as units of angle in a sexagesimal system, dividing hours and degrees into minutes and seconds.

3.2. NUMBERS IN EGYPT—THE FIRST DECIMAL SYSTEM IN HISTORY

Symbols for writing numbers appeared in Egypt at about the same time as in Mesopotamia, about 3000 BCE. As was the case in Mesopotamia, Egyptian mathematics developed out of practical needs. Mensuration; redistributing land after Nile floods; planning irrigation channels, pyramids, and temples; computing wages and taxes—all these tasks became so complex that human memory and verbal means alone became insufficient, and the need arose to record in written form words, orders, accounts, inventories, censuses, and so on. The Egyptian symbols were called *grammata hierogluphika* ("carved sacred signs") by the Greeks, from which the common name "hieroglyphs" is derived. Initially, hieroglyphs were pictograms or ideograms (symbols representing a word or idea) and later evolved into a representation of sounds (consonants). Hieroglyphs were either carved in stone monuments or written on papyrus, a paperlike material made from a grasslike plant (*Cyperus papyrus*) that grew to a height of three meters in the Nile Delta. In the dry climate of Egypt, papyrus lasts for a long time, and numerous documents have survived until today. We know about Egyptian mathematics essentially from a few papyri with mathematical content, written toward the end of the Middle Kingdom (about 1700 BCE) in hieratic script. The hieratic script consists of symbols that are late forms of hieroglyphs, obtained from them through a process of continuing simplification and schematization. The papyrus Rhind, written by the scribe Ahmose, contains eighty-five mathematical problems. This collection of exercises on geometry and arithmetic probably served to introduce other scribes to the art of mathematics and computation. Other famous papyri devoted to mathematics are the papyrus Moscow and the mathematical leather roll, which is now at the British Museum in London.

The Egyptians used a base-10 system. From the very beginning, they could write very large numbers, with special hieroglyphs for 10, 100, 1000, and so on, up to one million (see figure 3.4).

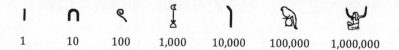

Figure 3.4: Hieroglyphs for powers of 10.

These symbols are for when writing from left to right. The symbols get flipped horizontally if the line containing the numeral is to be read from right to left.

It is very easy to understand how numerals were formed from these basic symbols. The numeral system was not a place-value system, but—very similar to the later Roman numeral system—based on addition. They simply repeated the corresponding symbol as often as needed. For example, the number 2578 would appear as in figure 3.5.

Figure 3.5: The number 2578 in hieroglyphs.

The Egyptians had no symbol for zero, and it was not needed to write numbers unambiguously.

3.3. ARITHMETIC IN EGYPT

The Egyptian way of doing calculations was based on reducing everything to addition. This is because addition of two numbers is so easy to do with this kind of numeral system. Consider for example, the computation in figure 3.6. In order to add the two numbers 2578 and 1859, you would just collect the symbols of the same kind and replace a group of ten symbols with the symbol of the next-higher order. The result 4437 can be easily read off:

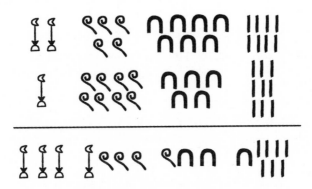

Figure 3.6: Adding two numbers.

Multiplying a number by 2 is easy because we only have to add the number to itself. Multiplying a number by 10 is even easier, because one only has to replace each symbol in the given number by the symbol of the next-higher order. Multiplying by any other number was reduced in an ingenious way to the task of addition and multiplication by 2. This is best explained by the following example:

If an Egyptian scribe wanted to multiply 12 by 58, he would create a table with two columns, starting with 1 and the multiplier 12, and double each entry in successive rows, as shown in table 3.1:

1	12	
2	24	✓
4	48	
8	96	✓
16	192	✓
32	384	✓

Table 3.1: Multiplying 12 by 58.

He would stop when reaching 32 in the first column, because the next doubling would lead to a number larger than 58 (the larger of the two numbers being multiplied) in the first column. The numbers in the second column are just the corresponding multiples of 12—for example, 192 = 16 × 12. Next, the scribe would mark all lines for which the sum of the numbers in the first column would give 58. This appears to be difficult at first sight but can be easily done as follows. We start from the bottom and mark the last line. Then we add the number 16 from the line above to 32, which gives 48. This is less than 58, hence we mark this line as well. We continue in this way, by adding the numbers in the first column and marking the lines, as long as the result is less than the desired multiplier. In our example, adding 8 to the previous result gives 56, which is still less than 58, therefore we mark this line. Adding 4 would give us 60, which would be too large. Therefore, we omit the line beginning with 4. Finally, adding 2 to 56 gives us the sought-after multiplier 58. Therefore, we mark the line beginning with 2. The first line is not needed. We have now marked the lines with 2, 8, 16, and 32 on the left side, and indeed 2 + 8 + 16 + 32 = 58. Now, to get the product of 12 × 58, you just have to add the numbers from the second column of the marked rows as follows:

$$12 \times 58 = 24 + 96 + 192 + 384 = 696.$$

You might wonder if this method always works. Is it indeed possible to write every multiplier as a sum of the numbers in the first column of the table? Yes it is! The reason is that the numbers in the first column are just powers of 2. Hence, what is effectively done with this method is to express the multiplier in the binary system, as a sum of powers of 2:

$$58 = 1 \times 2^5 + 1 \times 2^4 + 1 \times 2^3 + 0 \times 2^2 + 1 \times 2^1 + 0 \times 2^0 = 111010$$
(in base-2)

For example, if we want to know 12 × 45, we would have to express 45 as a sum of powers of 2, that is, as a sum of the numbers in the first column of that table. We find that 45 = 1 + 4 + 8 + 32.

We then take the corresponding numbers on the right side of the table, which are then added to give the result:

$$12 \times 45 = 12 + 48 + 96 + 384 = 540.$$

We see that multiplication can be done without memorizing any large multiplication tables, except doubling numbers.

Division can be done in an analogous way, as long as the division is without a remainder. For example, in order to divide 636 by 12, one would start with the same table as before (see table 3.2).

✓	1	12
	2	24
✓	4	48
	8	96
✓	16	192
✓	32	384

Table 3.2: Dividing 636 by 12.

However, now we would try to write 636 as a sum of the numbers on the *right* side. One finds, that, indeed, $636 = 384 + 192 + 48 + 12$. We then mark the corresponding lines, this time on the left side. Adding the numbers next to the check marks gives us the number of twelves that fit into 636, namely $1 + 4 + 16 + 32 = 53$. Hence $636 \div 12 = 53$.

This method only works when the division is without a remainder. The method is considerably more complicated in general cases and involves fractions. The Egyptians knew fractions, but only of the form $\frac{1}{n}$ with the numerator 1, the so-called unit fractions. The only exceptions were the very frequent expressions $\frac{2}{3}$ and $\frac{3}{4}$, for which they had their own symbols.

In order to denote a unit fraction, they wrote a mouth symbol ⬯ over the corresponding number (see figure 3.7 for examples).

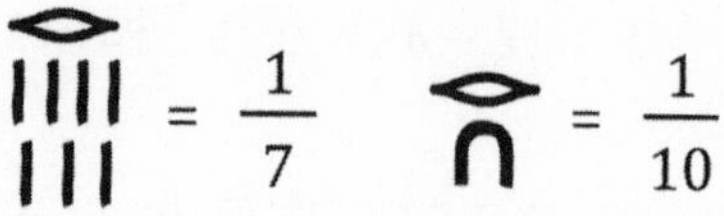

$$\overset{⬯}{\underset{|||}{||||}} = \frac{1}{7} \qquad \overset{⬯}{\cap} = \frac{1}{10}$$

Figure 3.7: Egyptian unit fractions.

The preference for fractions of the type $\frac{1}{n}$ is probably due to their method of division. For example, when they wanted to divide 23 by 16, they would try to write 23 by adding fractions of 16. A corresponding table is table 3.3:

✓	1	16
	$\frac{1}{2}$	8
✓	$\frac{1}{4}$	4
✓	$\frac{1}{8}$	2
✓	$\frac{1}{16}$	1

Table 3.3: Dividing 23 by 16.

Checked are the lines for which the sum of the numbers on the right yields 23. Adding the corresponding fractions on the left side provides the result

$$\frac{23}{16} = 1 + \frac{1}{4} + \frac{1}{8} + \frac{1}{16}.$$

This represents the result as a sum of unit fractions. Again, this is a special case and the Egyptians have found more complicated methods to deal with general cases. For all fractions of the form $\frac{2}{n}$ they used tables that expressed these as a sum of unit fractions. An interested reader may wish to investigate these extensions further.

3.4. NUMBERS IN CHINA

Egyptians, and similarly the Greeks and Romans, used an additive principle for writing numerals. That is, the symbols for one, ten, hundred, and so on are repeated as often as necessary to represent a number. Consider, for example, the Roman numeral MCCCXXIII. It has one symbol for "thousand" and repeats the symbol for "hundred" three times, and the symbols "ten" and "one" as often as it is necessary to represent 1323. The same number in Egyptian hieroglyphs is shown in figure 3.8, illustrating the same "additive" method of constructing numerals.

Figure 3.8: Egyptian and Roman additive system.

About three thousand years ago, the Chinese went a step further and developed a multiplicative-additive scheme. In its present form, this method of writing numerals has the number symbols shown in table 3.4:

1	2	3	4	5	6	7	8	9
一	二	三	四	五	六	七	八	九

10	100	1000	10000
十	百	千	万

Table 3.4: Chinese number symbols.

As with all other Chinese symbols, these signs represent words. They are just written forms of the spoken numbers, not separate kinds of symbols. Hence 七 corresponds to *seven* rather than *7*.

In spoken language, the Chinese use a numeral system, essentially following the construction principle shown in table 2.1 in section 2.9. The written numerals are just the translations into a written form of the spoken number words. Additionally, the verbal scheme uses multiplication and addition at the same time.

Placing one of the symbols for 1 to 9 after one of the symbols representing the powers of 10 implies addition:

$$十五 = 10 + 5 = 15, 千五 = 1000 + 5 = 1005.$$

Placing one of the symbols representing the numbers 1 to 9 before the higher units indicates multiplication:

$$五十 = 5 \times 10 = 50, 五千 = 5 \times 1000 = 5000.$$

You can see that the Chinese system is different from our written numeral system, but nevertheless is very similar to our way of pronouncing numerals.

For other Chinese numerals, the principles of addition and multiplication are combined, as we do in spoken language. Hence, it is easy to form longer number words. For example,

5724 = 五千七百二十四
(= five-thousand seven-hundred two-ten four).

For even higher numbers, the Chinese used the symbol for ten thousand as a new unit. Hence, the numeral for five million would have been (using the multiplicative principle)

5,000,000 = 五百万 (= five-hundred ten-thousands).

We would probably confuse this with 510,000, but it actually denotes a quantity of five hundred "ten-thousands"—that is, $500 \times 10,000 = 5,000,000$. Similarly, 一万万 would be $1 \times 10,000 \times 10,000$, which is one hundred million. This method of writing numbers is still in use today. We can see that a symbol for zero is absolutely not needed to represent a number unambiguously.

3.5. THE CHINESE PLACE-VALUE NOTATION

About two thousand years ago, during the Han dynasty, the Chinese developed another numeral system, based on the representation of numbers on a counting board. A counting board was an early calculator— a sort of checkerboard with square fields in which counting rods (little sticks made of bamboo and sometimes even of ivory) were arranged to symbolize numbers. It was easy to rearrange the rods in a field in order to represent different numbers, and this had to be done frequently during a calculation. Later, the counting rods found their way into writing, in two different, closely related forms: one in a vertical layout and one in a horizontal layout. The numerals in the vertical layout are shown in figure 3.9, the numbers in the horizontal layout are shown in figure 3.10.

Figure 3.9: Rod numbers in the vertical layout.

1	2	3	4	5	6	7	8	9

Figure 3.10: Rod numbers in the horizontal layout.

These rod-number symbols were either written or represented with counting rods placed in the square fields of a counting board. Here, the rightmost column symbolized the number of units, the next column to the left contained the tens, and then the hundreds went into the next column, and so on.

In principle, the number 2345 could now be represented, for example, by using rod numbers in the vertical layout, putting II III IIII IIIII into adjacent squares, but you can see the problem that this could cause, if by chance one of the sticks slid over into the next square. This would create the risk of confusion, changing the configuration, for example, to II III IIIII IIII, or 2354. The solution was simple and elegant—namely to arrange the sticks alternatingly in vertical and horizontal layout, as shown in figure 3.11. Typically, one would start with the vertical layout for the units, followed by the horizontal placement for the tens, and so on.

2	3	4	5
=	|||	≡	|||||

Figure 3.11: Rod number notation.

The rod numerals are, thus, written in a place-value system. The value of a "digit" depends on the column in which it is written. As long as the digits were placed within square fields, there was no need for a special symbol for zero because the square corresponding to a missing digit was simply left empty, as in figure 3.12

2	0	4	5
=		≡	|||||

Figure 3.12: Empty space instead of "zero."

The written notation often omitted the squares around the digits, and they were moved closer together. The missing symbol for zero usually was not a problem because two adjacent symbols in horizontal (or vertical) layout would indicate a "missing digit" in between. The symbol for zero was introduced to China in the eighth century through the influence of Indian scholars. Still, the alternating vertical-horizontal layout was kept as in ▮▦▮, which means 106929. Rod numbers in this style were in use for many centuries, not only in China, but also in Japan, Korea, and Vietnam. Figure 5.10 in section 5.9 shows a Chinese example of the use of rod numerals from the thirteenth century CE.

3.6. NUMBERS IN INDIA

In old India, about 1,500 years ago, a revolution took place that still influences our life today. Indian scholars invented a decimal place-value system with a concept of the number zero—not only as a symbol, but also as a quantity that could be used for counting and for calculations. It was not the first place-value system, and not the first decimal system, but it is the one that is still in use today.

India is a large country with many languages and subcultures; hence the language of the scholars has always been Sanskrit. Being scientists and poets at the same time, Indian scholars expressed everything in verses, even purely mathematical results—making use of the many verse meters available to them to reduce monotony. Moreover, they used an obscure and mystic way of speaking that is often unintelligible. It appears that the pure facts and methods were well known to them and were passed on by oral tradition. The written text just served to aid memory without providing too many details.

In particular, astronomy was a highly developed science, and much of present-day trigonometry was developed in the centers of astronomical research, in Ujjain and Pataliputra (Patna). Indian astronomers had a need for large numbers and developed a deep fascination for them.

At the beginning of the first millennium CE, the numeral system that was in use in India was not a place-value system. It was, actually, rather limited for scientific purposes and did not allow the representation of large numbers. It was also somewhat impractical because it contained too many different symbols. For example, they had special symbols for each of the numbers 10, 20, 30, . . . 90, 100, 200, 300, . . . 900, and so on. For us, this numeral system is of interest because it had digits for the numbers 1 to 9 that were "graphically designed" for easy distinguishability. There were numerous different writing styles for these symbols. A sample set (adapted from Ifrah) is shown in figure 3.13.[2] These symbols are predecessors of the digits that are in use today.

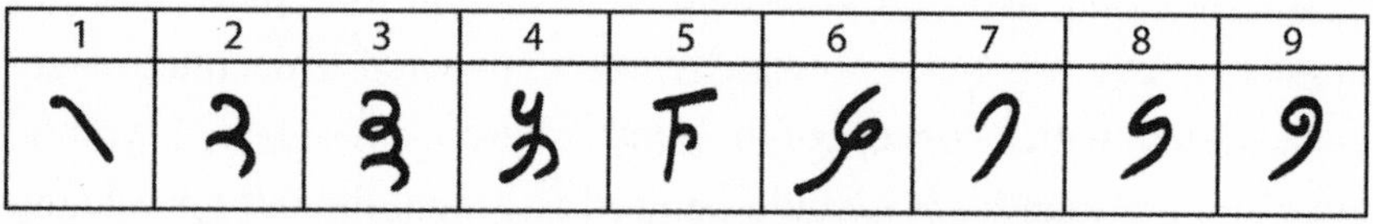

1	2	3	4	5	6	7	8	9
╲	३	३	ૠ	F	૭	૧	૬	૧

Figure 3.13: Digits derived from Brahmi writing.

In written text, however, Indian scholars did not use these symbols at all. Instead, they used the number words of Sanskrit to express numbers verbally. In addition to the number words for one to nine, Sanskrit has a number word for every power of 10 (see table 3.5).

éka	one		dasa	ten
dvi	two		sata	hundred
tri	three		sahasra	thousand
catúr	four		ayuta	ten thousand
pañca	five		laksha	hundred thousand
ṣáṣ	six		niyuta	one million
saptá	seven		krore	ten million
aṣṭá	eight		vyarbuda	hundred million
náva	nine		padma	one billion

Table 3.5: Sanskrit number vocabulary.

The sequence of number words for powers of 10 could even be continued to very high powers of 10, up to 10^{53}.

In the verbal description of a number, the scholars would name the units first and proceed to higher powers of 10, which is just the opposite of our present-day custom. For example, the numeral 4567 would have been expressed by saptá ṣáṣti pañcasata catúrsahasra (meaning "seven, sixty, five hundred, four thousand").

Very long numbers (for example, those used in astronomical texts) would therefore result in very long chains of words, which were difficult to incorporate into a verse. Probably at the beginning of the fifth century CE, a great idea evolved that helped to represent numbers more efficiently. In order to shorten the verbal expression of very long numbers, they did not pronounce the powers of 10 any longer and just kept the names of the units. In the example above, they would just say "seven-six-five-four." It was not necessary to say more, because the number words were always given in the same strict order of increasing powers of ten, and the Indian writers were, indeed, very conscious about this and had a special word, *sthanakramad* (meaning "in the order of position"), to describe this. So the spoken system was actually a place-value system.

In the written numeral place-value systems of Babylon and China, missing orders of magnitude had to be expressed by leaving a space. But this is not possible in a spoken system. So they put the word *śūnya*, which means "void," in place of the missing number. For example, they said *dvi-śūnya-tri* to distinguish 302 from *dvi-tri*, meaning 32. The oldest historical record of this "verbal positional system with zero" seems to be the Lokavibhāga, a treatise on cosmology, dated to 458 CE.

From the point of view of poetry, there was still a problem. Some results required a frequent repetition of one of the "digits." For example, one of the hypothetical cosmic cycles of Indian astronomy, the mahāyuga, lasted 4,320,000 years. This number would have to be represented by the following:

śūnya śūnya śūnya śūnya dvi tri catúr (void-void-void-void-two-three-four),

which is rather dull reading. So they had the idea to replace individual number words by other words, such as a verse metric often motivated by their aesthetic feeling for poetry. Each number word acquired many substitute words, which were also associated with that number. For example, the word *śūnya*, meaning "void" or "empty," could be replaced with the word for sky, atmosphere, or space, to name a few. As an example from a text of the year 629, Georges Ifrah reports the following:[3]

viyadambarakasaśūnyayamaramaveda = 4,320,000

This term consists of the words *sky-atmosphere-space-void-Yama-Rāma-Veda*, meaning

0-0-0-0-2-3-4.

In Hindu mythology, Yama, who later became the Hinduistic god of death, lived on Earth together with his twin sister, Yami, as the primordial couple. Thus, his name was associated with "two." Rāma was associated with "three" because of three famous persons with that name (two of them incarnations of Vishnu, and the third being the hero of the epic saga Rāmāyana). Veda is used as a word for *four* because the Veda, the main religious text of Hinduism, consists of four principal books.

There are many different words that could be used to express any number from zero to nine. In that way, even boring numbers could be turned into poetic verses. Texts written with that representation of numbers presented no problem for the Indian astronomers but were almost unreadable for the uninitiated. For the scholars, the verbal style even added to clarity because the Indian numeral symbols evolved in many different ways in various parts of India. Moreover, large differences in handwriting sometimes made the numeral symbols even more difficult to interpret than the verbal expression. Using the poetic representation, even little errors would have been noted because the substitution of a wrong word would probably have disturbed the rhythm of speech (that is, the metric of the verse).

3.7. SYMBOLIC NUMBER NOTATION AND THE ABACUS IN INDIA

The Indian verbal place-value system was not suitable for doing calculations. As with the Chinese, the Indians used the counting board, or an abacus, for practical calculations. The abacus is particularly important for the development of the place-value system. It was in use throughout the antique world. Long before it became the familiar counting frame with beads sliding on wires, it was just a board with vertical columns. Perhaps the simplest form is a sand abacus, a plane surface with fine sand on it, on which vertical lines are drawn with a pointed tool to separate the columns. Initially, numbers were represented by putting appropriate numbers of sticks or pebbles into the columns. (Recall the method of counting using pebbles, sticks, bones, shells, etc., as described in section 1.10.) The new idea here is that it is not necessary to use different kinds of objects in order to represent different orders of magnitude. Instead, one uses only one kind of object—for example, little pebbles (called *calculi* by the Romans). The order of magnitude is then represented by the column in which the pebbles are placed. Figure 3.14 shows, as an example, the same number as the one in figure 1.8, but this time realized on a sand abacus.

Figure 3.14: The number 1776 on a sand abacus.

The older method of figure 1.8 has the advantage that you can put everything into a bag and the number would still be preserved properly. The abacus, on the other hand, is made not for storing numbers, but for doing calculations. For example, when adding two numbers, you would simply add the pebbles in each column, and, as soon as the number in one column exceeds 10, you would remove 10 pebbles from that column and add one pebble to the next column.

The columns of an abacus, holding the ones, the tens, the hundreds, and so on, actually realize a place-value system: The first and rightmost column usually contains the units, the next column to the left represents the tens, the next column the hundreds, and so on. Quite early, the Indian scholars had the idea to write the symbols for the digits into the columns instead of placing pebbles or sticks. Thus, the number 1776 of figure 3.14 would then appear as in figure 3.15.

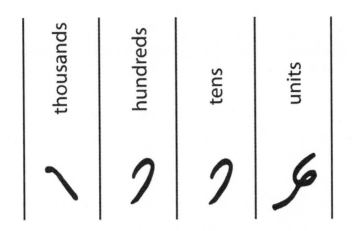

Figure 3.15: The number 1776 with Indian symbols written on an abacus.

Over time, the symbols used for calculations with an abacus also appeared in the writings of scholars. To make things more complicated, whenever the symbols appeared in a text (written from left to right), the order of the digits was reversed, in order to match the way the number words were written—starting with the units. Thus, the number 1776 would appear as 6771 in written form. An even more important difference between numbers on an abacus and numbers in writing has to do with the representation of zero. With an abacus, there was no need for a symbol denoting zero, because a missing order of magnitude would have been represented by an empty space. But in a written text, the symbols provided an abbreviation of the verbal number representation in Sanskrit, which frequently contained the word *śūnya* or one of its equivalents to denote the void, the absence of a "digit." Probably around the year 500, Indians had the idea to represent śūnya by a special symbol, a dot or a small circle (which later became our symbol 0). Still, the direction was reversed: 3200 would have been something like "sky-atmosphere-Yama-Rāma," or 0023, if written in symbols.

On an abacus, calculations were performed by writing digits in the columns and manipulating them according to very complicated rules that took a long time to learn. At the beginning of the sixth century, it was recognized that the symbol for zero could also be of use when performing calculations. The advantage was that the columns of the abacus were no longer needed, because the symbol for śūnya could be used for the empty column. The familiar calculations and manipulations of numbers could be done directly with the digits, without having to draw the columns first, because their value is uniquely determined by their position within the number. So mathematicians could do all computations in the same way as on the abacus. Hence, they became more and more used to numerals written in abacus-notation—with the rightmost digits representing the units. As a consequence, the direction of writing numerals was also changed to reflect the order of digits made familiar from the abacus. And now, as a result of a long development, the number 4,320,000 years in the cosmic cycle mahāyuga, finally, would have been written as shown in figure 3.16, which is the same way we do it today.

Figure 3.16: The number 4,320,000 in "modern" notation.

However, the notation enables a smooth transition from the abacus to written form. For example, the large repertoire of algorithms for doing calculations, which have been developed over centuries for the abacus, would then immediately be transformed into written form. The fact that the numbers and the calculations became independent of the abacus had important consequences. The digit zero gradually became a *number*. It was used not only as a symbol for an empty column of an abacus but also as something with which one can perform calculations, such as $5 - 5 = 0$, or $5 + 0 = 5$, or $5 \times 0 = 0$. Another important side effect of transferring calculations from an abacus to paper (or, rather, sheets of birch bark) was that intermediate results did not have to be erased, so it was easier to track down errors or rethink the applied methods in the search for simplifications. The complicated methods and rules used for calculations with an abacus could be gradually simplified; calculations could be done faster and more effectively. All this had an enormous impact on the mathematical sciences and astronomy, which in the following centuries experienced an unprecedented boom in India.

This was the state of art when Brahmagupta (598–668 CE), one of India's greatest mathematicians, was the head of the astronomical observatory of Ujjain. In the year 628 he wrote the famous book *Brahmasphutasiddhanta* ("Treatise of Brahma"). It used the decimal system, described the role of zero, and formulated, in particular, precise rules for doing calculations with zero. Moreover, Brahmagupta had rules for dealing with negative numbers and methods for computing square roots, solving equations, and much more. During the reign of the caliph al-Ma'mun (786–833 CE), Brahmagupta's text found its way to Baghdad and was translated into Arabic. Thus the information about the Indian numeral system spread from India into the expanding Arab world, where its ingenuity and importance was quickly recognized. In 825, Persian mathematician Muḥammad ibn Mūsā al-Khwārizmī (ca. 780–850), a scholar in the House of Wisdom, in Baghdad, wrote a book titled *On the Calculation with Hindu Numerals*. In Europe, the name al-Khwārizmī later developed into the word *algorithm*. Another book by al-Khwārizmī, titled *al-Kitab al-mukhtasar fi hisab al-jabr wa'l-muqabala* ("The Compendious Book on Calculation by Completion and Balancing") contained the word *al-jabr* in its title, which is the origin of the word *algebra*.

Thus, the Indian numeral system spread quickly throughout the Arab world, where culture and science were highly esteemed. Europe, at the same time, was in a period of economic decline and was culturally and scientifically backward. Thus it took another five hundred years until the Indian digits and numeral system finally reached Europe.

3.8. THE SLOW RECEPTION OF THE HINDU-ARABIC SYSTEM IN EUROPE

In medieval Europe, mathematics had no part in the general knowledge, not even of learned people. Thus, the performance of simple arithmetic tasks was a matter for specialists. It was done by professionals, who did calculations for a living with the help of an abacus in the Roman tradition. The results were communicated with the help of Roman numerals, which were predominantly used throughout the Middle Ages.

A first chance to introduce Hindu-Arabic numerals to Europe came toward the end of the first millennium. The French monk and mathematician Gerbert d'Aurillac (ca. 946–1003 CE) was an important scientist of his time, and during a long visit in Spain, where the medieval Moors had established a large Islamic cultural domain, he studied mathematics from the Arabic scholars. According to legend, he traveled to Seville and Cordoba, gaining access to Islamic universities in the disguise of an Islamic pilgrim.

Gerbert later became the teacher of Emperor Otto III. In the year 999 he was elected to succeed Pope Gregory V. As pope, he took the name Silvester II. It was the only time in history that a leading mathematician became the pope.

Gerbert made the symbols 1 through 9 known as symbols on an abacus, but despite his influence he did not succeed in popularizing the Hindu-Arabic algorithms or the use of zero and the place-value system. The reception of the Hindu-Arabic numeral system met with considerable resistance from the Catholic Church and the conservative accountants. In some places the resistance lasted until the fifteenth century. Thus, it was the conservatism of medieval Europe and the Church that effectively blocked the early introduction of Hindu-Arabic mathematics to Europe. The nine Hindu-Arabic digits became known as "Arabic digits" among professional calculators (albeit without the symbol for zero because the symbols were exclusively used on an abacus where the zero is not needed). But the next few centuries would change that. Through the returning crusaders and the development of commercial routes, more and more information about a vastly superior Islamic culture reached Europe, where interest in the achievements of Arabic science grew steadily.

An important proponent of the Hindu-Arabic numeral system in Europe was the Italian mathematician Leonardo da Pisa (ca. 1170–1240 CE), the most important European mathematician of his time. Today, he is better known under the name Fibonacci, probably evolving from the Italian "filius Bonacci," meaning "son of Bonacci." Fibonacci traveled throughout the Mediterranean and Islamic North Africa, where he learned about Arabian mathematics, and, in particular, about the Hindu-Arabic numeral system being used there. In 1202 he wrote the book *Liber Abaci* (usually translated as "Book of Calculation"), introducing the "modus Indorum," the method used by the Indians to write numbers. He thus made the advantages of the place-value system accessible to a larger audience in Europe. Fibonacci's word for 0 was *cephirum*, which turned into the Italian word *zefiro*, which later became the French *zéro* and the English *zero*. The Arabic word for zero was *Sifr*, which later developed into the English word *cipher*, as well as the German word *Ziffer* (meaning "digit"). This, then, takes us to our current number system, which is used extensively in today's technologically driven world.

CHAPTER 4

DISCOVERING PROPERTIES OF NUMBERS

4.1. THE SEARCH FOR MEANING

While everyone has an innate number sense, and while it is not too difficult to learn the syntax for counting, arithmetic usually poses problems of a new dimension. Learning to calculate is such an ordeal that it sometimes creates hostility toward mathematics and sympathy with those who fail. Stanislas Dehaene has formulated this as follows:

> Mental arithmetic poses serious problems for the human brain. Nothing ever prepared it for the task of memorizing dozens of intermingled multiplication facts, or of flawlessly executing the ten or fifteen steps of a two-digit subtraction. An innate sense of approximate numerical quantities may well be embedded in our genes; but when faced with exact symbolic calculation, we lack proper resources. Our brain has to tinker with alternate circuits in order to make up for the lack of a cerebral organ specifically designed for calculation. This tinkering takes a heavy toll. Loss of speed, increased concentration, and frequent errors illuminate the shakiness of the mechanisms that our brain contrives in order to "incorporate" arithmetic.[1]

While there are numerical prodigies who succeed very well with complicated arithmetic tasks, such as extracting the square root of a five-digit number mentally or multiplying two very large numbers mentally, most people fail miserably. Dr. Arthur Benjamin (1961–), a mathematics professor at Harvey Mudd College in California, for example, often demonstrates his mathematical talents for national audiences. Yet most mathematicians do not have this ability. Usually, mathematicians don't think of themselves as being particularly good at arithmetic, and the prospect of having to mentally evaluate 891×46 is not very appealing to them either. They are aware of the fact that the average human brain is ill-equipped for this kind of task.

Without doubt, the ability to perform exact computations is important in a developed society. Therefore, the insight that these tasks are rather difficult could well be considered a motivation and a historical reason for the emergence of "real mathematics." In a time before the advent of computers, one had to look for ways to achieve a deeper understanding that might relieve the trouble of tedious computational tasks. Instead of trying to perfect the mind to do error-free calculations, which is rather impossible, mathematicians looked for interesting properties of numbers and relationships between numbers. They preferred to play with numbers in search of logical structure and repeating patterns that might prove helpful. Finding meaning in the world of numbers makes life with them easier (or at least more entertaining).

Apart from the purely practical reasons for a deeper involvement with number properties, there is another reason that is more philosophical in nature. The human ability to create abstractions turns numbers into mathematical objects with a meaning of their own, which is independent of any concrete realization and thus applies to all kinds of different situations. The simple computation $5 + 3 = 8$ might refer to apples or days or persons. But the statement "$5 + 3 = 8$" can very well stand for itself. It need not refer to anything outside the world of numbers. It seems self-evident, objectively true, and independent of the human state of mind. Surely, it must have been true before there were any human beings—and will it not be true even after the extinction of the human race? It appears that in proven statements about natural numbers lies an a priori truth, a certainty and absoluteness, that is completely independent of human experience. If there exists an eternal and undisputable truth at all, isn't it here, in the arithmetic of natural numbers, where we come as close to it as we will ever get?

As soon as people started to think philosophically about meaning and truth, about illusion and reality, they also started to think about mathematics and the nature of numbers. Are numbers just an instrument for counting things, or is there more to them? It is perhaps a common human trait to suspect a deeper meaning and hidden truth under the surface. So the relationship between humans and numbers is shaped not only by the need to count, measure, and calculate for practical purposes, but also by the desire to understand numbers and their meaning from a more theoretical and philosophical point of view.

4.2. PYTHAGOREAN PHILOSOPHY OF NUMBERS

The origin of an elaborate philosophy of numbers lies in ancient Greece, where one made the distinction between *logistic* (counting and calculating for practical purposes) and *arithmetic* (philosophical number theory). Arithmetic as a philosophically motivated number theory is intimately connected with the school of Pythagoreans, who based their way of life upon the quasi-religious worship of numbers. Little is known about Pythagoras himself. He lived mostly in the second half of the sixth century BCE and was born on the Greek island Samos, close to the coast of Asia Minor, just a few kilometers from cities of Ephesus and Miletus. It is said that Pythagoras fled from the tyranny of Polycrates on Samos, and after travels to Mesopotamia and Egypt he settled in Croton around 530 BCE. Croton is today's Crotone in southern Italy, which then belonged to the Greek sphere of influence and had a considerable Greek population. In Croton he founded an influential secret order, which had the typical characteristics of a religious sect—secret conspirative meetings, a time of probation for new members, strict rules for nutrition and clothing, ascetic lifestyle, and its own cosmology. Later, after becoming too influential, the Pythagoreans were persecuted and Pythagoras left Croton. He died in Metapont (also southern Italy) in the early fifth century. His school continued its activities in the Greek cities of southern Italy for about one hundred years. One group of Pythagoreans were the *mathematikoi* (the learners), who engaged in developing the scientific aspects of Pythagoras's philosophy, while the *akousmatikoi* (the listeners) focused on the religious aspects of his teachings. Due to continuing political persecution, the school of Pythagoras dissolved in the late fifth century BCE. In the first century BCE, the Pythagoreans were revived in Rome, and most of the information about the original Pythagoreans stems from that later time.

What makes the Pythagoreans interesting and special for the history of mathematics is that, to them, numbers were the key to an understanding of the cosmos. The Pythagorean philosopher Philolaus of Croton (ca. 470–385 BCE) writes in fragment 4, "Indeed, everything that is known has number, for nothing is either conceived or known without this." And Aristotle (384–322 BCE), two generations later, even describes the Pythagorean doctrine as "All things are number." Aristotle writes the following, often-cited passage about the Pythagoreans' ideas in his book *Metaphysics*:

> The so-called Pythagoreans, who were the first to take up mathematics, not only advanced this study, but also having been brought up in it they thought its principles were the principles of all things. Since of these principles numbers are by nature the first, and in numbers they seemed to see many resemblances to the things that exist and come into being—more than in fire and earth and water . . . since, again, they saw that the modifications and the ratios of the musical scales were expressible in numbers—since, then, all other things seemed in their whole nature to be modeled on numbers, and numbers seemed to be the first things in the whole of nature, they supposed the

elements of numbers to be the elements of all things, and the whole heaven to be a musical scale and a number.[2]

This goes much further than the statement of Philolaos that everything *has* number. According to Aristotle, the Pythagoreans believed numbers to be the essence of all things. Numbers were not just an abstract construction of the human mind; for the Pythagoreans, numbers formed the basis, the principle, of all other things. And they came to this conclusion because they saw a variety of natural phenomena—from cosmic cycles to musical scales—that could be expressed through numbers and specific ratios of numbers.

Interestingly, Aristotle also states that the unit, the "One," is not itself a number; instead it is the fundamental principle that creates number and thus plays a very special role, philosophically. We find this also in Euclid's *Elements*, book 7, which starts with the definitions

- *A unit is that by virtue of which each of the things that exist is called* one.
- *A number is a multitude composed of units.*[3]

The "One" is the basic unit of which all numbers consist, and, as Aristotle explains, the unit is not a number in the same sense as a measure is not the things measured. Moreover, there was no need for a "number one" in counting, because if there was only one item, then there was no need to count. Counting, and hence numbers, therefore had to start with "two."

For the ancient Greek scholars, numbers were not just a useful tool. They regarded numbers as philosophical principles, as fundamental entities, as the essence of everything. Numbers had to be explored, as their properties would reveal the nature of all things. This quasi-religious state of mind was the driving force creating the Pythagorean tradition of systematically cultivating mathematics as a science. While the philosophical underpinning might appear obscure from today's perspective, the Pythagoreans' rational approach to the exploration of numbers by strictly logical reasoning nevertheless marks the historical origin of modern mathematics.

4.3. EVEN AND ODD NUMBERS

Early humans used little stones or pebbles as an aid for counting and simple arithmetic and as a means for exploring number properties. In Latin, a pebble used for counting was called *calculus* (meaning "little limestone")—the English word *chalk* refers back to this original meaning. Later, the Latin verb *calculare* meant "to reckon, to compute," which is also the modern meaning of the English verb "to calculate."

Probably our ancestors also had fun laying out calculi (plural of calculus) on a flat surface to produce regular shapes (such as triangles and squares), just as today's children have fun arranging their toy blocks or drawing points on a sheet of paper in regular patterns. When humans started to think about the world of numbers, they also started to explore the possible shapes and patterns that could be created with a certain number of calculi. The method of representing numbers as geometrically arranged objects or points is indeed very old and was certainly used by Pythagoras and the Pythagoreans around 500 BCE.

Figure 4.1: Even and odd numbers of objects.

One of the simplest observations that can be made by laying out a number of objects is about pairing. Depending on the total number of objects, we can either group all the objects in pairs, or precisely one object remains unpaired, as in figure 4.1. If it is possible to arrange a group of objects in pairs, we call the number *even*. Otherwise, when there is one object left over, it is called *odd*.

An understanding of even and odd might well have been achieved before the ability to count. Georges Ifrah, in his book *The Universal History of Numbers*, recounts the reports of early ethnologists about certain Australian and Oceanian aborigines who were only able to verbally express the numbers one and two.[4] They were able to count up to four using the numerals one, two, two-one, and two-two, calling all higher numbers "a lot." But even for larger numbers they appeared to have a clear feeling of even and odd to the extent that if two pins were removed from a set of seven, the aborigines hardly noticed it, but they saw immediately if only one pin was removed. To them, the numbers five, six, and seven were just "a lot," but obviously they could perceive the number six as a group of pairs, which makes it quite different from five and seven—thereby distinguishing between odd and even quantities.

Using today's abstract understanding of number, we describe a number as even if the collection contains precisely twice as many objects as it contains pairs. If the number of pairs is n, then the number of objects is $p = 2n$. We can thus characterize an even number p as follows:

A number p is even whenever $p = 2n$, where n is some natural number.

We have an odd number of objects if the attempt to arrange them in pairs results in one object remaining without a partner. An odd number q thus consists of n pairs and one remaining object:

A number q is odd whenever $q = 2n + 1$, where n is some natural number.

A special case remains with $q = 1$. From today's perspective, it makes sense to call 1 also an odd number, because if we have one object, we can form zero pairs and the one and only object is always left without a partner. This just means that we can let $n = 0$ in the formula $q = 2n + 1$. Likewise, we consider zero an even number (by allowing $n = 0$ in the formula $p = 2n$).

In history, this was not always considered obvious. Aristotle, in his *Metaphysics*, describes the views of the Pythagoreans on even and odd numbers, whose identification of number properties with philosophical principles is certainly not easy to understand. It was explained by Aristotle as follows:

> Evidently, then, these thinkers also consider that number is the principle both as matter for things and as forming both their modifications and their permanent states, and hold that the elements of number are the even and the odd, and that of these the latter is limited, and the former unlimited; and that the One proceeds from both of these (for it is both even and odd), and number from the One; and that the whole heaven, as has been said, is numbers.[5]

We talked already about the special role of the "One." The use of the words *limited* and *unlimited* in this context needs some explanation. Of course, the Greek scholars knew that there are infinitely many odd numbers. In what sense did the odd numbers represent to them the primary philosophical principle of "limitedness"? Giovanni Reale (1931–2014), in his book *A History of Ancient Philosophy*, gives the following interpretation (see figure 4.2). For an odd number, division into two parts ends with the one in the middle, which is perfectly meaningful. On the other hand, dividing an even number into two equal groups leaves an empty space in the middle, which, to the Pythagoreans, meant something without number and hence incomplete or "unlimited." This view of the unlimited nature of even numbers, as opposed to the limited nature of odd numbers, is illustrated in figure 4.2. In modern language we would just say: An even number can be divided by two without remainder, while an odd number divided by two always results with a remainder of 1.[6]

Figure 4.2: Even and odd: Division into two equal groups (demonstrating a remainder).

4.4. RECTANGULAR AND SQUARE NUMBERS

Experimenting a little, one sees that many numbers can be arranged to form a rectangle. For example, 15 dots can be arranged as 3 × 5, that is, as a rectangular array with 3 rows and 5 columns of points (see figure 4.3).

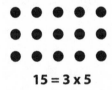

15 = 3 x 5

Figure 4.3: The rectangular number 15.

We do not regard a 5 × 3 rectangle as being different from a 3 × 5 rectangle, as one can be obtained from the other just by rotating the paper 90°. Sometimes there are several different ways to arrange rectangles. For example, 12 can be arranged as 2 × 6 or as 3 × 4 (see figure 4.4).

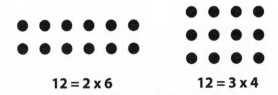

12 = 2 x 6   12 = 3 x 4

Figure 4.4: The rectangular number 12.

A number that can be represented geometrically by points arranged in the form of a rectangle is called a *rectangular number*. Of course, it is understood that a rectangle has more than one row and more than one column. Usually, 1 was not considered a rectangular number, because it is just a dot. Likewise, 2 is not rectangular, because two points just make a single line (row or column), which is not sufficient for a rectangle. Numbers like 3, 5, 7, and 11 are not rectangular numbers, because they cannot be arranged in rectangular form. A number is rectangular whenever it can be written as a product of two natural numbers other than itself and 1. Rectangular numbers are also often called *composite numbers*. The numbers greater than 1 that are not rectangular are called *prime numbers*. We will delve into these fascinating prime numbers later in the book.

A special case among the rectangular numbers is the *square numbers*. Square numbers can be represented by a rectangle, where the number of rows equals the number of columns. For example, $16 = 4 \times 4$ (see figure 4.5).

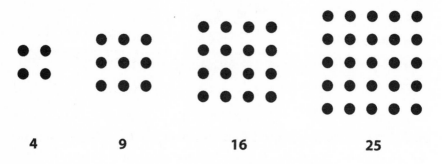

4   9   16   25

Figure 4.5: The first square numbers.

According to the definition, 1 is not a square number in the geometric sense, because 1 is not a rectangular number. And indeed, in Greek mathematics, 1 was not a number at all, as was explained earlier. Arithmetically, however, 1 is the square of 1, since $1 = 1 \times 1$, and today the sequence of square numbers usually includes the number 1. Sometimes it even contains 0, because zero is the square of zero: $0 = 0 \times 0$. The following is the sequence of nonnegative square numbers:

$$0, 1, 4, 9, 16, 25, 36, 49, 64, 81, 100, 121, 144, 169,$$
$$196, 225, 256, 289, \ldots$$

According to the online encyclopedia of integer sequences, this was the first sequence ever computed by an electronic computer, in the year 1949.

When playing with square numbers, people noticed that in order to produce the next larger square number, one always has to add an odd number of points. So in order to go from a 2×2 square to a 3×3 square, one has to add five points; and in order to produce the next larger square number, 4×4, one has to add another seven points, and so on, as shown in figure 4.6.

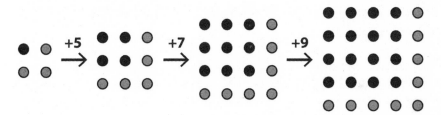

Figure 4.6: Adding odd numbers.

Even our first square number, 4, can be obtained as 1 plus 3. Thus we obtain the following beautiful pattern,

$$1 \qquad\qquad = 1 \times 1 = 1^2,$$
$$1 + 3 \qquad\qquad = 2 \times 2 = 2^2,$$
$$1 + 3 + 5 \qquad\quad = 3 \times 3 = 3^2,$$
$$1 + 3 + 5 + 7 \quad\; = 4 \times 4 = 4^2,$$
$$1 + 3 + 5 + 7 + 9 \; = 5 \times 5 = 5^2,$$

and so on.

Of course, the pattern can be extended indefinitely. The next line would be

$$1 + 3 + 5 + 7 + 9 + 11 = 6 \times 6 = 6^2,$$

which we can further appreciate by visualizing the geometric interpretation in figure 4.7.

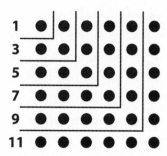

Figure 4.7: The sum of the first odd numbers gives a square number.

We see that in order to compute 5×5 we have to sum up the first five odd numbers. And 6×6 is the sum of the first six odd numbers. This leads to the amazing conjecture that

- the sum of the first n odd numbers equals $n \times n$, where n is any natural number.

When formulating a law that holds for all natural numbers, we have reached a new level of abstraction. The letter n in that statement does not refer to any particular number; rather it represents any number one wishes to substitute for n. It is a "placeholder" that can be replaced by any particular number (like 5, 6, or 273), which would turn the general statement into a statement about this particular number. For example,

- the sum of the first 273 odd numbers equals 273×273.

In mathematics, we call n a variable.

The sum of the first n even numbers can be constructed in a quite similar way. We do this in complete analogy to figure 4.7, with the only difference here being that we start with two dots at the beginning. Adding borders, we obtain a sequence of rectangular numbers, where the number of columns exceeds the number of rows by 1 (see figure 4.8). Notice that each of the borders contains an even number of dots.

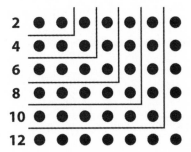

Figure 4.8: The sum of the first even numbers gives a number of the form $n \times (n + 1)$.

From this figure, we find that,

$$2 \qquad\qquad\qquad = 1 \times 2,$$
$$2 + 4 \qquad\qquad\quad = 2 \times 3 = 6,$$
$$2 + 4 + 6 \qquad\quad = 3 \times 4 = 12,$$
$$2 + 4 + 6 + 8 \quad\;\; = 4 \times 5 = 20,$$
$$2 + 4 + 6 + 8 + 10 = 5 \times 6 = 30,$$

and so on.

We can express this as a general rule:

- the sum of the first n even numbers equals $n \times (n + 1)$, where n is any natural number.

The British historian of Greek mathematics T. L. Heath (1861–1940) noted that the rectangles obtained as sums of even numbers in figure 4.8 on the basis of the even number 2 all have different proportions (3:2 is different from 4:3, which is different from 5:4, and so on), while the odd numbers all preserve the same quadratic form as shown in figure 4.7. This gives another interpretation as to why the even numbers were regarded by the Pythagoreans as unlimited and odd numbers as limited.

Obviously, the world of numbers is full of amazing regularities. To the Pythagoreans, the laws of numbers were the origin of the order of the cosmos.

4.5. TRIANGULAR NUMBERS

Another common shape is the triangle. In the days of Pythagoras, dots were arranged in a triangular shape, as shown in figure 4.9:

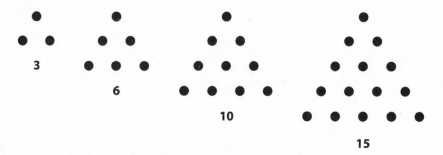

Figure 4.9: The first triangular numbers.

While the Greek philosophers would not have included the number 1, it is included today in the list of *triangular numbers*, as it is included in the list of square numbers. So the sequence of triangular numbers is

$$1, 3, 6, 10, 15, 21, 28, 36, 45, \ldots$$

Let's now try to find a rule that enables us to find a triangular number, given the previous one. The solution is in figure 4.10.

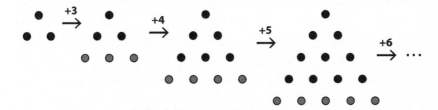

Figure 4.10: Generating triangular numbers by adding natural numbers.

We obtain the next triangular number just by adding a row at the bottom of the previous triangle. Every row has one dot more than the previous row. So the triangular numbers are simply created as sums of natural numbers:

$$
\begin{aligned}
1 &= 1, \\
1 + 2 &= 3, \\
1 + 2 + 3 &= 6, \\
1 + 2 + 3 + 4 &= 10, \\
1 + 2 + 3 + 4 + 5 &= 15,
\end{aligned}
$$

and so on.

The following formula states that the nth triangular number, which we conveniently call T_n, is given as the sum of all natural numbers up to n (always including the number 1):

$$T_n = 1 + 2 + \ldots + n, \text{ where } n \text{ is any natural number.}$$

The triangular figure for $10 = 1 + 2 + 3 + 4$ is just the familiar arrangement of bowling pins. For the Pythagoreans, this shape had a special meaning. It was called the *tetraktys* and was seen as a divine symbol of perfection representing the whole cosmos, including the sum of all possible dimensions. The first row, a single point, is the unity that generates all other dimensions. With the two points in the second row, they believed that one could represent a one-dimensional line. The third row, which consists of three points, can be arranged as a triangle in a two-dimensional plane, and the third row, which has four points, can be arranged to outline a three-dimensional figure, namely a tetrahedron. The sum of all these is ten, the Dekad, which was also the base of the number system that was already in use in ancient Greece. In the Attic numeral system that was used by the Athenians in the fifth century BCE, the numeral for ten was Δ, a Delta, which was the first letter of the word *Deka* (Δεκα), indicating ten, and one can't help noticing the similarity between Δ and the triangular outline of the tetraktys.

4.6. TRIANGULAR AND RECTANGULAR

By looking at figure 4.11, we see another interesting relationship for triangular numbers: Adding two consecutive triangular numbers obviously gives a square number. With just a little manipulation, we can see that happening geometrically.

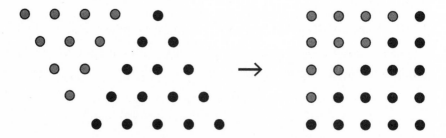

Figure 4.11: Triangular numbers and square numbers.

The relationship expressed in figure 4.11 means that every square number is the sum of two consecutive triangular numbers. In formulas, this statement can be elegantly and simply written as

$$T_{n-1} + T_n = n^2 \text{ (for all natural numbers } n \text{ greater than 1).}$$

A similar observation can be made with the help of figure 4.12. Taking the same triangular number twice obviously produces a rectangular number, where the number of columns exceeds the number of rows by one.

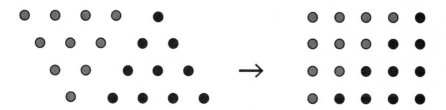

Figure 4.12: Twice a triangular number gives a rectangular number $n \times (n + 1)$.

This can be written as $2 \times T_n = n \times (n + 1)$. From this we obtain a useful formula,

$$T_n = \frac{n(n+1)}{2} .$$

With this formula, we can compute the nth triangular number immediately, without having to compute the sum of all natural numbers up to n. Note that either n or $n + 1$ must be an even number and can be easily divided by two. Thus, the formula really just requires us to evaluate a single multiplication. And this multiplication is equivalent to summing up all integers from 1 to n. Hence, for example, the sum of the first 100 natural numbers is easily obtained as

$$T_{100} = 1 + 2 + 3 + \ldots + 99 + 100 = \frac{100 \times 101}{2} = 50 \times 101 = 5050.$$

This particular triangular number often occurs in an anecdote about Carl Friedrich Gauss (1777–1855), one of the most important mathematicians of all time. Wolfgang Sartorius von Waltershausen, an early biographer, tells several stories of Gauss being a child prodigy with almost unbelievable skills in mental arithmetic. One of these stories (which Gauss himself often related in old age with amusement) is about nine-year-old Carl Friedrich at his elementary school, where a stern teacher confronted his students with the task of summing an arithmetic series. Much to the surprise of his teacher, Gauss produced the correct solution immediately, while all his classmates continued calculating for a very long time—producing wrong results in most cases. Later biographers beefed the story up with more mathematical details, claiming that the arithmetic series was the first 100 integers, and they also provided a method for how Gauss could have obtained the result 5050. Usually, this trick is explained as follows: In order to sum all integers between 1 and 100, young Carl Friedrich started by adding numbers from opposite ends of the sequence, that is, $1 + 100$, then $2 + 99$, $3 + 98$, and so on. He observed that in each case the sum is 101. The last sum in this sequence would be $50 + 51$, which shows that there is a total of fifty such sums, of 101. Hence the answer is $50 \times 101 = 5050$, as shown in figure 4.13.

1	+	2	+	3	+	4	+	...	+	48	+	49	+	50			
100	+	99	+	98	+	97	+	...	+	53	+	52	+	51			
101	+	101	+	101	+	101	+	...	+	101	+	101	+	101	=	50×101	= 5050

Figure 4.13: Gauss's trick for adding up all numbers from 1 to 100.

If you indeed try to solve the problem in the straightforward way—that is, by actually adding up all the numbers between 1 and 100—you will soon notice that this tends to be a rather tedious task. Obviously, a sudden flash of genius can really be helpful to solve a mathematical problem like this. But usually we cannot count on having this type of inspiration just when we need it. Much of mathematical research, thus, aims at developing methods that spare us the necessity of having ingenious ideas whenever we have to solve a mathematical problem. Our knowledge of triangular numbers obtained in this section would help us to add all natural numbers up to some value of n with similar speed as that of young Carl Friedrich Gauss. The formula for T_n effectively reduces this to a routine task—a simple multiplication.

4.7. POLYGONAL NUMBERS

If we take this discussion further, we enter a realm of numbers referred to as *polygonal numbers*, which get their name from the notion that these numbers can be placed in an arrangement that forms a regular polygon—one whose sides and angles are congruent. These polygonal numbers further enhance the appreciation of special numbers.

As we indicated earlier, three dots can be used to form an equilateral triangle, as can six dots. Therefore 3 and 6 are triangular numbers, as are 10 and 15 (see figure 4.9). We also recall that square numbers— such as 4, 9, 16, and 25—get their name from the fact that they can be arranged to form a square, as shown in figure 4.5.

Mathematicians soon started to think about possible generalizations to other regular polygons, for example, numbers that can be arranged to form regular pentagons. This is not too obvious and can be done in different ways. We learned how it was done in ancient Greece from the remaining fragments of a book on polygonal numbers written by Diophantus of Alexandria, who probably lived in the third century CE. Diophantus credits the definition of polygonal numbers to the Greek mathematician Hypsicles (who lived in the second half of the second century BCE).

The description given by Diophantus suggests that pentagonal numbers were constructed as shown in figure 4.14. These numbers begin with 5, 12, 22, and 35.

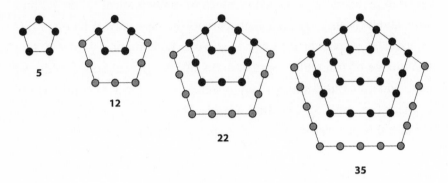

Figure 4.14: Pentagonal numbers.

The construction of these numbers follows a principle analogous to the case of triangles and squares, but it is not as elegant, because the larger figures do not have the same symmetry as the first one (since one of the corners plays a special role).

Next would be hexagonal numbers, which, similarly, represent the number of dots needed to form regular hexagons. These are the numbers 6, 15, 28, and 45 (see figure 4.15).

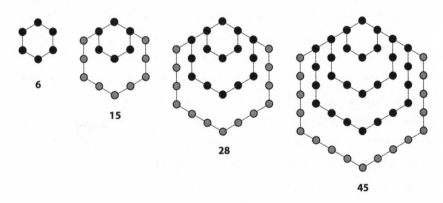

Figure 4.15: Hexagonal numbers.

Figure 4.16 lists the first polygonal numbers, again including 1 as the first element of each sequence. The numbers in between are the differences between adjacent polygonal numbers.

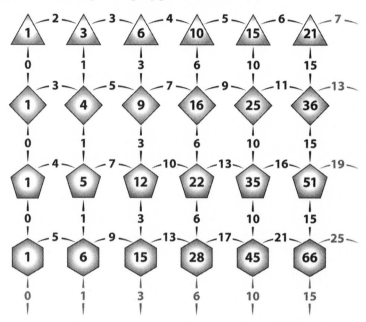

Figure 4.16: Relations between polygonal numbers.

Inspecting figure 4.16 more closely, we see that the polygonal numbers in the same column all have the same difference, and this difference is always a triangular number. For example, the fifth pentagonal number (35) has 10 dots more than the fifth square number (25) and 10 dots fewer than the fifth hexagonal number (45), and the difference 10 is just the fourth triangular number.

But this also means that any polygonal number can be obtained from triangular numbers. From the columns of figure 4.16 we can read off that the pentagonal numbers are obtained as

$$5 = 3 + 2 \times 1 = T_2 + 2T_1,$$
$$12 = 6 + 2 \times 3 = T_3 + 2T_2,$$
$$22 = 10 + 2 \times 6 = T_4 + 2T_3, \text{ and so on.}$$

This leads us to the general formula for the nth pentagonal number, which we denote by P_n:

$$P_n = T_n + 2T_{n-1}.$$

With our formula for T_n, which we obtained earlier, this can be rewritten as follows:

$$P_n = \frac{n(n+1)}{2} + 2\frac{(n-1)n}{2} = \frac{n^2}{2} + \frac{n}{2} + n^2 - n = \frac{3n^2 - n}{2} = \frac{n(3n-1)}{2}.$$

Formulas for all other polygonal numbers can be obtained in a similar way. The following formula, which Diophantus attributes to Hypsicles, gives the number of dots in a regular polygon (arranged in a way analogous to that of figure 4.14 and figure 4.15) where the polygon has k-vertices and its outer sides are made of n dots:

$$\frac{n^2(k-2) - n(k-4)}{2}, \text{ for } n = 1, 2, 3, \ldots, \text{ and } k = 3, 4, 5, 6, \ldots.$$

The reader may wish to check that this gives the right result for triangular numbers ($k = 3$), square numbers ($k = 4$), and pentagonal numbers ($k = 5$).

The construction of a polygonal number from the previous one involves the addition of a certain number of dots, which is also indicated in figure 4.16. This number, called the *gnomon*, is geometrically represented by the gray dots in figure 4.14 and figure 4.15 for the pentagon and the hexagon. Inspecting the lines of figure 4.16, we see that for triangular numbers the gnomon increases by steps of 1, and for square numbers it increases by steps of 2, which we have already established earlier. From the third line in figure 4.16 we see that the gnomon of pentagonal numbers increases in steps of 3, and for the hexagonal numbers in line 4 it increases in steps of 4.

We then see that every pentagonal number is a sum of integers differing by three; for example,

$$1 + 4 + 7 + 10 = 22, \text{ which is a pentagonal number.}$$

So, for example, if we want to know the sum of the first n integers from the sequence of numbers that starts at one and increases in steps of three, the answer is the nth pentagonal number:

$$1+4+7+10+13+16+\ldots \left(n \text{ summands} \right) = P_n = \frac{n\left(3n-1\right)}{2}.$$

You may now ask, what sort of truth about the universe can be obtained from considering polygonal numbers? What sort of philosophical insight into first principles did the Greek scholars of antiquity get from these considerations? Are there applications for polygonal numbers? Well, the consideration of polygonal numbers has certainly spurred the growth of mathematical knowledge. But it appears that the main motivation for this study is not its possible usefulness. It is much more the fascination with the unsuspected beauty and regularity that gives meaning to numbers that had no meaning before. It is the appearance of order in an initially hardly comprehensible variety of phenomena. Here, mathematics has become an art in itself that needs no external motivation.

4.8. TETRAHEDRAL NUMBERS

For the sake of completeness, we mention that the polygonal numbers have generalizations to higher dimensions. Points can be arranged in space to form regular polyhedrons. An example would be the stacking of cannon balls, as in figure 4.17, which realizes one of the Platonic solids, the tetrahedron.

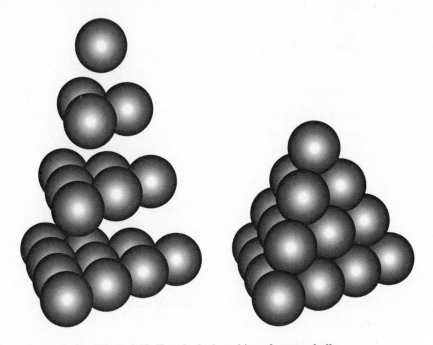

Figure 4.17: Tetrahedral stacking of cannon balls.

The left view shows the individual layers of the arrangement. We see that all layers are consecutive triangular numbers. Hence the number of cannonballs in the tetrahedral stack, where the side line of the base has four balls, is just the sum of the first four triangular numbers, $1 + 3 + 6 + 10 = 20$. The sequence of tetrahedral numbers obtained in this way starts with

$$1, 4, 10, 20, 35, 56, 84, 120, 165, 220 \ldots$$

We will encounter these numbers again in a completely different context in the next chapter.

CHAPTER 5

COUNTING FOR POETS

In the first chapter we discussed various aspects of numbers and counting. As long as one just thinks of counting marbles, this actually was "counting for children." But the abstraction principle, which states that you can count just about anything, even immaterial ideas, soon leads to counting tasks that are a whole lot more difficult. Consider, for example, the following problem: If at a New Year's Eve party everybody clinks glasses with everyone else, how many clinks will you hear? Problems like these can only be solved with elaborate and systematic thinking, and we give the answer to this and similar problems toward the end of this chapter. The need to learn more about these counting methods arose early in the history of humankind. Indeed, a very interesting solution can be traced back to the time of Vedic literature in old India, more than two thousand years ago.

5.1. VERSE METRIC—THE ROLE OF RHYTHM

The problem that we are going to consider in some detail shows how mathematical questions arise in quite unexpected ways. The problem has to do with the classification of verse meters in poetry. In order to understand why this problem (and its solution) arose first in old India, we have to understand what distinguishes Sanskrit poetry from modern English poetry.

The rhythm of speech is determined by the succession of stressed and unstressed syllables. A characteristic rhythmical structure is an important feature of most literature written in verses, which distinguishes poetry from prose. Poems are typically divided into lines that repeat a certain rhythmic pattern (with occasional variations, of course). The rhythmic structure of a line of verse is called its *meter*. In English (and many other languages, like, for example, German), most poetry uses highly regular meters, where stressed syllables occur in periodic intervals. Very often, each line of verse has a fixed number of syllables and stresses ("accentual-syllabic verse form"). This creates a rhythmic drive that in part is responsible for the fascination created by poems. As an example, observe the regular pattern of stressed and unstressed syllables in these lines from Shakespeare's *Macbeth*:

> **Dou**ble, **dou**ble, **toil** and **trou**ble;
> **Fire burn** and **cal**dron **bubb**le.

In many cases, the rhythm can be described as the repetition of a short and simple basic pattern, which is called *verse foot*. In this example, the foot of the verse consists of two syllables, the first is stressed, the second unstressed. This is called a *trochee*, and here it is repeated four times in each line, which gives a meter called *trochaic tetrameter*. Other common feet are the *iamb* (unstressed–stressed), the *dactyl* (stressed–unstressed–unstressed), and the *anapest* (unstressed–unstressed–stressed).

The English way of emphasizing syllables is called *accentuation*. This means that stressed syllables are typically spoken louder than unstressed syllables. Thus the verse meter is given by a characteristic succession of loud- and soft-spoken syllables. Of course, the loudness is not the only parameter that could vary. One could, for example, also change the duration (*quantity*) of a syllable or modulate the voice pitch (*intonation*), and indeed, all these methods will occur simultaneously. But it is characteristic of a language that one of these methods dominates and is most important for creating the rhythm of speech. In accentuating languages like English, loudness is mainly used for that purpose. There are other languages, like Japanese, where the duration of a syllable is most important and others, like Chinese, who use intonation even for distinguishing words.

The type of a language has an influence on which verse meters are common. Accentuating languages tend to have meters that repeat simple feet in a regular pattern, as explained above. Even the structure of verse feet has certain restrictions, making the appearance of two stressed (or three unstressed) syllables in succession rather unusual. As a rule, accentuating languages are therefore rather limited concerning the diversity of possible verse forms. Other languages show a much greater variety of common verse meters and are thus able to convey subtleties that cannot be expressed in English.

The antique Indo-European languages Greek, Latin, and Sanskrit are all *quantitative*, which means that they distinguish syllables by their quantity or duration. Thus a verse meter would be defined as a characteristic succession of long and short syllables in a line of verse. These languages tend to have a much greater variety of common meters than modern accentuating languages. For example, a website of the University of Heidelberg on Sanskrit language resources presently lists 1,352 different meters. It appears to be easier to arrange long and short syllables in an arbitrary succession without disturbing the flow of speech, than to arrange stressed and unstressed syllables. As speakers of an accentuating language, we cannot really appreciate the wealth of antique meters because we tend to change the antique pronunciation following an implicit rule according to which a long syllable becomes stressed and a short syllable unstressed. But this is rather difficult and wouldn't give the right impression for the original verse.

The enormous number of verse meters made possible by the quantitative character of the Sanskrit language led old Indian scholars to raise questions about the theoretically possible number of verse meters and their classification.

5.2. THE ORIGIN OF LINGUISTICS

The meter of a poem clearly makes it easier to learn the poem by heart. In a time without a written tradition, it was rather difficult to conserve important texts—hymns, songs, or the ritual texts spoken during religious ceremonies—and to pass them on to the next generation without errors. A text in verses helped to prevent errors because omissions or additions would change the meter and could thus be detected easily. Therefore, it is not astonishing that the earliest literature of humankind was created in verse form.

As the language of the antique Indian literature, Sanskrit has by far the greatest wealth of verse meters. It can be said that Sanskrit literature began with the Vedas, a huge collection of verses forming the basis of Hinduism and dating from the time between 1200 BCE and 800 BCE. The word *veda* means "knowledge," and the text aimed to represent the whole knowledge of that time about life, the universe, and everything.

The Vedas consist of thousands of hymns and mantras in verse form—the oldest part alone, the *Rigveda*, consists of 1,028 hymns comprising 10,600 verses. For many centuries, the only means to preserve the holy text was by oral tradition. And because the text was believed to be of divine origin, it was important to pass it on to the next generation without errors. Considering the enormous amount of text in the Vedas, learning it by heart was an impressive achievement, and it was certainly facilitated by the characteristic verse meters of the various hymns.

Over time, the meters of the ritual verses were associated with certain religious ceremonies and started to carry meaning themselves. This created the need to understand, investigate, and describe the effect and structure of verse meters. Hence, one began thinking about language in a rather abstract way—it was the birth of grammar and linguistics. Already in the first millennium BCE, a rich theory of meter was created. Prosody (that is, the study of meter) became one of the basic disciplines (*Vedangas*) of the Vedic science; other parts were ritual, phonetics, grammar, etymology, and astronomy.

The oldest scientific text about verse meters is the Chandahs Sutra of Pingala, who might have lived in the time between 400 and 200 BCE. Other than his name, almost nothing is known about Pingala, and only a linguistic analysis of the work gives some hints about the time of its writing. An important part of the Chandahs Sutra deals with mathematical questions about verse meters. Pingala not only was interested in describing the actually existing meters but also wanted to investigate all verse meters that are theoretically possible. Indeed, how many different meters can we think of? How can we find—in a systematic way—those meters that do not yet exist? These are genuine mathematical problems, although motivated by the study of literature and of a different kind than the problems arising from astronomy or geometry. Pingala was the first to give correct answers to these questions. However, Pingala's Chandahs Sutra is rather difficult to understand because it is written in a cryptic style in verse form. In order to interpret it, one often needs to refer to commentaries of later scholars, in particular, Halayudha, who lived in the late tenth century CE. Halayudha explained the mathematical content of Pingala's work and developed it further.

Pingala's work nevertheless hints at a treasure of highly developed mathematical knowledge in old India. In a time when science was not split into numerous isolated disciplines, mathematical knowledge and concern for mathematical questions was an inseparable part of every scholar's mind-set. In that way, the science of poetry and music could stimulate mathematics, and vice versa.

Pingala's ideas, which are intimately related to numbers and counting, constitute one of the roots of mathematical thinking. In the following, we embark on the adventure of describing those ideas in more detail. We will encounter the origin of a modern mathematical discipline that is important for many aspects of everyday life. Today it is taught in high schools and colleges throughout the world, but it has been long forgotten that its origin lies in poetry.

5.3. THE PROBLEM OF COUNTING METRIC PATTERNS

When we start counting verse meters, it is tempting to start analyzing poems, where each line contains a certain number of syllables. You could then ask how many different verse patterns with a certain number of syllables can be identified. However, for a quantifying language, like Sanskrit, we could also think of another approach, which emphasizes the time needed to recite a line of the poem. The linguistic unit for measuring the time is called *mora*. A mora is the time needed to pronounce a short, unstressed syllable. A long syllable is then said to have two moras (or morae) because it takes about twice as long to pronounce than a short syllable. Of course, a mora is not a physical time unit that can be measured in seconds, because verses can be pronounced with different speeds, depending on the individual interpretation. So instead of attempting to give a more precise definition of a mora, we stick with the sloppy definition of linguist James D. McCawley (1938–1999), who stated that a "mora is something of which a long syllable consists of two and a short syllable consists of one."[1]

- The duration of a short syllable is one mora.
- The duration of a long syllable is twice as long (two moras).

We will use the following definition of a verse meter:

- A meter is defined by a succession of short and long syllables.

We can represent meters in a graphical form, using the symbol – (called a *macron*) to denote a long syllable, and the symbol ⏑ (*breve*) to denote a short syllable. For example,

⏑ – ⏑ – ⏑ – ⏑ – ⏑ –

is a meter that consists of ten syllables (and has a duration of fifteen moras). It is called *iambic pentameter*. Another example from old India is a meter called *varatanu*, which means "woman with a beautiful body":

$$\smile \smile \smile - \smile \smile - \smile - \smile -$$

Varatanu is a meter with a duration of sixteen moras. It takes its name from a poem written in this meter, in which a young man greets his lover after a long night spent together.

When analyzing the possible number of different meters, we start by considering only verses where each line takes the same time to recite. Verse lines with the same total duration consists of a given number of, say, n, moras.

The following problem can, with some justification, be traced back to Pingala. We now state Pingala's problem in full generality. Remember that here we do not attempt to classify meters with a given number of syllables; instead, we are going to count the number of different meters with a given duration measured in moras.

Pingala's first problem:
How many different verse meters exist that have a total duration of n moras?

This problem is about counting the members of a particular set of objects—namely the set of all possible meters (arrangements of long and short syllables) with a fixed total duration. But it is of a different quality than the problem of counting the pebbles in figure 1.2 of chapter 1. Therefore, the usual counting method cannot be applied in a straightforward way. How do we put a finger on the "first" meter, and then on the second meter, and so on?

5.4. SOLVING PINGALA'S FIRST COUNTING PROBLEM IN SPECIAL CASES

A first step in a mathematical investigation is often to give a name to the object of interest. We are interested in the number of meters with a total duration of n moras, where n is an arbitrary natural number.

- The number of meters with a duration of n moras will be called $A(n)$.

For a mathematician, n would indeed be an arbitrary natural number. For the sake of completeness, a mathematician would also consider meters of verses with a duration $n = 1$ or 2 moras, which are absolutely irrelevant for poetry. But this has the advantage that the answer is very easy to find: Obviously, $A(1) = 1$, because the only meter with a duration of one mora is the meter that consists of exactly one short syllable. For a meter with a duration of two moras, there are already two possibilities because it could consist of two short syllables or one long syllable. Thus, we find $A(2) = 2$. Let us collect our results:

$$A(1) = 1, A(2) = 2.$$

Now, one could proceed in a systematic way and try to determine all possible meters of total duration $n = 3, 4, 5$, and so on. For example, if $n = 6$ (which is still too short for meaningful poetry), one would find the following list of thirteen different meters:

1:	— — —	6:	˘ — — ˘
2:	˘ ˘ — —	7:	— ˘ — ˘
3:	˘ — ˘ —	8:	— — ˘ ˘
4:	— ˘ ˘ —	9:	˘ ˘ ˘ — ˘
5:	˘ ˘ ˘ ˘ —	10:	˘ ˘ — ˘ ˘
		11:	˘ — ˘ ˘ ˘
		12:	— ˘ ˘ ˘ ˘
		13:	˘ ˘ ˘ ˘ ˘ ˘

Table 5.1: A list of verse meters with a duration of six moras.

But one soon has the impression that this procedure is not really helpful. For larger n, the number of possibilities gets very large and unmanageable. How can one be sure not to omit one of the possible patterns? It is probably wiser to look for another way of determining $A(n)$ for arbitrary n. As we will see, mathematicians do not always choose the direct approach. Sometimes, they attack a problem by working backward. In the next section, we give the general solution that results from this strategy.

5.5. A GENERAL SOLUTION TO PINGALA'S FIRST PROBLEM

One might obtain an idea for a solution when we look at the two groups of meters in table 5.1. The first group has 5; the second, 8 meters. What property distinguishes these two groups? Well, the first group contains all meters that end with a long syllable, while all meters of the second group end with a short syllable. Consider the first group. The parts that precede the long final syllable has a duration of four moras, and the group contains all possible meters with four moras combined with a long syllable. Hence the number of the meters of the first group is just $A(4)$. Similarly, the second group has all possible meters with a duration of five moras, combined with a short final syllable. Hence the number of meters in the second group is $A(5)$. From this we get the formula $A(4) + A(5) = A(6)$ for the number of possible meters with a duration of six moras.

Obviously, we can repeat that reasoning for any n. Any meter with n moras ends with either a long or a short syllable. Thus the set of all meters of length n can be divided into a set of meters that end with a long syllable and a second set of meters that end with a short syllable. The meters ending with a short syllable could start with an arbitrary part of length $n - 1$ at the beginning; hence there are $A(n - 1)$ such meters. And the meters that end with a long syllable start with an arbitrary part of length $n - 2$; hence there are $A(n - 2)$ such meters. We conclude that

$$A(n - 2) + A(n - 1) = A(n)$$

must hold for any n. Well, at least for n starting with at least 3, so that $n - 2$ is at least 1. We collect the results of our reasoning:

$A(1) = 1, A(2) = 2.$

$A(n - 2) + A(n - 1) = A(n)$, for all natural numbers n greater than 2.

Admittedly, this does not directly tell us what $A(n)$ is, but in a way, it solves Pingala's first problem. The formula describes how, starting with the "initial condition" for $A(1)$ and $A(2)$ we can easily compute step-by-step all the numbers $A(n)$:

$A(3) = A(1) + A(2) = 1 + 2 = 3,$

$A(4) = A(2) + A(3) = 2 + 3 = 5,$

$A(5) = 3 + 5 = 8, A(6) = 5 + 8 = 13, A(7) = 8 + 13 = 21$, and so on.

Every further number, $A(n)$ is the sum of the two preceding results. And now we can be sure that we haven't forgotten any of the possible meters of duration 6 in table 5.1, because with the new method we also find $A(6) = 13$.

With a little patience, we can easily compute the number $A(16)$ to find how many meters have a length of sixteen moras. The old Indian meter "woman with a beautiful body" mentioned earlier is just one of $A(16) = 1,597$ theoretically possible meters!

A discussion of the numbers $A(n)$ as the solution to Pingala's first problem in verse metrics can be found explicitly in the work of Hemachandra (1089–1172 CE). In the Western world, the sequence of numbers 1, 2, 3, 5, 8, 13, 21, 34 . . . , in which every number greater than 2 is the sum of the two preceding numbers, has been rediscovered quite often. Not knowing that these numbers were already known in old India more than one thousand years earlier, the French mathematician Édouard Lucas (1842–1891) called them *Fibonacci numbers* after Leonardo da Pisa, more popularly known today as Fibonacci (ca. 1170– ca. 1245). Fibonacci was the most important mathematician in medieval Europe, and he was largely responsible for the popularization of the Indo-Arabic numerals in Europe.

5.6. DISCOVERING COMMON TRAITS OF COUNTING PROBLEMS

One of the fascinating aspects of mathematics is that insight gained for one situation can be reapplied in quite different situations. The counting problem solved in the last section indeed occurs in many different contexts. It can be seen easily that the succession of long and short syllables in speech has much in common with the succession of long and short notes of a piece of music. Figure 5.1 shows all possible bars in six-eight time that consist only of quarter notes and eighth notes.

Figure 5.1: A list of all bars in 6/8 time containing only quarter notes and eighth notes.

We can see that the list of measures in figure 5.1 corresponds exactly to the list of all possible meters with a length of six moras, as shown earlier. And figure 5.2 shows one of the $A(16) = 1,597$ possible rhythms with a total length of 16 eighth notes, and consisting only of quarter notes and eighth notes:

Figure 5.2: Rhythm with a total length of 16 eighth notes.

The rhythm depicted in figure 5.2 occurs, for example, in Ludwig van Beethoven's Symphony no. 7, second movement. In the literature of India, this corresponds to the meter *rukmavati*, which has the following sequence of long and short syllables: − ⌣ ⌣ − − − − ⌣ ⌣ − − .

Here are two other examples:

The garden-path problem: You want to lay out a garden path with rectangular slabs. You can either place the slabs perpendicular or parallel to the direction of walking, as shown in figure 5.3. How many patterns are there to lay out sixteen slabs?

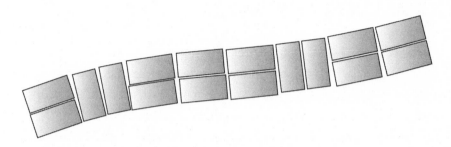

Figure 5.3: A garden path paved with sixteen slabs.

Do you recognize the similarity of this problem with Pingala's problem of arranging long and short syllables? A perpendicular slab would correspond to a single short syllable, and an element of two parallel slabs would correspond to a single long syllable. A long syllable takes the time of two short syllables and a parallel element consists of two slabs. The footpath is an arrangement of perpendicular (one slab) and parallel elements (two slabs), in very much the same way as a verse meter is an arrangement of short (one mora) and long (two moras) syllables. Thus, we conclude by analogy that the number of possible ways to lay out sixteen slabs in an arrangement of perpendicular and parallel elements is again $A(16) = 1,597$.

The problem of the postman: Every day, the driver of a parcel service has to use the same staircase with sixteen steps to deliver a parcel. Sometimes he climbs the stairs two at a time, sometimes he takes single steps. In order to add variety to his life, he decides to climb the stairs every day with another succession of single and double steps. How many ways of climbing the staircase are there?

A moment of thinking should reveal the similarity of this question with Pingala's first problem of counting rhythmic patterns of short and long syllables, or with the garden-path problem described previously. In all cases, the answer is $A(16) = 1,597$.

5.7. THE ART OF COUNTING SYLLABLES

Let's reconsider the problem of classifying verse meters. This was the original topic where the old Indian scholars, more than two thousand years ago, formulated for the first time typical questions of a scientific discipline nowadays called *mathematical combinatorics*.

In this section, we consider again the theoretical number of verse meters, but we slightly shift our point of view. Instead of asking for the number of meters with a given duration, we ask for the number of verse meters with a given length in syllables. Among the old Indian verse meters, it is the group of so-called Aksarachandas that are characterized by a fixed number of syllables. Again, every syllable has a fixed duration and is either long or short.

> Pingala's second problem:
>   How many different verse meters exist that have a total length of
> *n* syllables?

In order to realize one of the meters with a given number of *n* syllables, we have to distribute the long and short syllables in the verse. For the first syllable we can choose either a long or a short syllable. For each of these two beginnings, we have another two possibilities for the second syllable. This gives a total of four different possibilities for the first two syllables (namely $--$, $-\smile$, $\smile -$, and $\smile\ \smile$). For each one of these four forms, we have two possibilities to add a third syllable. Therefore, with every syllable we add, the number of possibilities is multiplied by two. We have

For 3 syllables: $2 \times 2 \times 2 = 2^3 = 8$ different meters
For 4 syllables: $2 \times 2 \times 2 \times 2 = 2^4 = 16$ different meters
For 5 syllables: $2 \times 2 \times 2 \times 2 \times 2 = 2^5 = 32$ different meters
. . .
For n syllables: $2 \times 2 \times \ldots \times 2$ (n factors) $= 2^n$ different meters

In this way we obtain, for example, in case of 24 syllables, more than 16 million possible verse meters—exactly 16,777,216. Very often, however, these verses consist of four similar parts (*padas*), and the first part alone determines the structure of the whole verse. But since Indian poetry has meters up to a length of more than one hundred syllables, we still obtain an enormous number of possible verse meters, of which only a few (actually several hundred) occur in practice.

It is interesting that old Indian scholars engaged in this kind of number game, which at first sight had little practical relevance. This could happen only because mathematics had already reached a fairly high level. Scholars had a deep knowledge about how to deal with numbers, and they were obviously proud to handle exceedingly large numbers. And they had already cultivated the ability to prove facts on the basis of logical arguments. In their examination of verse meters, we find a mathematical way of thinking that is visible in the ambition to understand all theoretically possible variants of a problem, even if not all the variants occur in reality.

5.8. THE ART OF COMBINATION

The Indian scholars also gave the answer to an even-more detailed and sophisticated problem: In how many ways can one combine a given number of short and long syllables into a verse? In honor of Pingala, we call this "Pingala's third problem."

Pingala's third problem:

How many different verse meters exist that have a total length of *n* syllables, among which are precisely *k* short syllables?

This is the same question as: In how many ways can we combine *k* short syllables and *n* − *k* long syllables? Here *n* is any natural number, but *k* can only have values between 0 and *n*. In the case of *k* = 0, the verse meter has only long syllables, and *k* = *n* means that it has only short syllables.

As in the case of Pingala's first problem, this one too is just a prototypical case of similar problems that are important in many different contexts. Indeed, big money is earned nowadays by exploiting the solution to this problem in lotto games. But more on this later.

We will approach Pingala's third problem step-by-step. As with Pingala's first problem, we start by giving the unknown number a name:

- The number of meters with *k* short syllables and a total of *n* syllables will be called $B(n,k)$.

It is often useful to approach a problem with a graphical illustration. Let us draw a single point on a piece of paper. This point should represent the beginning of the verse. From here, the verse may start either with a long or with a short syllable. These two alternatives will be shown in a "decision tree" by two arrows marked as L ("long") and S ("short"), as in figure 5.4.

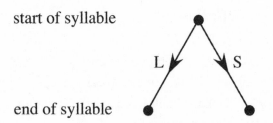

start of syllable

L   S

end of syllable

Figure 5.4: Two possibilities for the first syllable—long (L) or short (S).

What we have obtained so far, is $B(1,0) = 1$ and $B(1,1) = 1$. After the first syllable, the verse can continue in the same way. In figure 5.5 we see all possible steps that lead from the beginning of the verse to the end of the second syllable: LL, LS, SL, and SS. And this describes all possible verse meters consisting of four syllables.

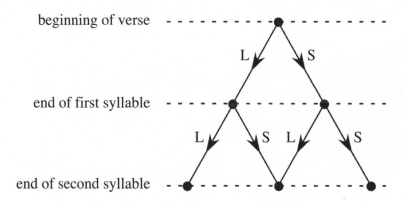

beginning of verse

end of first syllable

end of second syllable

Figure 5.5: Four possible paths lead to the end of the second syllable.

You can see that the two verses with the pattern LS and SL (trochee and iamb) end at the same point in the middle. This just indicates that both meters take the same time (three moras). If we count the time needed to reach the end of the second syllable by moras, as before, we find that it takes two moras (two short syllables) to reach the rightmost point, three moras to reach the point in the middle, and four moras to reach the leftmost point.

We collect our results for the second row of points:

$$B(2,0) = 1, B(2,1) = 2, \text{ and } B(2,2) = 1.$$

We continue to build the decision tree and add a line to the diagram for each additional syllable. Every path leading down from the top of the diagram determines a particular sequence of L and S syllables, and hence a particular verse meter. In figure 5.6, the marked path describes a seven-syllable meter with the pattern SLSLSSL = ⏑ – ⏑ – ⏑ ⏑ – (as in "I'd like to see you tonight").

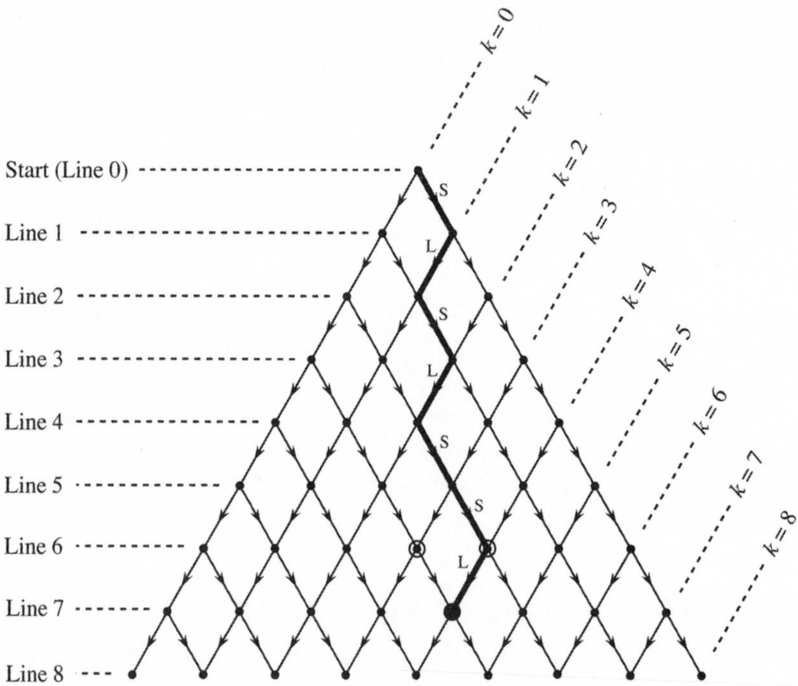

Figure 5.6: Decision diagram for verse meters with up to eight syllables.

Each point in the diagram can be identified by its line number and its place within that line. Note that line n marks the end of the nth syllable. The distance from the left side is described by the number k. In line n, the leftmost point has $k = 0$, and the rightmost point has $k = n$. Now every point in the diagram can be addressed by giving two numbers or "coordinates," the line number n and the (horizontal) distance k from the left: A general position in the diagram is (n, k), where k is between 0 and n. For example, the fat point in figure 5.6 has the coordinates (7,4) because it is in line 7 and 4 steps from the left border of the diagram. The top of the diagram has coordinates (0,0).

When you follow any path through the diagram that starts at the top, the number k is increased for every "S-step." In an "L-step," the number k is not changed. Any path in the diagram that ends at, say, $k = 4$, thus contains precisely four "k-steps" and corresponds to a verse with four short syllables. The coordinate k just counts the number of short syllables in the verse!

Thus, when we ask for the number $B(n,k)$ of verse meters with a total of n syllables of which k are short, we want to know the number of paths ending in line n at position k in the decision tree. And of course, the only allowed paths are those starting at the top of the diagram and going down in the direction of the arrows. A path going "upstairs" wouldn't mean anything. We have now found an important property of the number $B(n,k)$:

$B(n,k)$ = number of paths starting at (0,0) and ending at (n,k).

What have we gained now? Instead of counting possible verse meters, we now have to count the paths through a diagram, which doesn't appear much easier. At least we can now follow the paths with a finger and list all possible arrangements of long and short syllables. This works well for short verses. For example, we can find all verse meters ending in line 3 at position $k = 2$: LSS, SLS, and SSL (which today would be called *dactyl*, *amphibrach*, and *anapest*). This means, that $B(3,2) = 3$. We also notice that for all natural numbers $n = 1, 2, 3 \ldots$

- $B(n,0) = 1$, for all $n = 1, 2, 3 \ldots$ (only one path ends at the left border of line n).
- $B(n,n) = 1$, for all $n = 1, 2, 3 \ldots$ (only one path ends at the right border of line n).

If you try to count all possible paths ending at a particular point in the decision diagram, you will notice that this soon becomes rather difficult. Or can you easily find all thirty-five different paths that end at the point (7,4) in figure 5.6? Before we try to determine all the numbers $B(n,k)$ by actually counting all possible paths in the diagram, it is perhaps better to sit down thinking a bit more about their properties.

5.9. SOLVING PINGALA'S THIRD PROBLEM

In the section about Pingala's first problem, we were successful by "working backward." We were able to determine the number $A(n)$ with the help of the previous numbers $A(n-1)$ and $A(n-2)$. Perhaps a similar strategy will work here too?

Indeed, in figure 5.6, look at the fat point in line 7. How many paths lead to that point? All these paths obviously have to come from one of the two marked points in the previous line 6. Now we have just to add up all the paths leading to one of these two points in order to obtain the desired result. Thus we observe:

Number of paths leading to (7,4) = number of paths to (6,3) plus number of paths to (6,4).

As a formula, we can write this as follows:

$$B(7,4) = B(6,3) + B(6,4)$$

This observation is obviously true for all points in the diagram: The number of paths ending in a particular point is the sum of the paths leading to the immediately adjacent points in the line above. The points at the border are an exception to this rule, but we know already that there is only one path to each of the border points.

Now, let us write in the diagram the number of paths leading to each point. The prescription above gives us an easy method to determine the number of paths step-by-step. We write the results directly into the tree diagram, as in figure 5.7.

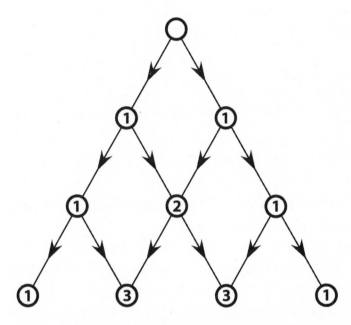

Figure 5.7: First lines of the decision tree with the numbers B(*n*,*k*) that count the number of paths leading from the top to that point.

It is now fairly easy to continue because we know that by adding two adjacent numbers, we obtain the number directly below. We can place "1" on all border points because we already know that only one path connects the border. For the sake of completeness, we put "1" even at the top—that is, we make the *definition* that $B(0,0) = 1$. The remaining places are then filled by simply adding the two adjacent numbers in the line above, as shown in figure 5.8.

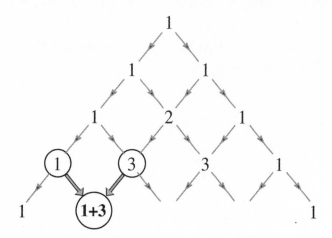

Figure 5.8: Constructing the numbers $B(n,k)$ step-by-step.

If we continue in this manner, by writing at each place the sum of the two nearest neighbors in the row above, we obtain a triangular table of numbers that can be continued indefinitely. Only our patience and the available space provide a natural limit (see figure 5.9). Here, for the sake of simplicity, we even omitted the arrows.

Indian astronomer and mathematician Varāhamihira (ca. 505) also knew the numbers $B(n,k)$, and he computed them in the way described above. The oldest record of the triangular arrangement seems to be in the writings of the Indian scholar Halayudha, who in the tenth century explained Pingala's findings. In Halayudha's commentary on Pingala's verse metric, this triangle is called the *Meru Prastara*, the "staircase to Meru." Halayudha saw in it a symbolic representation of the sacred Mount Meru, a mythological mountain in Hindu cosmology.

Like the Fibonacci numbers, the arithmetic triangle in figure 5.9 has been rediscovered often in other regions of the world. Shortly after its appearance in India, it was known in China to the palace eunuch and mathematician Jia Xian (1010–1070). Later, Yang Hui (about 1238–1298) explained its use in detail, quoting Jia Xian as his source.

Unfortunately, the original source was lost and the triangle became known as Yang Hui's triangle in China. Figure 5.10 shows an early drawing of Yang Hui's triangle. The numbers in the circles are written using so-called rod numerals.

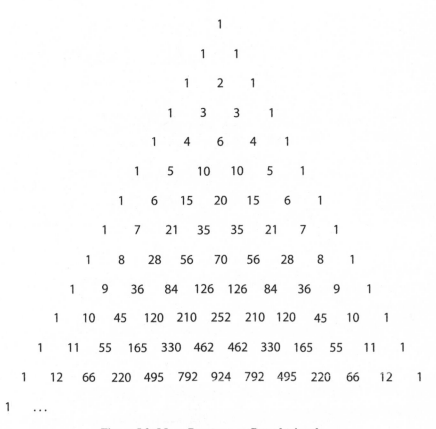

Figure 5.9: Meru Prastara, or Pascal triangle.

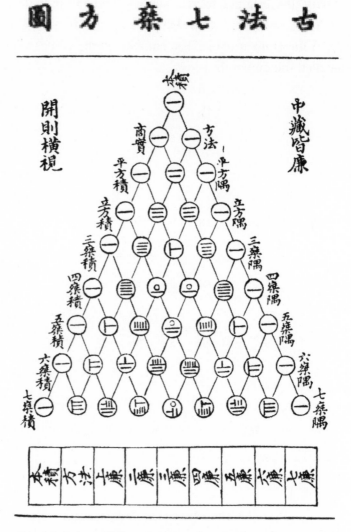

Figure 5.10: Yang Hui's triangle in a drawing from the year 1303. (From Wikimedia Commons, user Noe.)

About at the same time as in India, the arithmetic triangle was discovered by Al Karaji (ca. 953–1029) in Baghdad and later again by Omar Khayyām (1048–1131). Hence in Iran, it is known as the Khayyām triangle.

In Germany, the Renaissance scholar Peter Apian (1495–1552) was the first to publish the arithmetic triangle in Europe. In Italy, the *triangolo di Tartaglia* is named after Nicolo Tartaglia (1499–1557), a mathematician who is famous for finding a formula for solving cubic equations. In the modern Western mathematical literature, the triangle in figure 5.9 is known as the Pascal triangle after Blaise Pascal (1623–1662), who used the properties of the triangle to solve problems in probability theory. Pascal's *Traité du triangle arithmétique* ("Treatise on an Arithmetical Triangle") was published posthumously in 1665. His version of the triangle is shown in figure 5.11.

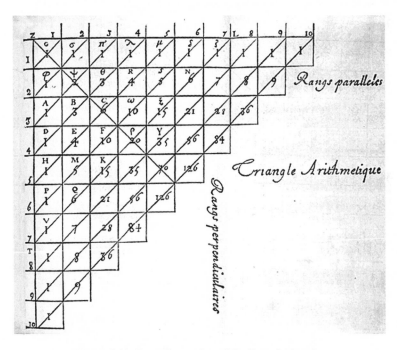

Figure 5.11: Pascal's version of the Pascal triangle.
(From B. Pascal, *Traité du Triangle Arithmétique*, 1654.)

5.10. THE PASCAL TRIANGLE AND PINGALA'S PROBLEMS

The Pascal triangle in figure 5.9 contains the full solution to Pingala's third problem. But not only this, the solutions to the Pingala's first and second problem are also hidden in there.

Pingala's second problem states that the number of all paths leading from the top of the decision diagram (figure 5.7) to a point in line n at position k is equal to the number of verse meters with n syllables of which k are short. This number has been called $B(n,k)$, and this is the number at position (n,k) in the Pascal triangle. In order to find the number of *all* meters with n syllables, we just have to sum up for this n all the numbers $B(n,k)$ with $k = 0$ up to $k = n$. This takes into account all meters with a length of n syllables and thus solves Pingala's second problem.

Indeed, we find that the sum of the numbers in each row is just a power of 2 (see figure 5.12). For example, the sum of the numbers in row number 5 (corresponding to five syllables) is

$$1 + 5 + 10 + 10 + 5 + 1 = 32 = 2^5,$$

and this is precisely the result that was obtained previously by a different reasoning.

$$1$$

1 syllable ..... $1 \quad + \quad 1 \quad = \quad 2$

2 syllables ..... $1 \quad + \quad 2 \quad + \quad 1 \quad = \quad 4$

3 syllables ..... $1 \quad + \quad 3 \quad + \quad 3 \quad + \quad 1 \quad = \quad 8$

4 syllables ..... $1 \quad + \quad 4 \quad + \quad 6 \quad + \quad 4 \quad + \quad 1 \quad = \quad 16$

5 syllables ..... $1 \quad + \quad 5 \quad + \quad 10 \quad + \quad 10 \quad + \quad 5 \quad + \quad 1 \quad = \quad 32$

6 syllables ..... $1 \quad + \quad 6 \quad + \quad 15 \quad + \quad 20 \quad + \quad 15 \quad + \quad 6 \quad + \quad 1 \quad = \quad 64$

$1 \quad + \quad 7 \quad + \quad 21 \quad + \quad 35 \quad + \quad 35 \quad + \quad 21 \quad + \quad 7 \quad + \quad 1$

Figure 5.12: Sum across lines in Meru Prastara.

Pingala's first problem asks that we determine the total number of meters with a given *duration*. Here duration is measured in moras, where a short syllable has one mora and a long syllable has two moras.

The solution to this problem is also hidden in the Pascal triangle. We only have to remember that all paths ending at a certain point (n,k) have precisely k short syllables and $n - k$ long syllables, hence they have the same duration. But there are other endpoints describing same duration. Consider figure 5.13, which shows some region of the decision tree. If we start at point "a," anywhere in the decision tree, it takes a long syllable to go to "b" and two short syllables to go to "c"—two moras in both cases. Hence the points "b" and "c" correspond to verse meters with the same duration (two moras longer than "a"). By the same reasoning, we find that all the points along a "shallow diagonal" (the dashed line in figure 5.13) correspond to verse meters of the same duration.

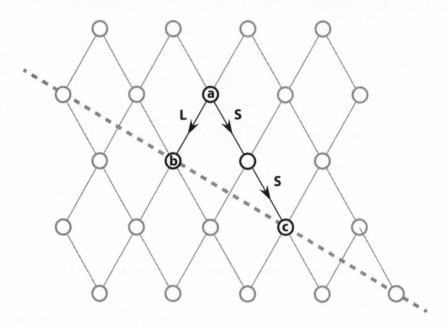

Figure 5.13: Points "b" and "c" correspond to verse meters with the same duration.

So if we want to know the total number of verse meters of a given duration, we have to sum up all entries in the Pascal triangle that sit along a shallow diagonal. This should give the solution to Pingala's first problem. Indeed, we find the Fibonacci numbers $A(n)$ as the sums along the shallow diagonals in the Pascal triangle, as illustrated in figure 5.14.

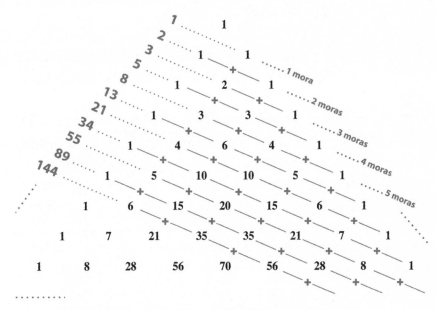

Figure 5.14: The sums along "shallow diagonals" are Fibonacci numbers.

Many other discoveries can be made within the Pascal triangle. You can see that the first diagonal, that is given by all numbers with $k = 1$, are just the natural numbers $1, 2, 3, 4 \ldots$,

$$B(n,1) = n.$$

The second diagonal is also a sequence familiar from the previous chapter; it is the sequence of triangular numbers $1, 3, 6, 10, 15, 21 \ldots$, characterized by the property

$$B(n + 1,2) = B(n,2) + B(n,1) = B(n,2) + n.$$

Moreover, the third diagonal sequence, the numbers with $k = 3$, are just the tetrahedral numbers $1, 4, 10, 20,$ and 35.

5.11. LOTTERY GAMES AND OTHER AMUSEMENTS

We find a wealth of applications for the results in the previous section when we contemplate the problem of counting meters from a slightly different perspective.

Consider, as an example, the seven-syllable meters with two short syllables. We can create these from a sequence of only long syllables by choosing two of them and replacing them with short syllables. This would give us one of the twenty-one different seven-syllable meters with two short syllables. If we ask for the number of all different seven-syllable meters containing just two short syllables, we could also ask in how many different ways we can select two elements from a collection of seven (see figure 5.16).

Figure 5.15: Choose any two of these numbers. You can do this in $B(7,2) = 21$ different ways.

Quite generally, $B(n,k)$ describes the number of possibilities to choose k objects out of n objects. Mathematicians are usually more precise when defining how to "choose." For example, the order doesn't matter. It makes no difference if you choose first object 6 and then object 2, or if you do it the other way around. Another condition is that you can choose every object only once. In a more mathematical way of speaking,

$B(n,k)$ describes the number of all k-element subsets of an n-element set.

The two-element subsets of the seven-element set $\{1,2,3,4,5,6,7\}$ are listed in the following table; there are $B(7,2) = 21$ different ways to choose two numbers out of seven (see table 5.2). The particular arrangement of these subsets in table 5.2 also explains why $B(7,2)$ is one of the triangular numbers.

$$\{1,2\}, \{1,3\}, \{1,4\}, \{1,5\}, \{1,6\}, \{1,7\},$$
$$\{2,3\}, \{2,4\}, \{2,5\}, \{2,6\}, \{2,7\},$$
$$\{3,4\}, \{3,5\}, \{3,6\}, \{3,7\},$$
$$\{4,5\}, \{4,6\}, \{4,7\},$$
$$\{5,6\}, \{5,7\},$$
$$\{6,7\}$$

Table 5.2: All two-element subsets of the seven-element set $\{1,2,3,4,5,6,7\}$.

We can now answer the following question from the beginning of this chapter:

Assume that there are n persons at a New Year's Eve party. At midnight, everybody clinks glasses with everyone else. How many clinks will you hear?

Think a moment before you read the solution in the next paragraph. Perhaps you can find the solution all by yourself?

One needs two persons to clink glasses. There will be as many clinks as you can choose two persons from the n persons at the party, that is, $B(n,2)$ clinks. If there are seven persons at the party, you will hear twenty-one clinks. Cheers!

If alcohol is not in short supply, people might toast somebody twice while forgetting others altogether. Hence, as it is often the case, mathematics would only provide an approximate solution for the real situation.

At a New Year's Eve party, you would wish luck to your friends, maybe hoping to finally make a big win yourself—a hope that allows casinos and lottery companies around the world a carefree existence. For example, in a popular lottery game, you can choose six numbers from a set of forty-nine when you buy a ticket. Then, in a public drawing, six numbers are randomly drawn from the pool of forty-nine numbers. You would win the jackpot prize if your ticket matched all the numbers in the drawing.

The number of possible outcomes of this game is the same as the number of all possible choices of six elements out of a set of forty-nine elements—that is, $B(49,6) = 13,983,816$. Only one of these choices wins. Hence the probability to win in that lottery game would be

$$\frac{1}{\text{number of possible outcomes}} = \frac{1}{13,983,816}.$$

This gives a chance of about 1 in 14 million. In order to visualize that probability, imagine a chain of domino tiles on the roadside, all along the street from New York City to Niagara Falls. As the travel distance is about 400 miles, and each domino is a little less than 2 inches long, this chain would contain about 14 million tiles. Assume that one of these tiles carries a mark on the bottom side. You are allowed to stop once during your trip and pick up one of the dominoes. Would you bet on finding the marked domino with one attempt (or even one hundred attempts)? The chance to win at the 6/49 lottery is about as good (or bad). And yet, every week, millions of people pay their wager.

CHAPTER 6

NUMBER EXPLORATIONS

6.1. THE FIBONACCI NUMBERS IN EUROPE

Pingala's numbers $A(n)$, introduced in chapter 5, section 4, in order to describe the number of verse meters with a given duration, are known today as *Fibonacci numbers* and are usually represented in an algebraic context with the symbol F_n representing the nth Fibonacci number. They are perhaps one of the most ubiquitous number sequences in mathematics. For historical reasons, one has $F_n = A(n-1)$ starting with $n = 2$, and the definition (which parallels the definition of the $A(n)$ in section 5.5) now reads

$F_1 = 1, F_2 = 1$.
$F_n = F_{n-1} + F_{n-2}$, for all natural numbers n greater than 2.

In other words, beginning with the numbers 1 and 1, each succeeding number is the sum of the two preceding numbers. They are named after Fibonacci, whose real name was Leonardo da Pisa and who is famous for promoting the use of the Hindu-Arabic numeral system in Europe during the early thirteenth century (see chapter 3, section 8). The numbers F_n appeared for the first time in the Western world in Fibonacci's book *Liber Abaci*, published in 1202. In chapter 12 of that book, he posed the following famous problem concerning the regeneration of rabbits:

161

Fibonacci's problem:

A man had one pair of newborn rabbits together in a certain closed place. He wishes to know how many pairs of rabbits will be created from the pair in one year, making the following assumptions concerning the nature of the rabbits:

- For a newborn pair, it takes two months to mature and afterward give birth to a new pair of rabbits.
- Mature pairs bear a new pair every month.
- No rabbit will die.

The page from *Liber Abaci* that contains the solution to this problem is shown in figure 6.1.

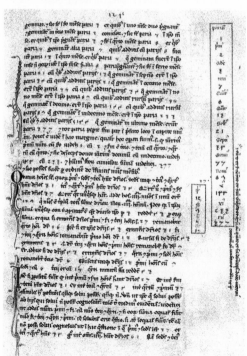

Figure 6.1: The page from *Liber Abaci* explaining the rabbit problem. (From Wikimedia Commons, user Otfried Liberknecht. Original from Fibonacci, *Liber Abaci*, 1202, located at Biblioteca Nazionale di Firenze, Florence, Italy.)

The situation described in Fibonacci's problem is shown in table 6.1. Notice that we will be counting pairs and not individual rabbits. Thus, we begin with a single pair, born at the beginning of the year. It needs two months to mature and reproduce. Thus, in the third month, there will be the first baby pair. At the end of this month, we then have two pairs. In the fourth month, the young pair is still growing, while the older one again gives birth to a baby pair. At the end of the fourth month, we therefore have three pairs. In month number five, these three pairs are still there because it is the assumption that the rabbits will not die. While the newborn pair from the fourth month has to wait to mature, two of the three pairs are now adults, and, hence, there will be two additional pairs in the fifth month.

month	number of rabbit-pairs	F_n
$n = 1$		1
2		1
3		2
4		3
5		5
6		8
7		13
8	. . .	21

Table 6.1: Proliferation of Fibonacci rabbits showing babies, young rabbits, and adults.

We see that the number of pairs obviously grows according to a rule that we call the *Fibonacci sequence*. The explanation is this: If the number of rabbit pairs in month n is denoted by F_n, then this number consists of the number (F_{n-1}) of rabbits in month $n - 1$, plus precisely F_{n-2} newborn pairs (children of all the rabbits that existed in month $n - 2$ because those are the adults in month n). This can be summarized as $F_n = F_{n-1} + F_{n-2}$, which (together with the initial condition) defines the Fibonacci sequence.

Now, we can easily compute the number of pairs produced in one year. It is $F_{12} = 144$. This is the solution of Fibonacci's famous problem. But this is just the beginning, as an incredible explosion of the population of rabbits would follow using this scheme. After two years, the number of pairs would be $F_{24} = 46,368$. After one hundred months (a little over eight years), the number of rabbits would be

$$F_{100} = 354,224,848,179,261,915,075.$$

The animals shown in table 6.1 are, in fact, not rabbits, but copies of Albrecht Dürer's famous painting *Young Hare* from the year 1502 CE. If these animals had reproduced according to Fibonacci's assumptions for rabbit reproduction, then today the number of pairs—about five hundred years, or six thousand months, later—would be described by a number with 1,254 digits:

$$F_{6000} = 377,013,149,387,799 \ldots (\textit{1,224 digits omitted}) \ldots$$
$$475,233,419,592,000.$$

The whole mass of the observable universe, if converted into hares, would by far not be sufficient to create that many pairs. Obviously, the rules defined by Fibonacci for the proliferation of rabbits are not, in the long run, very realistic. But, of course, it was not Fibonacci's goal to give a realistic model of population growth; he just wanted to provide the reader with an intellectually stimulating and entertaining mathematics problem.

6.2. GENERATIONS OF RABBITS

Another stimulating question about the rabbit problem would be the following: How many pairs of rabbits belong to a particular generation of rabbits in a given month? Table 6.2 provides the answer. It shows the number of rabbit pairs for each month, ordered according to the generation to which they belong. Originally, we started with a single pair constituting the first generation. This pair is always there because Fibonacci rabbits never die. The children of the first pair are listed in column 2; they belong to the second generation. Starting with month 2, the original pair gives birth to a new pair each month; consequently, the number of second-generation pairs increase by one each month. The third-generation pairs are the children of the second-generation rabbits, or the grandchildren of the original pair.

month	generation							sum
	1	2	3	4	5	6	7	
1	1							1
2	1							1
3	1	1						2
4	1	2						3
5	1	3	1					5
6	1	4	3					8
7	1	5	6	1				13
8	1	6	10	4				21
9	1	7	15	10	1			34
10	1	8	21	20	5			55
11	1	9	28	35	15	1		89
12	1	10	36	56	35	6		144

Table 6.2: Number of Fibonacci rabbits per generation.

Let us consider the number of pairs of the kth generation in month m. We denote this number by $S(m,k)$. This number consists of

- all rabbits of kth generation that existed already in the month before ($= S(m-1,k)$), plus
- the number of the newborn pairs of the kth generation.

However, the number of newborn pairs is exactly equal to the number of parent pairs (belonging to generation $k-1$) that existed two months earlier, that is, $S(m-2,k-1)$. Therefore,

$$S(m,k) = S(m-1,k) + S(m-2,k-1).$$

That is, the numbers $S(m,k)$ are obtained as the sums of two previously obtained numbers, as indicated for the numbers 9, 10, and 56 in table 6.2. With this rule you could easily compute further entries to the chart in table 6.2.

Perhaps you have already recognized the similarity of the numbers in table 6.2 with the numbers in another arrangement, the Meru Prastara, or Pascal triangle, detailed in section 5.9. In fact, the chart in table 6.2 is just a distorted version of the Pascal triangle, whose entries have now received a new interpretation, namely as the number of pairs of Fibonacci rabbits belonging to a certain generation in a certain month. The columns in table 6.2 are the steep diagonals of the Pascal triangle; the rows are the shallow diagonals. This interpretation of the Pascal triangle is shown in figure 6.2, where the black lines correspond to the rows and the gray lines to the columns in table 6.2.

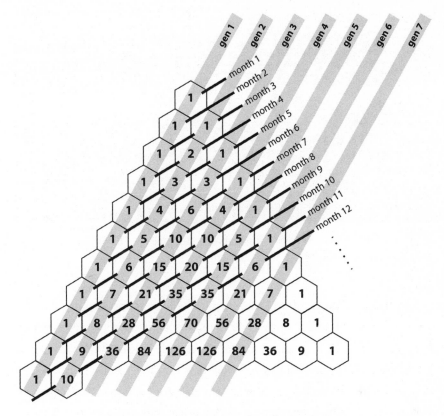

Figure 6.2: Pascal triangle and rabbit generations.

The consideration above offers an alternative to the approach described in chapter 5 that led us to Meru Prastara, the old Indian version of the Pascal triangle. However, this path was not taken by Fibonacci, and in Europe it took several more centuries until the triangle was finally rediscovered by Blaise Pascal in the seventeenth century.

Likewise, the Fibonacci numbers were not identified as anything special during the time Fibonacci wrote *Liber Abaci* in 1202. Centuries passed, and the numbers still went unnoticed. Then in the 1830s, C. F. Schimper and A. Braun noticed that the numbers appeared as the number of spirals of bracts on a pinecone (see figure 6.3).

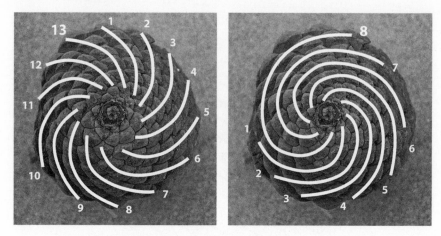

Figure 6.3: The number of spirals of a pinecone are Fibonacci numbers 8 and 13.

In the mid-1800s, the Fibonacci numbers began to capture the fascination of mathematicians. They took on their current name (*Fibonacci numbers*) from François-Édouard-Anatole Lucas (1842–1891), the French mathematician usually referred to as Edouard Lucas. Lucas is well known for his invention of mathematical recreations like the Tower of Hanoi puzzle, which appeared in 1883 under the name of N. Claus de Siam, which is an anagram of Lucas d'Amiens. His four-volume work on recreational mathematics (1882–1894) has become a classic. Lucas died as the result of a freak accident at a banquet when a plate was dropped and a piece flew up and cut his cheek. He died of erysipelas a few days later.

Lucas devised his own sequence by following the pattern set by Fibonacci. The Lucas numbers form a sequence of numbers much like the Fibonacci numbers, with which they share many properties. The Lucas numbers differ from the Fibonacci numbers in that they begin with different initial numbers:

$L_1 = 1, L_2 = 3.$
$L_n = L_{n-1} + L_{n-2}$, for all natural numbers n greater than 2.

Hence, the sequence of Lucas numbers starts with

$$1, 3, 4, 7, 11, 18, 29 \ldots$$

At about this time, French mathematician Jacques-Philippe-Marie Binet (1786–1856) developed a formula for finding any Fibonacci number, given its position in the sequence. That is, with Binet's formula we can find the 118th Fibonacci number without having to list the previous 117 numbers. The Binet formula for the nth Fibonacci number is

$$F_n = \frac{1}{\sqrt{5}} \left[\left(\frac{1+\sqrt{5}}{2} \right)^n - \left(\frac{1-\sqrt{5}}{2} \right)^n \right].$$

It should be noted that numerous other mathematicians prior to Binet came up with a formula analogous to this one; however, over time, the formula took on the name of Binet.

Today, these celebrated numbers still hold the fascination of mathematicians around the world. The Fibonacci Association was created in 1963 to provide enthusiasts an opportunity to share discoveries about these intriguing numbers and their applications. Through the Fibonacci Association's *The Fibonacci Quarterly*, many new facts, applications, and relationships about these numbers can be shared worldwide. According to its official website, *The Fibonacci Quarterly* is meant to serve as a focal point for interest in the Fibonacci numbers and related ideas, especially with respect to new results, research proposals, challenging problems, and innovative proofs of already-known relationships.

The Fibonacci numbers seem to crop up in countless botanical structures, such as the number of spirals of bracts on a pineapple or on a pinecone. They also appear when counting branches of various trees, and they make themselves omnipresent in architecture and art, as they are closely tied to the golden ratio. We recommend an excursion through these marvelous numbers, which can be found in *The Fabulous Fibonacci Numbers* by A. S. Posamentier and I. Lehmann (Amherst, NY: Prometheus Books, 2007). There you will find the enormous variety of appearances of the Fibonacci numbers.

6.3. MORE ABOUT THE PASCAL TRIANGLE

Let us now return to the Pascal triangle and explore some of its properties. This triangular arrangement provides us with a wealth of unusual, and quite amazing, relationships. For example, in any row where the second element (i.e., number) is a prime number, then every other number in that row will be a multiple of that prime number. For example in the eleventh row, the second number is 11, and therefore every other number in that row (55, 165, 330, 462) is a multiple of 11—of course with the exception of the 1s on either end of the row.

Another curious property of this triangular arrangement of numbers is shown in figure 6.4. Starting at the border, go down along one diagonal up to an arbitrary point. At the terminal point, look in the other direction, as indicated in figure 6.2, to see the number that will be the sum of the numbers that you passed through to get to that point. In figure 6.4, the circled numbers are the sum of the numbers in the shaded part of the diagonal.

A special case of this observation is that the numbers in the third diagonal are just the sums of the numbers in the second diagonal. As an example, consider the shaded part of the second diagonal in figure 6.5. It indicates the natural numbers from 1 to 7. The sum of these numbers may be found by simply looking to the number below and to the left of the number 7. It is $28 = 1 + 2 + 3 + 4 + 5 + 6 + 7$.

The second diagonal is just the sequence of natural numbers. In section 4.5, we identified the triangular numbers as the sum of consecutive natural numbers beginning with 1. Hence, the third diagonal is the sequence of triangular numbers, marked in bold in figure 6.5.

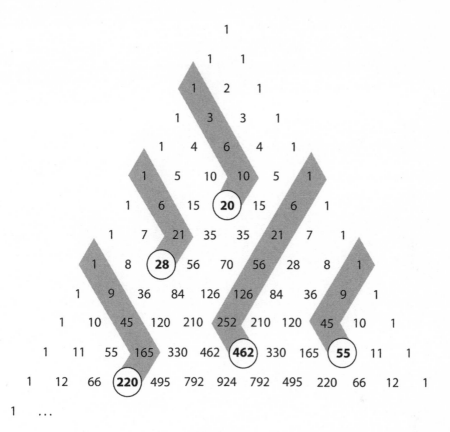

1

1   1

1   2   1

1   3   3   1

1   4   6   4   1

1   5   10   10   5   1

1   6   15   (20)   15   6   1

1   7   21   35   35   21   7   1

1   8   (28)   56   70   56   28   8   1

1   9   36   84   126   126   84   36   9   1

1   10   45   120   210   252   210   120   45   10   1

1   11   55   165   330   462   (462)   330   165   (55)   11   1

1   12   66   (220)   495   792   924   792   495   220   66   12   1

1   ...

Figure 6.4

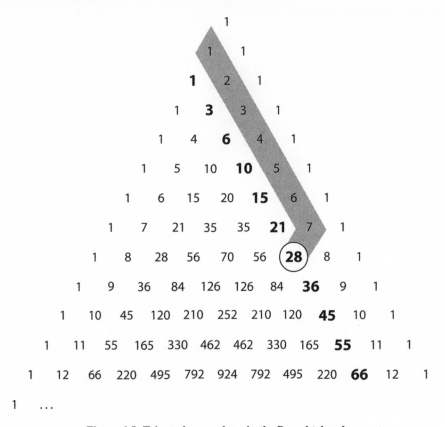

Figure 6.5: Triangular numbers in the Pascal triangle.

There are lots of interesting patterns to be found in the Pascal triangle. Consider the following in figure 6.6. Here we have marked the numbers encircling the number 10. Take the product of the numbers in circles, namely, $5 \times 6 \times 20 = 600$, and then the product of the numbers in polygonal shapes, $4 \times 10 \times 15 = 600$, and you will find a surprising result—the products are equal.

We can repeat this by considering similar shapes anywhere in the Pascal triangle. Figure 6.6 also shows two other examples: The numbers encircling 9 illustrate the products $1 \times 8 \times 45 = 360$ and $1 \times 10 \times 36 = 360$. Once again, notice that the two products are equal. Just to drive a point home, we will repeat this one more time with the shaded cells that encircle the number 210. Here the two products are 84 × 252 × 330 = 6,985,440 and 120 × 126 × 462 = 6,985,440.

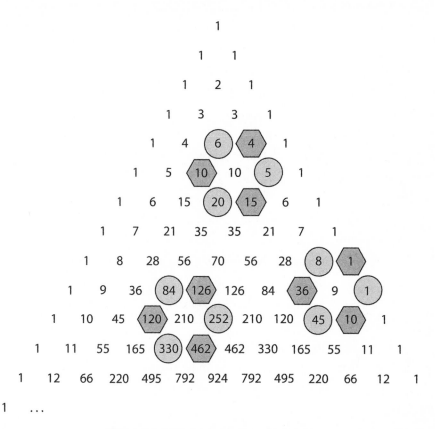

Figure 6.6: Patterns in the Pascal triangle.

6.4. COMBINATORIAL GEOMETRY

What makes the Pascal triangle so truly outstanding is the many fields of mathematics it touches or involves. We have seen in chapter 5, section 11, how the Pascal triangle arises in the theory of lotto games (probability theory). For sheer enjoyment, we shall consider some more applications of the Pascal triangle now.

In table 6.3 we have listed circles in the first column with an increasing number of points on the circle. With the circles that have two or more points, we have drawn all possible connecting segments. Each column lists the number of polygons that one can count in the figure shown in the first column of that row. You will notice that we have essentially created the Pascal triangle with the exception of the first 1 at the left (except for the top 1). The reason for the appearance of the Pascal triangle in table 6.3 can be understood as follows: In chapter 5, we denoted the number in the Pascal triangle that is n steps down from the top and k steps from the left border by $B(n,k)$ (e.g., $B(7,2) = 21$). According to chapter 5, section 11, the number $B(n,k)$ equals the number of all k-element subsets of an n-element set. Consider, for example, the number of line segments connecting two of the vertices of a heptagon. Obviously, this is the same as the number of 2-vertex subsets of the set of seven vertices, that is, $B(7,2) = 21$ (which is indeed the number in line 7 and column 2 of table 6.3). Similarly, each triangle formed from the vertices of a heptagon corresponds to a choice of three vertices out of the given seven vertices. Hence the number of those triangles is equal to the number of 3-vertex subsets, that is, $B(7,3) = 35$.

The Pascal triangle can also help us predict into how many pieces a plane can be divided by a given number lines, where no two are parallel and no three are concurrent (i.e., containing a common point of intersection). Figure 6.7 shows that three lines can divide the plane into seven regions, and four lines can divide the plane into eleven regions. The condition is that no two lines are parallel, and not more than two lines intersect at the same point.

Figure	Points	Segments	Triangles	Quadrilaterals	Pentagons	Hexagons	Heptagons
	1						
	2	1					
	3	3	1				
	4	6	4	1			
	5	10	10	5	1		
	6	15	20	15	6	1	
	7	21	35	35	21	7	1

Table 6.3: Numbers of cyclic polygons.

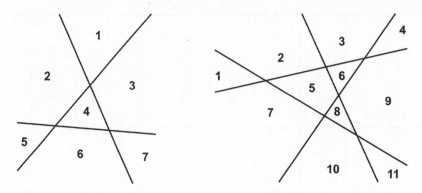

Figure 6.7: Regions formed by three or four lines in a plane.

Among other information, these numbers are listed in the chart in table 6.4. Just wait a bit and you will see how this ties in with the Pascal triangle.

Values of n	Number of Segments Formed by n Points on a Line	Number of Regions Formed by n Lines in a Plane	Number of Space-Regions Formed by n Planes in Space
0	1	1	1
1	2	2	2
2	3	4	4
3	4	7	8
4	5	11	15
5	6	16	26

Table 6.4: Number of geometric partitions.

For example, when there are three points on the line, the line is partitioned into four line segments. Or, consider the case with five lines in a plane intersecting in such a way that no two lines are parallel and no three lines are concurrent, we find that there are sixteen regions determined. This should give you some idea as to how the numbers in the chart have been generated. But now we shall see how the Pascal triangle could also have generated these numbers.

The columns of the chart can be obtained from the Pascal triangle by taking the sum of the shaded elements in each of the rows, as shown in figure 6.8.

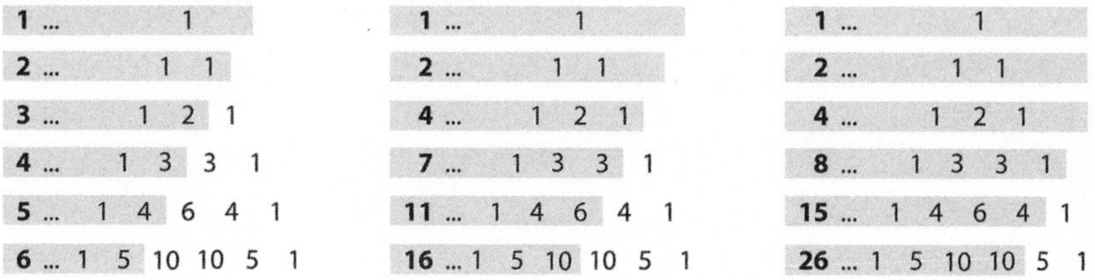

```
 1 ...          1              1 ...          1             1 ...          1
 2 ...         1  1            2 ...         1  1           2 ...         1  1
 3 ...       1  2  1           4 ...       1  2  1          4 ...       1  2  1
 4 ...     1  3  3  1          7 ...     1  3  3  1          8 ...     1  3  3  1
 5 ...   1  4  6  4  1        11 ...   1  4  6  4  1        15 ...   1  4  6  4  1
 6 ... 1  5 10 10  5  1       16 ... 1  5 10 10  5  1       26 ... 1  5 10 10  5  1
```

Figure 6.8: Partial sums of the rows in the Pascal triangle.

6.5. THE BINOMIAL EXPANSION

The Pascal triangle contains many unexpected number relationships beyond the characteristics that Pascal intended in his original use of the triangle. Pascal's original use was to exhibit the coefficients of successive terms of a binomial expansion.

We shall now consider the binomial expansion—that is, taking a binomial such as $(a + b)$ to successively higher powers. By now, you ought to be able to recognize the pattern formed by the coefficients of the terms shown in figure 6.9. The coefficients of each of the binomial-expansion lines is also represented as a row of the Pascal triangle.

$$(a + b)^0 = 1$$

$$(a + b)^1 = 1a + 1b$$

$$(a + b)^2 = 1a^2 + 2ab + 1b^2$$

$$(a + b)^3 = 1a^3 + 3a^2b + 3ab^2 + 1b^3$$

$$(a + b)^4 = 1a^4 + 4a^3b + 6a^2b^2 + 4ab^3 + 1b^4$$

$$(a + b)^5 = 1a^5 + 5a^4b + 10a^3b^2 + 10a^2b^3 + 5ab^4 + 1b^5$$

$$(a + b)^6 = 1a^6 + 6a^5b + 15a^4b^2 + 20a^3b^3 + 15a^2b^4 + 6ab^5 + 1b^6$$

$$(a + b)^7 = 1a^7 + 7a^6b + 21a^5b^2 + 35a^4b^3 + 35a^3b^4 + 21a^2b^5 + 7ab^6 + 1b^7$$

$$(a + b)^8 = 1a^8 + 8a^7b + 28a^6b^2 + 56a^5b^3 + 70a^4b^4 + 56a^3b^5 + 28a^2b^6 + 8ab^7 + 1b^8$$

Figure 6.9: The binomial expansion.

This allows us to expand a binomial without actually multiplying it by itself many times to get the end result. There is a pattern also among the variables' exponents: one descends while the other ascends in value—each time keeping the sum of the exponents constant—that is, the sum is equal to the exponent of the power to which the original binomial was taken.

The formulas in figure 6.9 can be written in compact form in a single line, which we provide here because of its beauty. This formula is called the *binomial theorem*.

$$\left(a + b\right)^n = \sum_{k=0}^{n} \binom{n}{k} a^{n-k} b^k$$

The symbol $\binom{n}{k}$ is pronounced "n over k." This is the modern notation for the numbers in the Pascal triangle, the so-called binomial coefficients, which we denoted by $B(n,k)$ in chapter 5.

Actually, there is a formula that allows us to compute any binomial coefficient given its position n,k in the Pascal triangle, without having to evaluate the binomial coefficients above. This formula uses the notation

$$n! = 1 \times 2 \times 3 \times \ldots \times (n-1) \times n$$

for the product of all natural numbers from 1 to n. The expression $n!$ is pronounced "n-factorial." With this abbreviation, the binomial coefficient is given by

$$\binom{n}{k} = \frac{n!}{k!\,(n-k)!} = B(n,k).$$

For example,

$$\binom{7}{3} = \frac{7!}{3!\,(7-3)!} = \frac{1 \times 2 \times 3 \times 4 \times 5 \times 6 \times 7}{1 \times 2 \times 3(1 \times 2 \times 3 \times 4)} = \frac{5 \times 6 \times 7}{1 \times 2 \times 3} = \frac{210}{6} = 35$$

gives the binomial coefficient $B(7,3) = 35$. According to chapter 5, it describes in how many ways we can choose three elements out of a set of seven elements. It also tells us how many ways seven coin tosses come up with three heads.

For $a = 1$ and $b = 1$, the expressions in figure 6.9 can be simplified, as all products of powers of a and b can be replaced by 1. We obtain, for example,

$$(1 + 1)^6 = 2^6 = 1 + 6 + 15 + 20 + 15 + 6 + 1,$$

which is just the sum of the entries in the corresponding row of the Pascal triangle. In that way, we can reproduce the result of figure 5.12: The sums across the lines of the Pascal triangle are the powers of 2.

For $a = 10$ and $b = 1$, all factors $a^n b^k$ in figure 6.9 simply become 10^n. This leads to an interesting observation. For example,

$$(10 + 1)^3 = 11^3 = \mathbf{1} \times 10^3 + \mathbf{3} \times 10^2 + \mathbf{3} \times 10^1 + \mathbf{1} = 1{,}331.$$

The binomial formula, in this case, provides the representation of a number in our decimal numeral system. And the digits of this number are the entries of the Pascal triangle in the corresponding row. This shows us that if one reads a row of the Pascal triangle as a single number whose digits are the elements of that row, we get a power of 11. Indeed, $121 = 11^2$, and the numbers 1, 4, 6, 4, 1 in the fourth row lead to $14{,}641 = 11^4$. However, beginning with the fifth row, we would have to regroup the digits:

$$11^5 = 1 \times 10^5 + 5 \times 10^4 + 10 \times 10^3 + 10 \times 10^2 + 5 \times 10^1 + 1 = 161{,}051.$$

In this case, some of the elements in the Pascal triangle have more than one digit, and one has to carry the leading digits over to the next order. So, for example, the sixth power of 11 has to be computed as follows:

$$
\begin{array}{r}
11^6 = 1 \\
6 \\
1\ 5 \\
2\ 0 \\
1\ 5 \\
6 \\
1 \\
\hline
1\ 7\ 7\ 1\ 5\ 6\ 1
\end{array}
$$

Figure 6.10: How to determine a power of 11 from the Pascal triangle.

The occurrence of Fibonacci numbers, powers of 2, and powers of 11 in the Pascal triangle are once again illustrated in figure 6.11.

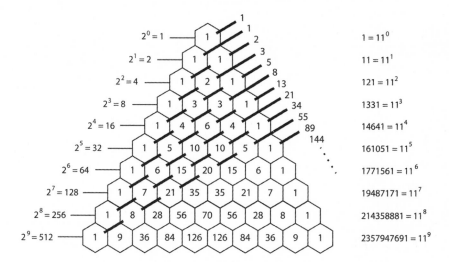

Figure 6.11: Amazing number relationships in the Pascal triangle

There are many number relationships present in the Pascal triangle. The turf is fertile. The opportunity to find more gems in this triangular arrangement of numbers is practically boundless! We encourage the motivated reader to search for more hidden treasures embedded in this number arrangement.

CHAPTER 7

PLACEMENT OF NUMBERS

7.1. MAGIC SQUARES

For some people, the expression "recreational mathematics" might be an oxymoron—a contradiction in terms, like "living dead" or "dark light." And yet there are many people, amateurs and professionals alike, who play around with numbers and mathematical objects just for fun. And often they make very interesting discoveries. One of the mathematicians well known for his activities in the field of recreational mathematics was Edouard Lucas, who, among other things, popularized the Fibonacci numbers in the nineteenth century (see chapter 5). In the twentieth century, a well-known writer with an intense interest in recreational mathematics was American author Martin Gardner (1914–2010). During a period of twenty-five years, he published a column called "Mathematical Games" in the journal *Scientific American* and wrote many books on recreational mathematics.

Mathematical puzzles form an important part of recreational mathematics. They are often similar to crossword puzzles, but the entries are numbers instead of words. A famous number-related puzzle is Sudoku (Japanese for "single number"), a logic-based number-placement puzzle that originated in Japan and gained worldwide popularity in 2005. A similar, less popular puzzle, but with a closer relation to arithmetic operations than Sudoku, is known under the name Kenken or KenDoku.

One of the first puzzles in the history of mathematics was the magic square, which is as fascinating today as it was ages ago. The task is to find a square arrangement of numbers so that the sum of the numbers in each row and each column is the same as the sum of the numbers in each of the two diagonals. The first known example of a magic square is the Lo Shu square, with numbers arranged as shown in figure 7.1. It was known to Chinese mathematicians as early as 650 BCE, and it became important in feng shui, the art of placing objects to achieve harmony with the surrounding environment.

Figure 7.1: Lo Shu square and the magic turtle.

A legend from that time tells us that once there was a huge flood on the Lo River in China, and the people tried to placate the river's god. But each time they offered a sacrifice, a turtle emerged from the river and walked around the offering, until a child noticed a strange pattern of dots on the turtle's shell. After studying these markings, the people realized that the correct amount of sacrifices to make would be 15. And after they did so, the river god was satisfied and the flood receded.

The number 15 is the sum of numbers in each row, column, and diagonal of the Lo Shu magic square.

Magic squares appear throughout history; they became popular among Arabian mathematicians in Baghdad, who even designed 6 × 6 magic squares and published them in an encyclopedia in 983 CE. In the tenth century, a famous magic square, called *Chautisa Yantra*, appeared in India. The 4 × 4 magic square, shown in figure 7.2, is found on the Parshvanath temple in Khajuraho, India. Here the sum of each row, each column, and the diagonals is 34.

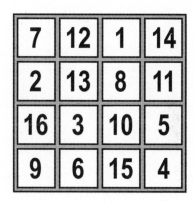

Figure 7.2: Chautisa Yantra.

There is one magic square, however, that stands out from the rest for its beauty and additional properties—not to mention its curious appearance. This particular magic square has many properties beyond those required for a square arrangement of numbers to be considered "magic." This magic square even comes to us through art, and not through the usual mathematical channels. It is depicted in the background of the famous 1514 engraving by the renowned German artist Albrecht Dürer (1471–1528), who lived in Nürnberg, Germany. (See figure 7.3.)

Figure 7.3: *Melencolia I*, engraving by Albrecht Dürer (1514).

Remember, a magic square is a square matrix of numbers, where the sum of the numbers in each of its columns, rows, and diagonals is the same. As we begin to examine the magic square in Dürer's etching, we should take note that most of Dürer's works were signed by him with his initials, one over the other, and with the year in which the work was made included. Here we find it in the dark-shaded region near the lower right side of the engraving (figures 7.3 and 7.4). We notice that it was made in the year 1514.

Figure 7.4: Initials AD of Albrecht Dürer and the year 1514.
(Detail of *Melencolia I*.)

The observant reader may notice that the two center cells of the bottom row of the Dürer magic square depict the year as well. Let us examine this magic square more closely. (See figure 7.5.)

Figure 7.5: Dürer's magic square.
(*Left*: Detail of *Melencolia I*.)

First let's make sure that it is, in fact, a true magic square. When we evaluate the sum of each of the rows, columns, and diagonals, we always get the result 34. That is all that is required for this square matrix of numbers to be considered a "magic square." However, this Dürer magic square has lots of properties that other magic squares do not have. Let us now marvel at some of these extra properties.

• The four corner numbers have a sum of 34:

$$16 + 13 + 1 + 4 = 34$$

• Each of the four corner two-by-two squares has a sum of 34:

$$16 + \ 3 + \ 5 + 10 = 34$$
$$2 + 13 + 11 + \ 8 = 34$$
$$9 + \ 6 + \ 4 + 15 = 34$$
$$7 + 12 + 14 + \ 1 = 34$$

- The center two-by-two square has a sum of 34:

$$10 + 11 + 6 + 7 = 34$$

- The sum of the numbers in the diagonal cells equals the sum of the numbers in the cells not in the diagonal:

$$16+10+7 + 1 + 4 + 6 +11+13=$$
$$3 + 2 + 8 + 12 + 14+15+ 9 + 5 =68$$

- The sum of the squares of the numbers in both diagonal cells is

$$16^2 + 10^2 + 7^2 + 1^2 + 4^2 + 6^2 + 11^2 + 13^2 = 748.$$

This number is equal to

  □ the sum of the squares of the numbers not in the diagonal cells:
  $$3^2 + 2^2 + 8^2 + 12^2 + 14^2 + 15^2 + 9^2 + 5^2 = 748$$

  □ the sum of the squares of the numbers in the first and third rows:
  $$16^2 + 3^2 + 2^2 + 13^2 + 9^2 + 6^2 + 7^2 + 12^2 = 748$$

  □ the sum of the squares of the numbers in the second and fourth rows:
  $$5^2 + 10^2 + 11^2 + 8^2 + 4^2 + 15^2 + 14^2 + 1^2 = 748$$

  □ the sum of the squares of the numbers in the first and third columns:
  $$16^2 + 5^2 + 9^2 + 4^2 + 2^2 + 11^2 + 7^2 + 14^2 = 748$$

  □ the sum of the squares of the numbers in the second and fourth columns:
  $$3^2 + 10^2 + 6^2 + 15^2 + 13^2 + 8^2 + 12^2 + 1^2 = 748$$

- The sum of the cubes of the numbers in the diagonal cells equals the sum of the cubes of the numbers not in the diagonal cells:

$$16^3 + 10^3 + 7^3 + 1^3 + 4^3 + 6^3 + 11^3 + 13^3 =$$
$$3^3 + 2^3 + 8^3 + 12^3 + 14^3 + 15^3 + 9^3 + 5^3 = 9{,}248$$

- Notice the following beautiful symmetries:

$$2 + 8 + 9 + 15 = 3 + 5 + 12 + 14 = 34$$
$$2^2 + 8^2 + 9^2 + 15^2 = 3^2 + 5^2 + 12^2 + 14^2 = 374$$
$$2^3 + 8^3 + 9^3 + 15^3 = 3^3 + 5^3 + 12^3 + 14^3 = 4624$$

- Adding the first row to the second, and the third row to the fourth, produces a pleasing symmetry:

16 + 5 = **21**	3 + 10 = **13**	2 + 11 = **13**	13 + 8 = **21**
9 + 4 = **13**	6 + 15 = **21**	7 + 14 = **21**	12 + 1 = **13**

- Adding the first column to the second, and the third column to the fourth, produces a pleasing symmetry:

16 + 3 = **19**	2 + 13 = **15**
5 + 10 = **15**	11 + 8 = **19**
9 + 6 = **15**	7 + 12 = **19**
4 + 15 = **19**	14 + 1 = **15**

A motivated reader may wish to search for other patterns in this beautiful magic square. Remember, this is not a typical magic square, where all that is required is that all the rows, columns, and diagonals have the same sum. This Dürer magic square has many more properties. Likewise, it is worthwhile to explore the Chautisa Yantra in figure 7.2 in order to find additional properties.

7.2. GENERAL PROPERTIES OF MAGIC SQUARES

You might wonder how it could be that both the Chautisa Yantra and the Dürer magic square have 34 as their "magic number." But, actually, this would necessarily be the case for any 4 × 4 magic square that uses the numbers from 1 to 16. The sum of these numbers is $1 + 2 + 3 + \ldots + 16 = 136$. In a magic square, every row of numbers contributes exactly a quarter of this sum because there are four rows and all rows are required to have the same sum. Therefore, the sum across each row is a quarter of 136, which is 34. By the definition of a magic square, the sum of the numbers in each column and each diagonal must also be 34.

In that way, we can even obtain a formula for the magic number of any $n \times n$ magic square. For this, we remind the reader of the discussion in chapter 4 about the sum of the first n natural numbers. Such a number is called a *triangular number*, T_n, and was determined by the formula

$$T_n = 1 + 2 + 3 + \ldots + \left(n-1\right) + n = \frac{n}{2}\left(n+1\right).$$

A magic square of size $n \times n$ contains all the natural numbers from 1 to n^2. Applying the formula above for this situation, we find that the sum of natural numbers from 1 to n^2 is

$$T_{\left(n^2\right)} = \frac{n^2}{2}\left(n^2 + 1\right).$$

However, if it is required that each of the n rows must have the same sum, S_n, then the sum of each row must be

$$S_n = \frac{T_{(n^2)}}{n} = \frac{n}{2}\left(n^2+1\right).$$

And, in a magic square, this number must be the sum of any row, column, or diagonal.

For $n = 3$, this formula indeed gives the magic number of the Lo Shu square:

$$S_3 = \frac{3}{2}\left(9+1\right) = 15.$$

Here, we will consider magic squares consisting of all numbers from 1 to n^2, where n, the number of row or columns, is called the *order* of the magic square. However, if one adds a constant number k to all numbers in a magic square, one would obtain another magic square, with numbers ranging from $k + 1$ to $k + n^2$, and with magic number $kn + S_n$. Similarly, *multiplying* each number of a magic square with a constant k would give a magic square with magic number kS_n.

The question that would logically be asked is, how does one construct a magic square? How did Dürer come up with this special magic square? According to their order, we distinguish three types:

(1) magic squares of odd order (n is an odd number),
(2) magic squares of doubly-even order (n is a multiple of 4),
(3) magic squares of singly-even order (n is a multiple of 2, but not 4).

The Dürer magic square is a doubly-even magic square.

7.3. HOW TO CONSTRUCT A DOUBLY-EVEN MAGIC SQUARE

Since we have the Dürer square at hand, we will begin by discussing the construction of the doubly-even magic squares. Let us begin with the smallest of these—namely, those with four rows and columns. We begin our construction of this doubly-even magic square by first placing the numbers in the square in numerical order, as shown in the first square of figure 7.6.

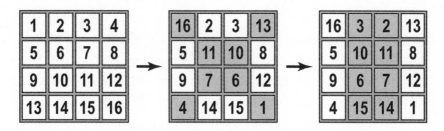

Figure 7.6: Constructing Dürer's square in three steps.

This is not yet a magic square, because all the small numbers are in the first row and the large numbers are in the last row. But a quick inspection shows that the sum along each diagonal already has the required value of 34. Any rearrangement of the numbers within a diagonal will not change their sum. In the next step, we must try to get some of the large numbers into the upper part of the square. To do this we will exchange within the first diagonal (the "main diagonal") the numbers 1 and 16 and the numbers 6 and 11. Similarly, in the secondary diagonal, we will exchange the numbers 13 and 4, as well as the numbers 10 and 7. The cells to be changed are shaded in the second square of figure 7.6. We now have been able to get some large numbers in the first row, and, indeed, the sum of the first row is 34! Quickly checking the remaining rows and columns reveals that this square is indeed a magic square. (No need to check the diagonals again, because exchanging numbers within a diagonal does not change its sum!) Thus, we have constructed our first magic square! However, the magic square we have obtained is *not* the same as the one that Dürer pictured in his *Melencolia I* etching. Dürer apparently interchanged the positions of the two middle columns to allow his square to show the date that the picture was made, 1514, in the middle of the bottom row. This resulting arrangement of numbers is shown as the last square in figure 7.6, which is Dürer's magic square, and it has many more properties than the magic square constructed in the first step.

Once you have obtained a magic square, you can try to generate a new one by starting with the given magic square. Any change you apply to an existing magic square should not change the sums of rows, columns, and diagonals. For example, if you exchange the second and the third column, as Dürer did in the second step described above in figure 7.6, this had no influence on the sums along the rows. But it might change the sum along the diagonals because this step exchanges numbers between the diagonals. In general, this can be repaired by switching the second and third *row*, which does not change the sum along the columns but restores the numbers in the diagonals. You might want to try this for the Chautisa Yantra of figure 7.2. This one would not remain a magic square if only the two central columns were exchanged. There you would have to exchange the central rows as well. Dürer's square is again special in that it remains a magic square when columns 2 and 3 are exchanged (and also if you exchange columns 1 and 4, or rows 1 and 4, or rows 2 and 3).

In general, exchanging columns 1 and 4 (or for that matter, columns 2 and 3) and then exchanging the corresponding rows would preserve the "magic property" of a square.

Another general method to create a new magic square from an existing one is to replace each number by its complement. The complement of a number a in a magic square is a number b, such that $a + b$ is 1 greater than the number of cells. In a square of order 4, two numbers are complementary if their sum is 17. (The first step in figure 7.6 can also be described as the replacement of the numbers in the diagonals by their complements).

You may wish to generate new magic squares using this technique. There are a total of 880 possible magic squares of order 4. By the way, there is no magic square of order 2, and there is essentially only one magic square of order 3—the Lo Shu square of figure 7.1—because all other magic squares of order 3 can be obtained from the Lo Shu square by rotation or reflection.

The next larger doubly-even magic square is of order 8—that is, with eight rows and columns. Once again we place the numbers in the cells in proper numerical order, as shown in figure 7.7.

1	2	3	4	5	6	7	8
9	10	11	12	13	14	15	16
17	18	19	20	21	22	23	24
25	26	27	28	29	30	31	32
33	34	35	36	37	38	39	40
41	42	43	44	45	46	47	48
49	50	51	52	53	54	55	56
57	58	59	60	61	62	63	64

Figure 7.7: First step in the construction of a magic square of order 8.

This time we will once again replace the numbers in the diagonals by their complement—in this case, the complement of a number is the number that will produce a sum of 65. However, the diagonals in this case are the diagonals of each of the 4 × 4 squares included in the 8 × 8 square—here they are the shaded numbers. The completed magic square with all the appropriate cell changes is shown in figure 7.8.

64	2	3	61	60	6	7	57
9	55	54	12	13	51	50	16
17	47	46	20	21	43	42	24
40	26	27	37	36	30	31	33
32	34	35	29	28	38	39	25
41	23	22	44	45	19	18	48
49	15	14	52	53	11	10	56
8	58	59	5	4	62	63	1

Figure 7.8: A magic square of order 8.

7.4. CONSTRUCTION OF A MAGIC SQUARE OF ORDER 3

A systematic construction of all possible 3 × 3 magic squares would begin by considering the matrix of letters representing the numbers 1 to 9 shown in figure 7.9. Here the sums of the rows, columns, and diagonals are denoted by r_j, c_j, and d_j, respectively. In a magic square of order 3, all these number sums would be equal to the magic number 15.

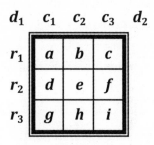

Figure 7.9: A map of a general magic square

In a magic square, we would thus have

$$r_2 + c_2 + d_1 + d_2 = 15 + 15 + 15 + 15 = 60.$$

However, this sum can also be written as

$$r_2 + c_2 + d_1 + d_2 = (d + e + f) + (b + e + h) + (a + e + i) + (c + e + g) =$$
$$3e + (a + b + c + d + e + f + g + h + i) = 3e + 45$$

Therefore, $3e + 45 = 60$, and $e = 5$. Thus it is established that the center position of a magic square of order 3 must be occupied by the number 5.

Recall that two numbers of an nth order magic square are said to be complementary if their sum is $n^2 + 1$. In a 3×3 magic square, two numbers are complementary if their sum is $9 + 1 = 10$. We can now see that numbers on opposite sides of 5 are complementary. For example, $a + i = d_1 - e = 15 - 5 = 10$, and, therefore, a and i are complementary. But so are the pairs g and c, b and h, and d and f.

Let us now try to put 1 in a corner, as shown in figure 7.10. Here $a = 1$, and therefore i must be 9, so that the diagonal adds up to 15. Next we notice that 2, 3, and 4 cannot be in the same row (or column) as 1, since there is no natural number less than 9 that would be large enough to occupy the third position of such a row (or column). This would leave only the two shaded positions in figure 7.10 to accommodate these three numbers (2, 3, and 4). Since this cannot be the case, our first attempt was a failure: the numbers 1 and 9 may occupy only the middle positions of a row (or column).

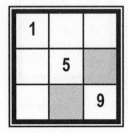

Figure 7.10: A non-magic-square construction—false start.

Therefore, we have to start with one of the four possible, positions remaining for 1, for example, as we show in the first square of figure 7.11. We note that the number 3 cannot be in the same row (or column) as 9, for the third number in such a row (or column) would again have to be 3 to obtain the required sum of 15. This is not possible, because a number can be used only once in the magic square. Additionally, we have seen above that 3 cannot be in the same row (or column) as 1. This leaves only the two shaded positions in figure 7.11 for the number 3. The number opposite 3 is always 7, because then 3 + 5 + 7 = 15.

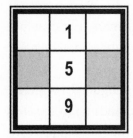 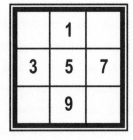

Figure 7.11: The development of one of several possible magic squares.

We continue with the second square in figure 7.11, showing one of two possibilities for the placement of 3 and 7 (the other possibility has 3 and 7 exchanged). It is now easy to fill in the remaining numbers. There is only one such possibility, shown in the third square of figure 7.11.

How many different squares are there? We could start by putting the number 1 in any of the four positions in the middle of a side. We then have two possibilities for placing 3. After that, the construction is unique. This produces the eight magic squares shown in figure 7.12.

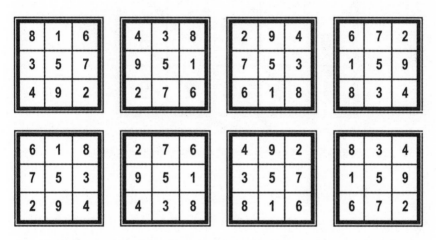

Figure 7.12: There are precisely eight magic squares of order 3.

7.5. CONSTRUCTING ODD-ORDER MAGIC SQUARES

You might now want to extend this technique to construct other odd-order magic squares. However, it can become somewhat tedious. The following is a rather mechanical method for constructing an odd-order magic square.

Begin by placing a 1 in the first position of the middle column. Continue by placing the next consecutive numbers along the diagonal line, as in figure 7.13.

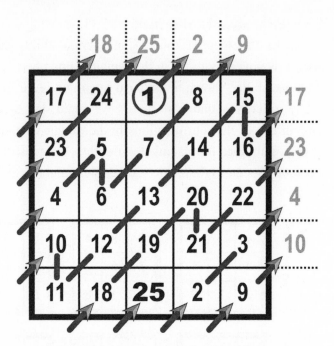

Figure 7.13: Construction of an odd-order magic square.

Whenever you drop off the square on one side, you enter again on the opposite side. So the gray number 2 in figure 7.13 (which fell off the grid) must be placed in the last row. Analogously, the gray number 4 will be placed in the first column. The process continues by consecutively filling each new cell along the diagonal line until an already-occupied cell is reached (as is the case with the number 6). Rather than placing a second number in an already-occupied cell, the number is placed below the previous number. The process continues until the last number is reached. After some practice with this procedure, you will begin to recognize certain patterns (e.g., the last number always occupies the middle position of the bottom row). This is just one of many ways of constructing odd-order magic squares. Not counting rotations and reflections, there are 275,305,224 different 5 × 5 magic squares. Exact numbers for higher-order magic squares are unknown.

7.6. CREATING SINGLY-EVEN MAGIC SQUARES

A different scheme is used to construct magic squares of singly-even order (i.e., where the number of rows and columns is even but not a multiple of 4). Any singly-even order (say, of order n) magic square may be separated into quadrants (figure 7.14). For convenience, we will label these quadrants as A, B, C, and D.

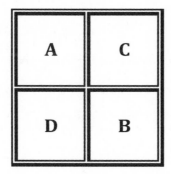

Figure 7.14: Quadrants of a magic square of singly-even order.

The order of the square should be n, a singly-even number, hence the order of each of the quadrants must be odd. We denote the order of the quadrant by $k = 2m + 1$ (which is always an odd number for $m = 1, 2, 3 \ldots$). As there is no magic square of order 2, the smallest singly-even magic square will have order 6—in which case $m = 1$ and $k = 3$. We have

$$n = 2k = 2\,(2m + 1) = 6, 10, 14, 18 \ldots \text{(for } m = 1, 2, 3, 4 \ldots)$$

Each of the four quadrants contains k^2 different numbers. We start by creating a magic square of odd-order k according to the method described earlier. For $n = 6$ and $k = 3$, the starting point will thus be one of the variants of the Lo Shu square. We will choose the first of the magic squares shown in figure 7.12.

We begin by entering this magic square into quadrant A. The magic squares in quadrants B, C, and D will be obtained as shown for the case $n = 6$ in figure 7.15.

8	1	6	8+18	1+18	6+18
3	5	7	3+18	5+18	7+18
4	9	2	4+18	9+18	2+18
8+27	1+27	6+27	8+9	1+9	6+9
3+27	5+27	7+27	3+9	5+9	7+9
4+27	9+27	2+27	4+9	9+9	2+9

Figure 7.15: First step of the construction of a singly-even magic square of order 6.

Here, square B is obtained by adding $k^2 = 9$ to all numbers of square A. Square C is obtained by adding k^2 to all numbers of square B, and square D is obtained by adding k^2 to all numbers of square C.

Notice that adding a fixed number to all numbers of a magic square does not change the magic property: The sum of rows, columns, and diagonals would still remain the same. Thus the squares B, C, and D are also magic squares, only they do not use the numbers from 1 to k^2. For example, square B uses instead the numbers from $k^2 + 1$ to $2k^2$ (for $n = 6$, the numbers 10 to 18). The square of order n obtained in this way is shown in figure 7.16. Although it has magic squares in its quadrants, it is not yet a magic square itself.

8	1	6	26	19	24
3	5	7	21	23	25
4	9	2	22	27	20
35	28	33	17	10	15
30	32	34	12	14	16
31	36	29	13	18	11

Figure 7.16: Second step of the construction of a singly-even-order magic square of order 6.

Continuing along with our construction of the singly-even-order magic square, we have to make some adjustments to the square we have developed to this point. Recall that the integer m determines the order through the formula $n = 2 (2m + 1)$.

In general, the adjustments will be the following: We first take the numbers in the first m positions in each row of quadrant A, except the middle row, where we will skip the first position and then take the next m positions. Then we will exchange the numbers in these positions with the correspondingly placed numbers in square D below. We then take the last $m - 1$ cells in each row of square C and exchange them with the numbers in the corresponding cells of square B.

For $n = 6$ and $m = 1$, the positions in squares A and D that will be changed during that procedure are shaded in figure 7.16. Since, in this case, $m - 1 = 0$, the squares B and C on the right side remain unaltered. The resulting square is shown in figure 7.17. You may verify that it is indeed a magic square.

35	1	6	26	19	24
3	32	7	21	23	25
31	9	2	22	27	20
8	28	33	17	10	15
30	5	34	12	14	16
4	36	29	13	18	11

Figure 7.17: The singly-even-order magic square obtained from figure 7.16.

We illustrate this procedure once again with the next-larger singly-even magic square, which is of order $n = 10$, and in this case $m = 2$.

1. Starting with a magic square of order 5, we take the one created by the method explained previously (figure 7.13).
2. We fill the four quadrants of the $n \times n$ square. We create square B by adding 25 to all numbers of square A, and then continue as we did earlier. The result is shown as the first square in figure 7.18.
3. Take the first two positions of each row of quadrant A, except the middle row, where you skip the first cell and then take the next two positions. Exchange the numbers in these cells with the

numbers in the corresponding cells of square D. Figure 7.18 has the corresponding positions shaded.

17	24	1	8	15	67	74	51	58	65
23	5	7	14	16	73	55	57	64	66
4	6	13	20	22	54	56	63	70	72
10	12	19	21	3	60	62	69	71	53
11	18	25	2	9	61	68	75	52	59
92	99	76	83	90	42	49	26	33	40
98	80	82	89	91	48	30	32	39	41
79	81	88	95	97	29	31	38	45	47
85	87	94	96	78	35	37	44	46	28
86	93	100	77	84	36	43	50	27	34

92	99	1	8	15	67	74	51	58	40
98	80	7	14	16	73	55	57	64	41
4	81	88	20	22	54	56	63	70	47
85	87	19	21	3	60	62	69	71	28
86	93	25	2	9	61	68	75	52	34
17	24	76	83	90	42	49	26	33	65
23	5	82	89	91	48	30	32	39	66
79	6	13	95	97	29	31	38	45	72
10	12	94	96	78	35	37	44	46	53
11	18	100	77	84	36	43	50	27	59

Figure 7.18: Construction of a higher-order singly-even magic square.

4. To complete the magic square, we take last $m - 1$ positions (here the last positions, since $m - 1 = 1$) in each row of the squares C and B and interchange them. This gives us the magic square shown as the second square in figure 7.18.

We now have a procedure for constructing each of the three types of magic squares: the odd-order magic square and both the singly-even and the doubly-even magic squares.

We end this discussion about magic squares with a curiosity, just for entertainment. You can verify that the first square in figure 7.19 is a magic square. The sum of its rows, columns, and diagonals is 45.

12	28	5
8	15	22
25	2	18

twelve	twenty eight	five
eight	fifteen	twenty two
twenty five	two	eighteen

6	11	4
5	7	9
10	3	8

Figure 7.19: An alphamagic square.

However, this square has an additional property that makes it a so-called alphamagic square. Replace the numbers by their written words. The number of letters in each word generates a new magic square—the third square in figure 7.19. You can convince yourself of its magic property either by computing all sums of the rows, columns, and diagonals, or by noticing that it also can be obtained from the Lo Shu square by adding two to all its numbers. (Remember, adding a constant number to all numbers of a magic square generates a new magic square.)

7.7. PALINDROMIC NUMBERS

There are certain categories of numbers that have particularly strange characteristics so that we can consider them for their common curious property. And sometimes a playful approach leads to difficult mathematical problems and interesting questions. Here we consider numbers that read the same in both directions: left to right or right to left. These are called *palindromic numbers*. First note that a palindrome can also be a word, phrase, or sentence that reads the same in both directions. Figure 7.20 shows a few amusing palindromes.

A
EVE
RADAR
REVIVER
ROTATOR
LEPERS REPEL
MADAM I'M ADAM
STEP NOT ON PETS
DO GEESE SEE GOD
PULL UP IF I PULL UP
NO LEMONS, NO MELON
DENNIS AND EDNA SINNED
ABLE WAS I ERE I SAW ELBA
A MAN, A PLAN, A CANAL, PANAMA
A SANTA LIVED AS A DEVIL AT NASA
SUMS ARE NOT SET AS A TEST ON ERASMUS
ON A CLOVER, IF ALIVE, ERUPTS A VAST, PURE EVIL; A FIRE VOLCANO

Figure 7.20: Palindromes

There is a well-known Latin palindromic sentence that stems from the second century CE and has an additional amazing property. It reads: *"Sator arepo tenet opera rotas,"* which commonly translates to "Arepo the sower (farmer) holds the wheels with effort." (See figure 7.21.) This so-called Templar magic square—named after the Order of the Templars—places these letters in a five-by-five square arrangement. Now you can read the sentence in all directions. This is quite astonishing! The Templar magic square is very old—it has been found in excavations of the Roman city of Pompeii, which had been buried in the ashes of Vesuvius. In medieval times, people attributed magical properties to it and used it as a spell to protect against witchcraft. Five examples of it were discovered in Mesopotamia in 1937, and there are some specimens of it in Britain, Cappadocia, Egypt, and Hungary.

Figure 7.21: Templar magic square.

A palindrome in mathematics would be a number such as 666 or 123321 that reads the same in either direction. For example, the first four powers of 11 are palindromic numbers:

$11^0 = 1$
$11^1 = 11$
$11^2 = 121$
$11^3 = 1331$
$11^4 = 14641$

It is interesting to see how a palindromic number can be generated from other given numbers. All you need to do is to continually add a number to its reversal (that is, the number written in the reverse order of digits) until you arrive at a palindrome. For example, a palindrome can be reached with a single addition with the starting number 23: the sum 23 + 32 = 55, a palindrome.

Or it might take two steps, such as with the starting number 75: the two successive sums 75 + 57 = 132 and 132 + 231 = 363 have led us to a palindrome.

Or it might take three steps, such as with the starting number 86:

$$86 + 68 = 154, 154 + 451 = 605, 605 + 506 = 1111.$$

The starting number 97 will require six steps to reach a palindrome; while the starting number 98 will require twenty-four steps to reach a palindrome.

Be cautioned about using the starting number 196; this one has not yet been shown to produce a palindrome number—even with over three million reversal additions. We still do not know if this one will ever reach a palindrome. If you were to try to apply this procedure with 196, you would eventually—at the sixteenth addition—reach the number 227,574,622, which you would also reach at the fifteenth step of the attempt to get a palindrome from the starting number 788. This would then tell you that applying the procedure to the number 788 has also never been shown to reach a palindrome. As a matter of fact, among the first 100,000 natural numbers, there are 5,996 numbers for which we have not yet been able to show that the procedure of reversal additions will lead to a palindrome. Some of these are: 196, 691, 788, 887, 1675, 5761, 6347, and 7436.

Using this procedure of reverse and add, we find that some numbers yield the same palindrome in the same number of steps, such as 554, 752, and 653, which all produce the palindrome 11011 in three steps. In general, all integers in which the corresponding digit pairs symmetric to the middle 5 have the same sum will produce the same palindrome in the same number of steps. However, there are other integers that produce the same palindrome, yet in a different number of steps, such as the number 198, which with repeated reversals and additions will reach the palindrome 79497 in five steps, while the number 7299 will reach this number in two steps.

For a two-digit number ab with digits $a \neq b$, the sum $a + b$ of its digits determines the number of steps needed to produce a palindrome. Clearly, if the sum of the digits is less than 10, then only one step will be required to reach a palindrome—for example, $25 + 52 = 77$. If the sum of the digits is 10, then $ab + ba = 110$, and $110 + 011 = 121$, and two steps will be required to reach the palindrome. The number of steps required for each of the two-digit sums 11, 12, 13, 14, 15, 16, and 17 to reach a palindromic number are 1, 2, 2, 3, 4, 6, and 24, respectively.

We can arrive at some lovely patterns when dealing with palindromic numbers. For example, some palindromic numbers when squared also yield a palindrome. For example, $22^2 = 484$ and $212^2 = 44944$. On the other hand, there are also some palindromic numbers that, when squared, do not yield a palindromic number, such as $545^2 = 297,025$. Of course, there are also nonpalindromic numbers that, when squared, yield a palindromic number, such as $26^2 = 676$ and $836^2 = 698,896$. These are just some of the entertainments that numbers provide. You may want to search for other such curiosities.

Numbers that consist entirely of 1s are called *repunits*. All the repunit numbers with fewer than ten 1s, when squared, yield palindromic numbers. For example,

$$1111^2 = 1234321.$$

There are also some palindromic numbers that, when cubed, yield again palindromic numbers.

To this class belong all numbers of the form $n = 10^k + 1$, for $k = 1$, 2, 3 . . . . . When n is cubed, it yields a palindromic number that has $k - 1$ zeros between each consecutive pair of 1,3,3,1.

$k = 1, n = 11$: $11^3 = 1331$
$k = 2, n = 101$: $101^3 = 1030301$
$k = 3, n = 1001$: $1001^3 = 1003003001$

We can continue to generalize and get some interesting patterns, such as when n consists of three 1s and any even number of 0s symmetrically placed between the end 1s when cubed will give us a palindrome. For example,

$111^3 = 1367631,$

$10101^3 = 1030607060301,$

$1001001^3 = 1003006007006003001,$ and

$100010001^3 = 1000300060007000600030001.$

Taking this even a step further we find that when n consists of four 1s and 0s in a palindromic arrangement, where the places between the 1s do not have same number of 0s, then n^3 will also be a palindrome, as we can see with the following examples:

$11011^3 = 1334996994331$ and

$10100101^3 = 1030331909339091330301.$

However, when the same number of 0s appears between the 1s, then the cube of the number will not result in a palindrome, as in the following example: $1010101^3 = 1030610121210060301.$ As a matter of fact, the number 2201 is the only nonpalindromic number that is less than 280,000,000,000,000, and that, when cubed, yields a palindrome: $2201^3 = 10662526601.$

However, just for amusement, consider the following pattern with palindromic numbers:

$$12321 = \frac{333333}{1+2+3+2+1}$$

$$1234321 = \frac{44444444}{1+2+3+4+3+2+1}$$

$$123454321 = \frac{5555555555}{1+2+3+4+5+4+3+2+1}$$

$$12345654321 = \frac{666666666666}{1+2+3+4+5+6+5+4+3+2+1}$$

and so on.

An ambitious reader may search for other patterns involving palindromic numbers.

7.8. NAPIER'S RODS

Here, we introduce a calculating system that depends on the placement of numbers. The Scottish mathematician John Napier (1550–1617), who is perhaps best known for having invented logarithms and for using the decimal point in his calculations, also introduced a mechanical system for multiplication known as *Napier's Rods*. The method is based on a technique invented by the Arabs in the thirteenth century. When it finally arrived in Europe, it became known as multiplication "per gelosia" ("by jealousy"). It was a system for performing multiplication using only addition. Napier significantly improved the system through the use of specially constructed strips, as shown in figure 7.22. The rods can be made out of cardboard, wood, or, as John Napier did when he invented this system, bone, thus providing us with another name for this method: *Napier's Bones*. Before reading what follows, you may want to spend a little time examining figure 7.22 and trying to understand the logic of the construction.

The arrangement in figure 7.22 is a multiplication table. There are ten vertical rods, each of which has a specific column from the multiplication table written on it in a peculiar manner. Notice how the rod marked at the top with the digit "5" continues downward, with each of the multiples of 5 (10, 15, 20, etc.) written such that the tens digit is above the diagonal line and the ones digit is below it. The same principle can be observed in the other rods: the fifth entry on the number 7 rod is 35, which is the same as the product $5 \times 7 = 35$. (Observe also that we put a 0 above the slash in entries where the product is less than 10.)

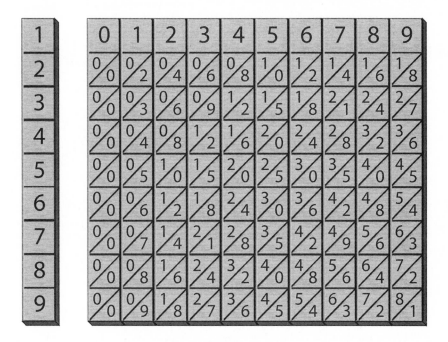

Figure 7.22: Napier's Rods.

These rods can be rearranged freely, permitting us to construct the numbers we want to multiply, and then to perform the computation using only addition. How is this possible? Let's look at an example to learn about the method Napier devised.

We will choose two numbers at random, in this case 284 and 572, and then select the rods whose top digits will allow us to construct our number. It doesn't matter which of these two numbers we choose to represent first. Thus, in this example we will construct 572, selecting the rods numbered 2, 5 and 7, and then putting them in the correct order. (See figure 7.23.)

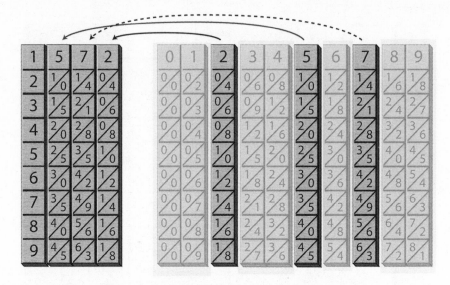

Figure 7.23: Multiplication of 572 × 284—First step.

We have written the digits 1 through 9 along the left-hand side in a single column. In Napier's original construction, these numbers were written or engraved along the side of a shallow box inside which the rods fit snugly. If you choose to re-create this example on your own, writing the numbers on a sheet of paper will work just fine, as long as you make sure to line up the tops of your own rods as you place them.

As you may have already guessed, the next step is to identify the rows that we will need to construct our second number. With physical rods it would not be possible to extract these rows, but for our illustration we will rearrange them as indicated by the arrows to form the number 284, again maintaining proper alignment. (See figure 7.24.)

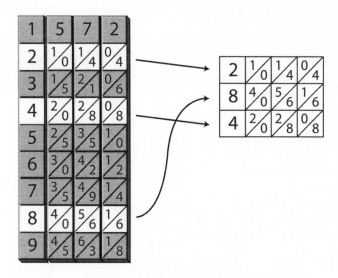

Figure 7.24: Multiplication of 572 × 284—Second step.

In order to illustrate the next step, we will de-emphasize the boundaries between the rods, while highlighting the diagonal lines. At the end of each diagonal we have created a space where our sum can be written, as indicated by the dashed arrows. It looks like our product will be a six-digit number, as there are six diagonals in our final computation. (See figure 7.25.)

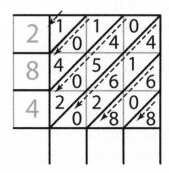

Figure 7.25: Multiplication of 572 × 284—Third step.

Starting in the lower right corner, we find the sum of each diagonal and, whenever that sum is greater than 9, write the digit to be carried in a slightly smaller font inside the box, as well as at the head of the next diagonal. Looking at the second diagonal, you can see the sum: $6 + 0 + 8 = 14$, which means the tens digit of our final product will be 4, while the 1 is carried to the top of the third diagonal and added to the other numbers there, as shown in figure 7.26.

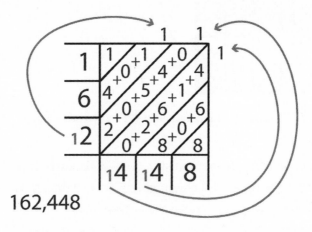

Figure 7.26: Sums of the diagonals. Addition with carrying.

Proceeding along each diagonal, we see the sums are 8, 14, 14, 12, 6, and 1. Reading these in order from the top down and from left to right, without the carried digits, we get 1 6 2 4 4 8, which indicates that our final product is 162,448. You might want to verify that this is the correct product—using our modern-day device, the electronic calculator!

Let's see how this method works. Normally, the multiplication of two numbers is performed by successive digit multiplications and positional arithmetic. When you do multiplication according to the method typically taught in elementary school, you place one number above the other with a line underneath and multiply pairs of digits. As you do so, you write the ones digit of each product below the line, carrying the tens digits when necessary, and taking the sum of the partial products at the end of the process. To illustrate how Napier's Rods work, we will break this process down step-by-step.

The first step is to multiply 572 by 4. The products of these multiplications are $4 \times 2 = 8$, $4 \times 7 = 28$, and $4 \times 5 = 20$. Carrying the 2 from the second multiplication and adding these together, we get a partial total of 2,288, which is the same result obtained by adding the diagonals of row four of the second figure in this section. (See figure 7.27.)

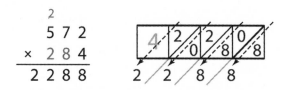

Figure 7.27: Intermediate step 1—multiplication of 572 by 4.

Repeating this process for the second digit, 8, we get $8 \times 572 = 4{,}576$, which again is the same result we get from adding the terms in the diagonals of the eighth row. According to the algorithm we know from elementary school, we insert a 0 in the ones column, leaving us with 45,760 in the new final row. (See figure 7.28.)

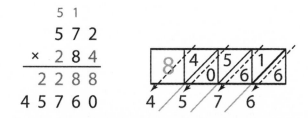

Figure 7.28: Intermediate step 2—multiplication of 572 by 8.

Next, we multiply 2 × 572 and insert two 0s, giving 114,400, the first four digits of which we recognize from the second row of Napier's Rods (figure 7.29).

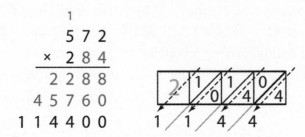

Figure 7.29: Intermediate step 3—multiplication of 572 by 2

Finally, we add these three numbers, again finding a result of 162,448, which is indeed the correct product, as shown in figure 7.30.

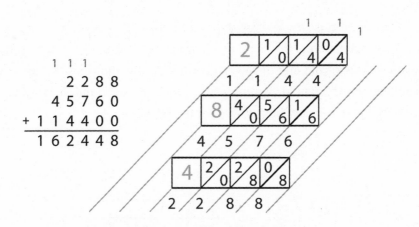

Figure 7.30: Adding the intermediate results.

To complete our illustration of this method, we will do one final alteration: instead of adding the digit products as we go, we will instead write the products as we did when constructing Napier's Rods, using a leading 0 for any number less than 10. Each product will be written with the appropriate offset, but in the same order we used when performing the previous operations.

Alongside this, we will draw the relevant portion of our Napier's Rods, this time rotated one-quarter turn as we have in figure 7.31.

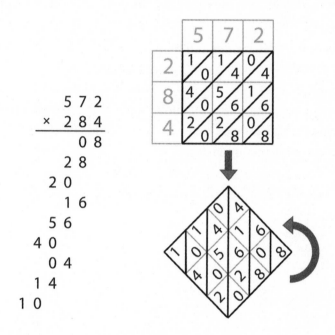

$$\begin{array}{r} 5\ 7\ 2 \\ \times\ \ 2\ 8\ 4 \\ \hline 0\ 8 \\ 2\ 8 \\ 2\ 0 \\ 1\ 6 \\ 5\ 6 \\ 4\ 0 \\ 0\ 4 \\ 1\ 4 \\ 1\ 0 \end{array}$$

Figure 7.31: Comparison of Napier's Rods with the common method.

Do you notice anything interesting? That's right—each digit we produce using the traditional method of multiplication is also present in the Napier's Rods representation, and in the proper column! Also, if you look closely at the bold-outlined rows, you will notice that there is an exact correspondence between these rows and the respective digit products. So, for instance, the final three rows on the left are 10, 14, and 04, and these same numbers are in the top column in the figure on the right.

As we have observed, the method of Napier's Rods is mechanically identical to our elementary-school algorithm, but it can make keeping track of the positions of each digit much easier. As an added advantage, it helps us avoid multiplication errors—after all, most of us can do addition much more accurately than we can multiply!

In this chapter, we have noticed that not only do numbers demonstrate peculiarities in and of themselves but also the placement of numbers can be significant. There are surprising relationships found in magic squares, and we can arrange numbers to assist in calculations, such as with Napier's Rods. Selective number placement can also provide some curious recreations, such as the generation of palindromic numbers. Thus we can see that the position of a number can also open up some interesting vistas.

CHAPTER 8
SPECIAL NUMBERS

8.1. PRIME NUMBERS

We begin by reviewing the definition of a *prime number*—a number greater than 1 that has only two divisors: the number itself and the number 1. For example, the first few prime numbers are: 2, 3, 5, 7, 11, 13, 17, and 19. We can notice that there is only one even prime number, 2, and all the other prime numbers are odd. In chapter 4 we characterized prime numbers as nonrectangular numbers greater than 1. It is impossible to represent a prime number by a rectangular array of objects with more than one row or column (see figure 8.1).

Figure 8.1: The number 11 is a prime number.

A *rectangular number*, on the other hand, can be arranged in rectangular form, which means that it can be written as a product of at least two numbers greater than 1, such as $45 = 3 \times 15 = 9 \times 5 = 3 \times 3 \times 5$. Such numbers are also called *composite numbers*.

In fact, every composite number can be written as a product of prime numbers. For example, $297 = 3 \times 3 \times 3 \times 11$, and $9{,}282 = 2 \times 3 \times 7 \times 13 \times 17$. It is important to note that the prime numbers that occur in the factorization of a composite number are uniquely determined.

> Every natural number greater than 1 either is a prime number itself or has *exactly one* factorization into primes.

This famous theorem was already known to Euclid. Because of its importance, it is called the *fundamental theorem of arithmetic*.

An early method for finding prime numbers was developed by the Greek scholar Eratosthenes (276–194 BCE). Using his method, we can list all the numbers—at least as many as we wish—beginning with the number 2, and cancel every successive multiple of 2. This leaves us with all the odd numbers. We then continue along with the next uncanceled number, in this case the number 3, and once again we cancel out all the successive multiples of 3. The next uncanceled number is the number 5, and once again all successive multiples of 5 become canceled. What remains uncanceled as we continue this process are all the prime numbers. Figure 8.2 shows this "sieve of Eratosthenes," a table of the odd numbers, with straight lines striking out the multiples of 3, 5, 7, 11, and 13. The remaining encircled numbers are the prime numbers between 3 and 309. A listing of all prime numbers less than 10,000 is in the appendix, section 2.

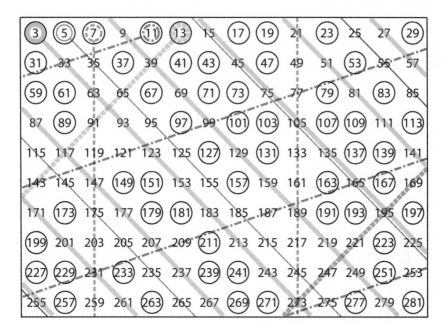

Figure 8.2: Sieve of Eratosthenes.

Does the sequence of prime numbers continue indefinitely? Or will the process of canceling multiples eventually stop, when all numbers have been struck out? Euclid gave an ingenious proof that there are, indeed, infinitely many prime numbers. The argument is as follows: Suppose we know only a finite number of prime numbers—say, for example, 2, 3, 5, 7, 11, 13, 17, and 19. Then we can show that there must be another one. That one would be found by forming the product of all known prime numbers, and adding 1 to obtain a larger number, which for our example would be as follows:

$$2 \times 3 \times 5 \times 7 \times 11 \times 13 \times 17 \times 19 + 1 = 9{,}699{,}691.$$

This number is not necessarily a prime number—as is the case here, where $347 \times 72{,}953 = 9{,}699{,}691$. If this number were a prime number, then we would have found another prime. The product of successive known primes plus 1 may be a prime (e.g., $2 \times 3 + 1 = 7$) or may not be a prime (e.g., $3 \times 5 + 1 = 16$). Now assume that we did not generate a new prime—as was the case above. Then we will look for the smallest number $q > 1$ that divides this number exactly. The number q cannot be in our list of known prime numbers, because none of these divides 9,699,691 exactly (as there will always be a remainder 1). Then q must be a prime number itself, because otherwise, q would be a product of smaller numbers, which are also factors of 9,699,691, and then q would not be the smallest factor. So we have found a new prime number q that is not in our list of known prime numbers. For every finite list of prime numbers, we can find a new one using this procedure. Therefore, the list of all prime numbers cannot be finite.

8.2. SEARCHING FOR PRIMES

Mathematicians have spent years trying to find a general formula that would generate primes. There have been many attempts, but none have succeeded. We can test the expression $n^2 - n + 41$ as a possible formula for generating prime numbers by substituting various positive values for n. As we proceed, we begin to notice that as n ranges in value from 1 through 40, only prime numbers are being produced. When we let $n = 41$, the value of $n^2 - n + 41$ is $(41)^2 - 41 + 41 = (41)^2$, which is not a prime number. A similar expression, $n^2 - 79n + 1601$, produces primes for all values of n up to 80. But for $n = 81$, we have $(81)^2 - 79 \times 81 + 1601 = 1763 = 41 \times 43$, which is, again, not a prime number. You might now wonder if it is possible to have a polynomial expression in n with integral coefficients whose values would be primes for every positive integer n. Don't waste your time, since the Swiss mathematician Leonhard Euler (1707–1783) proved that no such formula can exist. Euler showed that any proposed expression will produce at least one nonprime. The idea of Euler's proof is rather simple. It goes this way: First assume that such a polynomial expression exists, and we represent it in general form as $a + bx + cx^2 + dx^3 + \ldots$ (understanding that some of the coefficients may be zero). Let the value of this expression be a prime number s when $x = m$ (a positive integer). Therefore, $s = a + bm + cm^2 + dm^3 + \ldots$. Similarly, let t be the value of the expression when $x = m + ns$, then

$$t = a + b\,(m + ns) + c\,(m + ns)^2 + d\,(m + ns)^3 \ldots.$$

This may be transformed to

$$t = (a + bm + cm^2 + dm^3 + \ldots) + A,$$

where A represents the remaining terms, all of which are multiples of s. (Moreover, it can be shown that $A \neq 0$ for a suitable choice of n.) But the expression within the parentheses is, by hypothesis, equal to s. This makes the whole expression a multiple of s, that is, $t = ks$ (with $k > 1$) and the number produced is, therefore, not a prime. Every such expression will produce at least one prime, but not necessarily more than one. Consequently, no polynomial expression can generate primes exclusively. Mathematicians continued to conjecture about forms of numbers that generated only primes, despite the fact that the above argument had already been accepted.

The French mathematician Pierre de Fermat (1601–1665), who made many significant contributions to the study of number theory, conjectured that all numbers of the form $F_n = 2^{(2^n)} + 1$, where $n = 0$, 1, 2, 3, 4 \ldots , were prime numbers. If you try this for F_n for values of $n = 0, 1, 2$, you will see that the first three numbers derived from this expression are 3, 5, and 17. For $n = 3$, you will find that $F_n = 257$, and $F_4 = 65,537$. We notice these numbers are increasing in size at a very rapid rate. For $n = 5$, $F_n = 4,294,967,297$, and Fermat could not find any factor of this number. Encouraged by his results, he expressed the opinion that all numbers of this form are probably also prime. Unfortunately, he stopped too soon, for in 1732, Euler showed that $F_5 = 4,294,967,297 = 641 \times 6,700,417$, and it is, therefore, not a prime! It was not until 150 years later that the factors of F_6 were found:

$$F_6 = 18,446,744,073,709,551,617 = 247,177 \times 67,280,421,310,721.$$

Many more numbers of this form have been found, but, as far as it is now known, *none* of them are prime numbers. It seems that Fermat's conjecture has been completely turned around, and one now wonders if any primes beyond F_4 exist.

The largest known prime as of early 2013 is $2^{57,885,161} - 1$, which has 17,425,170 digits. This prime number is of the form $2^k - 1$, with some natural number k. Such primes are called *Mersenne primes*, as they were discovered by the French monk Marin Mersenne (1588–1648). The first few such Mersenne primes are obtained for $k = 2, 3, 5, 7, 13, 17, 19, 31, 61, 89, 107$, and 127.

You may have noticed that all the numbers k in this list are prime numbers themselves. Indeed, $2^k - 1$ can only be prime when k is prime. But this does not guarantee that when k is prime, $2^k - 1$ will also be prime. For example, $k = 11$ is prime, but $2^{11} - 1 = 2,047 = 23 \times 89$ is not a prime, and so $k = 11$ does not qualify.

While, according to Euclid's proof, there are infinitely many prime numbers, it is not known whether there are infinitely many Mersenne primes.

There are many interesting relationships regarding prime numbers. For example, the famous *Goldbach conjecture* made in a letter in 1742 to Leonhard Euler by German mathematician Christian Goldbach (1690–1764), which states that

- *every even integer greater than 2 can be expressed as the sum of 2 prime numbers.*

The first few of these are: $4 = 2 + 2$, $6 = 3 + 3$, $8 = 3 + 5$, $10 = 5 + 5$, $12 =$ $5 + 7$, $14 = 7 + 7$, $16 = 5 + 11$, and so on. Remember, this is a conjecture, which implies that we have not yet proved that is true for all numbers; however, to date no one has ever proved it to be wrong. This is one of the ongoing challenges for number theorists.

There is also the twin-prime conjecture that states that there are infinitely many pairs of prime numbers whose difference is 2, such as the numbers 3 and 5, or the numbers 5 and 7, or the numbers 17 and 19 (adjacent encircled numbers in figure 8.2).

We also have a relationship among "three-primes," where the difference between any two consecutive members of the triple is not greater than 4. For example, the simplest triple of such primes is (2, 3, 5). Another one would be (2, 3, 7). Table 8.1 lists a sampling of these prime triples.

2, 3, 5	107, 109, 113
3, 5, 7	191, 193, 197
5, 7, 11	193, 197, 199
7, 11, 13	223, 227, 229
11, 13, 17	227, 229, 233
13, 17, 19	277, 281, 283
17, 19, 23	307, 311, 313
37, 41, 43	. . .
41, 43, 47	7,873; 7,877; 7,879
67, 71, 73	247,603; 247,607; 247,609
97, 101, 103	5,037,913; 5,037,917; 5,037,919
101, 103, 107	88,011,613; 88,011,617; 88,011,619
103, 107, 109	1,385,338,061; 1,385,338,063; 1,385,338,067

Table 8.1: Some triples of primes.

8.3. PLAYING WITH PRIMES

There are some patterns among prime numbers. For example, there are reversible prime numbers—that is, those prime numbers whose reverse is also a prime number: 13 and 31, 17 and 71, 37 and 73, 79 and 97, 107 and 701, 113 and 311, 149 and 941, 157 and 751.

Referring back to our previous discussion of palindromic numbers, there are also palindromic numbers that are prime numbers, such as: 2, 3, 5, 7, 11, 101, 131, 151, 181, 191, 313, 353, 373, 383, 727, 757, 787, 797, 919, 10301, 10501, 10601, 11311, 11411, 12421, 12721, 12821, and 13331.

Moreover, there are repunit numbers that are also prime, such as 11, 1111111111111111111, and 11111111111111111111111, with the next two such prime repunit numbers having large numbers of 1s, specifically, 317 and 1,031 1s.

There are prime numbers that have the characteristic that any other arrangement of the digits will also produce a prime number. The first few of these are: 2, 3, 5, 7, 11, 13, 17, 31, 37, 71, 73, 79, 97, 113, 131, 199, 311, 337, 373, 733, 919, and 991. It is believed that larger such primes are repunit primes.

There are also prime numbers that remain prime numbers even when their digits are moved in a circular fashion. For example, the prime number 1,193 can have its digits "rotated" to form the following numbers: 1,931; 9,311; 3,119. Since all of these rotated-digit variations yield a prime number, we call the number 1,193 a *circular prime number*. Other such circular prime numbers are: 2; 3; 5; 7; 11; 13; 17; 31; 37; 71; 73; 79; 97; 113; 131; 197; 199; 311; 337; 373; 719; 733; 919; 971; 991; 1,193; 1,931; 3,119; 3,779; 7,793; 7,937; 9,311; 9,377; 11,939; 19,391; 19,937; 37,199; 39,119; 71,993; 91,193; 93,719; 93,911; and 99,371.

Continuing this exploration of numerical oddities, we arrive at the amazing prime number 193,939. This is also a circular prime number because all of its rotated-digits numbers are prime numbers:

193,939; 939,391; 393,919; 939,193; 391,939; and 919,393.

But in this case there is another arrangement of these digits that results in another circular prime number:

199,933; 999,331; 993,319; 933,199; 331,999; and 319,993.

On the other hand, there are some prime numbers that, when any one of their digits is changed to another value, will always result in a composite (nonprime) number. Some of these are:

294,001; 505,447; 584,141; 604,171; 971,767; 1,062,599; 1,282,529; 1,524,181; 2,017,963; 2,474,431; 2,690,201; 3,085,553; 3,326,489; and 4,393,139.

There are also prime numbers that are the sum of two consecutive squares—that is, $n^2 + (n + 1)^2$. The first few of these so-called sum-of-consecutive-squares primes are: $1 + 4 = 5$, $4 + 9 = 13$, and $16 + 25 = 41$, and the rest of the first few are: 61; 113; 181; 313; 421; 613; 761; 1,013; 1,201; 1,301; 1,741; 1,861; 2,113; 2,381; 2,521; 3,121; 3,613; 4,513; 5,101; 7,321; 8,581; 9,661; 9,941; 10,531; 12,641; 13,613; 14,281; 14,621; 15,313; 16,381; 19,013; 19,801; 20,201; 21,013; 21,841; 23,981; 24,421; and 26,681.

Then there are additive primes, which are prime numbers where the sum of their digits is also a prime number. Some of these are: 2, 3, 5, 7, 11, 23, 29, 41, 43, 47, 61, 67, 83, 89, 101, 113, and 131.

In recreational mathematics we often seek strange characteristics of numbers for entertainment. For example, there is a set of prime numbers that is called *minimal primes*. It is believed that there are only twenty-six such prime numbers, with the characteristic that any subsequence of the digits will be a composite number (a nonprime number). To better understand what is meant by "minimal primes," let us consider one example. For the prime number 6,949 we can establish the following numbers, taking the digits in sequence as follows: 6, 9, 4, 69, 94, 49, 64, 99, 694, 699, and 949, all of which are composite numbers. The list of the twenty-six known minimal primes is as follows: 2; 3; 5; 7; 11; 19; 41; 61; 89; 409; 449; 499; 881; 991; 6,469; 6,949; 9,001; 9,049; 9,649; 9,949; 60,649; 666,649; 946,669; 60,000,049; 66,000,049; and 66,600,049.

8.4. UNSOLVED QUESTIONS

There are a number of conjectures about prime numbers that have evolved over the years. Some conjectures have been proved, and some still remain open, such as the Goldbach conjecture and the twin primes conjecture mentioned earlier. Here are some other "facts" about prime numbers that have not been proved or disproved yet:

- There are infinitely many prime numbers of the form $n^2 + 1$, where n is a natural number.
- There is always a prime number between n^2 and $(n + 1)^2$.
- There is always a prime number between n and $2n$.
- There is an arithmetic progression of consecutive prime numbers for any given finite length, such as 251, 257, 263, 269, which has a length of 4. So far, the largest such length is 10.
- If n is a prime number, then $2^n - 1$ is not divisible by the square of a prime number.
- There are infinitely many prime numbers of the form $n! - 1$.
- There are infinitely many prime numbers of the form $2^n - 1$ (i.e., Mersenne primes).
- Every Fermat number $2^{2^n} - 1$ is a composite number for $n > 4$.
- The Fibonacci numbers (see chapter 6, section 1) contain an infinite number of prime numbers. Here are some of these: 2; 3; 5; 13;

89; 233; 1,597; 28,657; 514,229; 433,494,437; 2,971,215,073; and 99,194,853,094,755,497.

The study of primes is boundless; we have merely shown some of the peculiarities that can be discovered among the prime numbers. Yet there are many other peculiarities that readers may want to discover on their own.

8.5. PERFECT NUMBERS

Most mathematics teachers probably told you often enough that everything in mathematics is perfect. While we would then assume that everything in mathematics is truly perfect, might there still be anything more perfect than something else? This brings us to numbers that hold such a title: *perfect numbers*. This is an official designation by the mathematics community. In the field of number theory, we have an entity called a perfect number, which is defined as a number equal to the sum of its proper divisors (i.e., all the divisors except the number itself).

The smallest perfect number is 6, since $6 = 1 + 2 + 3$, which is the sum of all its divisors, excluding the number 6 itself. By the way, 6 is the only number that is both the sum *and* product of the same three numbers: $6 = 1 \times 2 \times 3 = 3!$ Also, $6 = \sqrt{1^3 + 2^3 + 3^3}$. It is also fun to notice that $\frac{1}{1} = \frac{1}{2} + \frac{1}{3} + \frac{1}{6}$ and that both 6 and its square, 36, are triangular numbers (see chapter 4).

The next-larger perfect number is $28 = 1 + 2 + 4 + 7 + 14$, which is the sum of all the divisors of 28, excluding 28 itself. The next perfect number is 496, since again, $496 = 1 + 2 + 4 + 8 + 16 + 31 + 62 + 124 + 248$, which is the sum of all of the divisors of 496, excluding 496 itself. The first four perfect numbers were known to the ancient Greeks. They are 6; 28; 496; and 8,128. It was Euclid who came up with a theorem to generalize a procedure to find a perfect number. He said that for an integer, k, if $2^k - 1$ is a prime number, then one can construct a perfect number using the formula $2^{k-1}(2^k - 1)$. That is, every Mersenne prime (see section 2 of this chapter) gives rise to a perfect number. As noted earlier, $2^k - 1$ can only be prime when k is a prime number. It should be noted that any perfect number obtained through Euclid's formula is an even perfect number. Leonhard Euler finally proved that *every* even perfect number can be obtained in this way. It is not known whether there are any odd perfect numbers. None has yet been found.

Using Euclid's method for generating perfect numbers, we get table 8.2, where for the values of k we get $2^{k-1}(2^k - 1)$ as perfect numbers when $2^k - 1$ is a Mersenne prime number. All presently known Mersenne primes are listed in the appendix, section 3.

k	Mersenne Prime $2^k - 1$	Perfect Number $2^{k-1}(2^k - 1)$
2	3	6
3	7	28
5	31	496
7	127	8,128
13	8,191	33,550,336
17	131,071	8,589,869,056
19	524,287	137,438,691,328
31	2,147,483,647	2,305,843,008,139,952,128
61	2,305,843,009,213,693,951	2,658,455,991,569,831,744, 654,692,615,953,842,176

Table 8.2: The first perfect numbers.

As of early 2013, there are forty-eight known Mersenne primes, and, therefore, only forty-eight known perfect numbers. A complete table of these is given in the appendix, section 4. Just to show some of them in their complete form, look at these perfect numbers:

For $k = 61$: $2^{60}(2^{61} - 1) =$
$\qquad$ 2,658,455,991,569,831,744,654,692,615,953,842,176
For $k = 89$: 2^{88} $(2^{89} - 1) =$ 191,561,942,608,236,107,294,793,378,
$\qquad$ 084,303,638,130,997,321,548,169,216

Presently, the largest known perfect number is obtained for $k = 57{,}885{,}161$. This number has 34,850,340 digits.

By observation, we notice some additional properties of perfect numbers. For example, they all seem to end in either a 6 or a 28, and these are preceded by an odd digit. They also appear to be triangular numbers, which are the sums of consecutive natural numbers; for example,

$$496 = 1 + 2 + 3 + 4 + \ldots + 30 + 31 = T_{31}.$$

(See chapter 4 for a definition of the triangular numbers T_n). Indeed, if p is a Mersenne prime, then the corresponding perfect number is the triangular number with index p, that is, the sum of the first p integers.

From the work of the Italian mathematician Franciscus Maurolycus (1494–1575) we know that every even perfect number is also a hexagonal number. In general, the nth hexagonal number is given by $H_n = 2n^2 - n = n(2n - 1)$ (see chapter 4, section 7). Inserting $n = 2^{k-1}$, this formula gives for the 2^{k-1}th hexagonal number the expression $2^{k-1}(2^k - 1)$. We also know that every even perfect number has this form, when $2^k - 1$ is prime.

To take this a step further, every perfect number *after* the number 6 is the partial sum of the series:

$$1^3 + 3^3 + 5^3 + 7^3 + 9^3 + 11^3 + \ldots.$$

For example,

$$28 = 1^3 + 3^3,$$
$$496 = 1^3 + 3^3 + 5^3 + 7^3,$$
$$8{,}128 = 1^3 + 3^3 + 5^3 + 7^3 + 9^3 + 11^3 + 13^3 + 15^3.$$

This connection between the perfect numbers greater than 6 and the sum of the cubes of consecutive odd numbers is far more than could ever be expected! You might try to find the partial sums for the next few perfect numbers—another challenge for the motivated reader.

8.6. KAPREKAR NUMBERS

There are other numbers that have unusual peculiarities as well. Sometimes these peculiarities can be understood and justified through an algebraic representation, while at other times a peculiarity is simply a quirk of the base-10 number system. In any case, these numbers provide us with some rather entertaining amusements that ought to motivate us to look for other such peculiarities or oddities.

Consider, for example, the number 297. When we take the square of that number, we get $297^2 = 88{,}209$, and, strangely enough, if we were to split it up into two numbers, the sum of the two numbers results in the original number: $88 + 209 = 297$. Such a number is called a *Kaprekar number*, named after the Indian mathematician Dattaraya Ramchandra Kaprekar (1905–1986) who discovered such numbers. Here are a few more examples:

$$9^2 = 81 \ldots 8 + 1 = 9$$
$$45^2 = 2025 \ldots 20 + 25 = 45$$
$$55^2 = 3025 \ldots 30 + 25 = 55$$
$$703^2 = 494{,}209 \ldots 494 + 209 = 703$$
$$2{,}728^2 = 7{,}441{,}984 \ldots 744 + 1{,}984 = 2{,}728$$
$$4{,}879^2 = 23{,}804{,}641 \ldots 238 + 04{,}641 = 4{,}879$$
$$142{,}857^2 = 20{,}408{,}122{,}449 \ldots 20{,}408 + 122{,}449 = 142{,}857$$

A more comprehensive table is given in the appendix, section 5. Some higher Kaprekar numbers are: 38,962; 77,778; 82,656; 95,121; 99,999; ... 538,461; 857,143 ....

There are also further variations, such as the number 45, which we would consider a *Kaprekar triple*, since it behaves as follows: $45^3 = 91,125 = 9 + 11 + 25 = 45$. Other Kaprekar triples are: 1, 8, 10, 297, and 2322. Curiously enough, the number 297, which we previously demonstrated as a Kaprekar number, is also a Kaprekar triple, since $297^3 = 26,198,073$, and $26 + 198 + 073 = 297$. Readers may choose to find other Kaprekar triples.

8.7. THE KAPREKAR CONSTANT

An oddity that is apparently a quirk of the base-10 number system is the *Kaprekar constant*, which is the number 6,174. This constant arises when one takes a four-digit number with at least two different digits, forms the largest and the smallest number from these digits, and then subtracts these two newly formed numbers. Continuously repeating this process with the resulting differences will eventually result in the number 6,174. When the number 6,174 is reached and the process is continued—that is, creating the largest and the smallest number and then taking their difference ($7,641 - 1,467 = 6,174$)—we will always get back to 6,174. This is called the *Kaprekar constant*. To demonstrate this with an example, we will carry out this process with a randomly selected number. When choosing the number, avoid numbers with four identical digits, such as 3,333. For numbers with fewer than four digits, you obtain four digits by padding the number with zeros on the left, such as 0012. For our example, we will choose the number 2,303:

- The largest number formed with these digits is 3,320.
- The smallest number formed with these digits is 0,233.

- The difference is 3,087.
- The largest number formed with these digits is 8,730.
- The smallest number formed with these digits is 0,378.
- The difference is 8,352.
- The largest number formed with these digits is 8,532.
- The smallest number formed with these digits is 2,358.
- The difference is 6,174.
- The largest number formed with these digits is 7,641.
- The smallest number formed with these digits is 1,467.
- The difference is 6,174.

And so the loop is formed, since you will continue to get the number 6,174. Remember, all of this began with an arbitrarily selected four-digit number whose digits are not all the same, and will always end up with the number 6,174, which then gets you into an endless loop (i.e., continuously getting back to 6,174). It should never take more than seven subtractions to reach 6,174. If it does, then there must have been a calculating error.

Incidentally, another curious property of 6,174 is that it is divisible by the sum of its digits:

$$\frac{6174}{6+1+7+4} = \frac{6174}{18} = 343.$$

By the way, were we to apply this continuous subtraction scheme with arbitrary three-digit numbers (not all the same), we would reach the number 495, which would then result in a similar loop returning to the number 495.

8.8. THE MYSTICAL NUMBER 1,089

The number 1,089 has a number of oddities attached to it. One characteristic of this number can be seen by taking its reciprocal and getting the following:

$$\frac{1}{1089} = 0.\overline{0009182736455463728191}.$$

With the exception of the first three zeros and the last 1, we have a palindromic number—918,273,645,546,372,819—since it reads the same in both directions. Furthermore, if we multiply 1,089 by 5, we also get a palindromic number (5,445); and if we multiply 1,089 by 9, we get 9,801, the reverse of the original number. By the way, the only other number of four or fewer digits whose multiple is the reverse of the original number is 2,178, since 2,178 × 4 = 8,712.

Let us now do multiplication by 9 of some numbers that are modifications of 1,089—say 10,989; 109,989; 1,099,989; 10,999,989; and so on—and then marvel at the results:

$$10,989 \times 9 = 98,901$$
$$109,989 \times 9 = 989,901$$
$$1,099,989 \times 9 = 9,899,901$$
$$10,999,989 \times 9 = 98,999,901$$

and so on.

Returning to the number 1,089, we find that it has embedded in it a very entertaining oddity. Suppose you select any three-digit number whose units digit and hundreds digit are not the same, and then reverse that number. Now subtract the two numbers you have (obviously, the larger minus the smaller). Once again reverse the digits of this arrived-at difference, and add this new number to the difference. The result will always be 1,089.

To see how this works, we will choose any randomly selected three-digit number—say 732. We now subtract $732 - 237 = 495$. Reversing the digits of 495, we get 594, and then we add these last two numbers: $495 + 594 = 1,089$. Yes, this will hold true for all such three-digit numbers—amazing! This is a cute little "trick" that can be justified with simple algebra. We offer here an exercise in elementary algebra for the interested reader who is curious why this surprising trick works the way it does. We begin by representing an arbitrarily selected three-digit number, htu, as $100h + 10t + u$, where h represents the hundreds digit, t represents the tens digit, and u represents the units digit (see chapter 1, section 12). The number with the digits reversed is then $100u + 10t + h$. We will let $h > u$. Therefore, the original number is larger than the number with the reversed digits. Next, we subtract the reversed number uth from the original number htu (the minuend) with the usual algorithm, that is, by subtracting the digits at the units place. In this case, we have $u - h < 0$. Therefore, following the usual method of subtraction, we have to take 1 from the tens place (of the minuend), in order to make this subtraction possible. This makes the units place $u + 10$. Next, consider the subtraction at the tens place. The tens digits of the two numbers to be subtracted were equal, but now 1 was taken from the tens digit of the minuend, and the value of this digit became $10(t - 1)$. In order to enable subtraction in the tens place, 1 has to be taken away from the hundreds digit of the minuend. The hundreds digit of the minuend then becomes $h - 1$, making the value of the tens digit $10(t - 1) + 100 = 10(t + 9)$.

When we do the first subtraction, we actually subtract the two numbers in the following form:

$$
\begin{array}{lll}
100\,(h-1) & +\,10\,(t+9) & +\,u+10 \\
-\,100\,u & -\,10\,t & -\,h \\
\hline
100\,(h-u-1) & +\,10\,(9) & +\,u-h+10
\end{array}
$$

Therefore the subtraction $htu - uth$ gives

$$100(h - u - 1) + 10 \times 9 + (u - h + 10).$$

Reversing the digits of this difference gives us

$$100(u - h + 10) + 10 \times 9 + (h - u - 1).$$

By adding these last two expressions, we obtain

$$100(h - u - 1) + 10 \times 9 + (u - h + 10) + 100(u - h + 10) +$$
$$10 \times 9 + (h - u - 1) = 1000 + 90 - 1 = \textbf{1089}.$$

This algebraic justification enables us to inspect the general case of this arithmetic process, thereby allowing us to guarantee that this process holds true for all numbers.

8.9. SOME NUMBER PECULIARITIES

Number oddities need not necessarily be restricted to a single number. There are times when these oddities appear with partner numbers. Consider the addition of the two numbers 192 + 384 = 576. You may ask, what is so special about this addition? Look at the outside digits (bold): **192** + **384** = **576**. They are in numerical sequence left to right (1, 2, 3, 4, 5, 6) and then reversing to get the rest of the nine digits (7, 8, 9). You might have also noticed that the three numbers we used in this addition problem have a strange relationship, as you can see from the following:

$$192 = 1 \times 192,$$
$$384 = 2 \times 192,$$
$$576 = 3 \times 192.$$

The representation of all nine digits often fascinates the observer. Let's consider a number of such situations.

One such unexpected result happens when we subtract the symmetric numbers consisting of the digits in consecutive reverse order and in numerical order: $987,654,321 - 123,456,789$ to get $864,197,532$. This symmetric subtraction used each of the nine digits exactly once in each of the numbers being subtracted, and, surprisingly, resulted in a difference that also used each of the nine digits exactly once.

Here are a few more such strange calculations—this time using multiplication—where on either side of the equals sign all nine digits are represented exactly once: $291,548,736 = 8 \times 92 \times 531 \times 746$, and $124,367,958 = 627 \times 198,354 = 9 \times 26 \times 531,487$.

Another example of a calculation where all the digits are used exactly once (not counting the exponent), is $567^2 = 321,489$. This also works for the following: $854^2 = 729,316$. These are, apparently, the only two squares that result in a number that allow all the digits to be represented once.

When we take the square and the cube of the number 69, we get two numbers that together use all the ten digits exactly once. $69^2 = 4,761$, and $69^3 = 328,509$. That is, the two numbers 4,761 and 328,509 together represent all ten digits.

A somewhat convoluted calculation that results in a surprise ending begins with the following: $6,667^2 = 44,448,889$. When this result, 44,448,889, is multiplied by 3 to get 133,346,667, we notice that the last four digits are the same as the four digits of the number we began with, namely, 6,667. We use this example to bring us to a more general number oddity, which occurs when we take the number 625 to any power. We notice that the resulting number will always end with the last three digits being 625 (see figure 8.3).

625^1	=	**625**
625^2	=	390,**625**
625^3	=	244,140,**625**
625^4	=	152,587,890,**625**
625^5	=	95,367,431,640,**625**
625^6	=	59,604,644,775,390,**625**
625^7	=	37,252,902,984,619,140,**625**
625^8	=	23,283,064,365,386,962,890,**625**
625^9	=	14,551,915,228,366,851,806,640,**625**
625^{10}	=	9,094,947,017,729,282,379,150,390,**625**
. . .		

Table 8.3: Powers of 625.

There are only two such numbers of three digits that have this property. The other is 376, which we can see from the list in table 8.4.

376^1	=	**376**
376^2	=	141,**376**
376^3	=	53,157,**376**
376^4	=	19,987,173,**376**
376^5	=	7,515,177,189,**376**
376^6	=	2,825,706,623,205,**376**
376^7	=	1,062,465,690,325,221,**376**
376^8	=	399,487,099,562,283,237,**376**
376^9	=	150,207,149,435,418,497,253,**376**
376^{10}	=	56,477,888,187,717,354,967,269,**376**
. . .		

Table 8.4: Powers of 376.

If one questions whether there are two-digit numbers that have this property, the answer is clearly yes, and they are 25 and 76.

Number oddities are boundless. Some of these seem a bit far-fetched but nonetheless can be appealing to us from a recreational point of view. For example, consider taking any three-digit number that is multiplied by a five-digit number, all of whose digits are the same. When you add its last five digits to the remaining digits, a number will result where all digits are the same. Here are a few such examples:

$$237 \times 33,333 = 7,899,921, \text{ then } 78 + 99,921 = 99,999;$$
$$357 \times 77,777 = 27,766,389, \text{ then } 277 + 66,389 = 66,666;$$
$$789 \times 44,444 = 35,066,316, \text{ then } 350 + 66,316 = 66,666;$$
$$159 \times 88,888 = 14,133,192, \text{ then } 141 + 33,192 = 33,333.$$

These amazing number peculiarities, although entertaining, allow us to exhibit the beauty of mathematics so as to win over those individuals who have not had the experience of seeing mathematics from this point of view. We offer some more of these here to further entice the reader.

Armstrong Numbers

As we continue to expose some of the most celebrated numbers in mathematics, we come to those that are often referred to as *Armstrong numbers* or narcissistic numbers. In 1966, Michael F. Armstrong, while teaching a course in Fortran and general computing, came across these numbers as an exercise for his students. These numbers were named Armstrong numbers and were popularized in an article by Tim Hartnell in the February 23, 1988, issue of the *Australian* newspaper; and in the April 19, 1988, edition, the author formally named them "the Armstrong numbers." The Armstrong numbers have the property that each number is equal to the sum of its digits, when each is taken to the power equal to the number of digits in the original number. For example, we have the three-digit Armstrong number 153, which is equal to the sum of its digits, each taken to the third power as $1^3 + 5^3 + 3^3 = 1 + 125 + 27 = 153$.

The nine-digit number $472{,}335{,}975 = 4^9 + 7^9 + 2^9 + 3^9 + 3^9 + 5^9 + 9^9 + 7^9 + 5^9$ is, therefore, also an Armstrong number. All Armstrong numbers are shown in the appendix, section 6, where we notice that there are no Armstrong numbers for $k = 2, 12, 13, 15, 18, 22, 26, 28, 30$, and 36 (and $k > 39$). In fact, there are only eighty-nine Armstrong numbers in the decimal system. The largest Armstrong number is thirty-nine digits long, and it is equal to the sum of its digits, each of which is taken to the thirty-ninth power:

$$1^{39} + 1^{39} + 5^{39} + 1^{39} + 3^{39} + 2^{39} + 2^{39} + 1^{39} + 9^{39} + 0^{39} + 1^{39} + 8^{39} + 7^{39} +$$
$$6^{39} + 3^{39} + 9^{39} + 9^{39} + 2^{39} + 5^{39} + 6^{39} + 5^{39} + 0^{39} + 9^{39} + 5^{39} + 5^{39} + 9^{39} +$$
$$7^{39} + 9^{39} + 7^{39} + 3^{39} + 9^{39} + 7^{39} + 1^{39} + 5^{39} + 2^{39} + 2^{39} + 4^{39} + 0^{39} + 1^{39} =$$
$$115{,}132{,}219{,}018{,}763{,}992{,}565{,}095{,}597{,}973{,}971{,}522{,}401.$$

The following is a list of the *consecutive* Armstrong numbers.

$k = 3$: 370; 371
$k = 8$: 24,678,050; 24,678,051
$k = 11$: 32,164,049,650; 32,164,049,651
$k = 16$: 4,338,281,769,391,370; 4,338,281,769,391,371
$k = 25$: 3,706,907,995,955,475,988,644,380;
    3,706,907,995,955,475,988,644,381
$k = 29$: 19,008,174,136,254,279,995,012,734,740;
    19,008,174,136,254,279,995,012,734,741
$k = 33$: 186,709,961,001,538,790,100,634,132,976,990;
    186,709,961,001,538,790,100,634,132,976,991
$k = 39$: 115,132,219,018,763,992,565,095,597,973,971,522,400;
    115,132,219,018,763,992,565,095,597,973,971,522,401

Incidentally, our first Armstrong number, 153, has some other amazing properties as well. It is also a triangular number, where

$$1 + 2 + 3 + 4 + 5 + 6 + 7 + 8 + 9 + 10 + 11 + 12 + 13 + 14 + 15 + 16 + 17 = 153.$$

The number 153 is not only equal to the sum of the cubes of its digits, but it is also a number that can be expressed as the sum of consecutive factorials $1! + 2! + 3! + 4! + 5! = 153$.

Can you discover any other properties of this ubiquitous number 153?

CHAPTER 9

NUMBER RELATIONSHIPS

9.1. BEAUTIFUL NUMBER RELATIONSHIPS

Having observed some numbers that exhibit special characteristics, we now consider noteworthy relationships among other numbers. There are pairs of numbers that partner in rather unexpected relationships. There are partnerships that lead us to refer to them as amicable numbers or friendly numbers. There are numbers that partner as triples, where Pythagorean triples are a prime example. In this chapter, we will consider these and other number relationships that add some unexpected dimensions to the beauty of numbers as they relate to one another.

It is hard to imagine that there are certain number pairs that yield the same product even when both numbers are reversed. For example, $12 \times 42 = 504$ and, if we reverse each of the two numbers, we get $21 \times 24 = 504$. The same thing is true for the number pair 36 and 84, since $36 \times 84 = 3,024 = 63 \times 48$.

At this point you may be wondering if this will happen with any pair of numbers. The answer is that it will only work with the following fourteen pairs of numbers:

$12 \times 42 = 21 \times 24 = 504$

$12 \times 63 = 21 \times 36 = 756$

$12 \times 84 = 21 \times 48 = 1{,}008$

$13 \times 62 = 31 \times 26 = 806$

$13 \times 93 = 31 \times 39 = 1{,}209$

$14 \times 82 = 41 \times 28 = 1{,}148$

$23 \times 64 = 32 \times 46 = 1{,}472$

$23 \times 96 = 32 \times 69 = 2{,}208$

$24 \times 63 = 42 \times 36 = 1{,}512$

$24 \times 84 = 42 \times 48 = 2{,}016$

$26 \times 93 = 62 \times 39 = 2{,}418$

$34 \times 86 = 43 \times 68 = 2{,}924$

$36 \times 84 = 63 \times 48 = 3{,}024$

$46 \times 96 = 64 \times 69 = 4{,}416$

A careful inspection of these fourteen pairs of numbers will reveal that in each case the product of the tens digits of each pair of numbers is equal to the product of the units digits. We can justify this algebraically as follows: For the numbers z_1, z_2, z_3, and z_4, we have

$$z_1 \times z_2 = (10a + b) \times (10c + d) = 100ac + 10ad + 10bc + bd \text{, and}$$
$$z_3 \times z_4 = (10b + a) \times (10d + c) = 100bd + 10bc + 10ad + ac.$$

Here a, b, c, d represent any of the ten digits: 0, 1, 2, . . . 9, where $a \neq 0$ and $c \neq 0$.

Let us assume that $z_1 \times z_2 = z_3 \times z_4$. Then,

$$100ac + 10ad + 10bc + bd = 100bd + 10bc + 10ad + ac,$$

that is, $100ac + bd = 100bd + ac$, and $99ac = 99bd$, or $ac = bd$, which is what we wanted to prove.

There are times when the numbers speak more effectively for themselves than does any explanation. Here the numbers are related in a rather unusual way. A visual inspection of this relationship is far better than a written one. Simply enjoy!

$$1^1 + 6^1 + 8^1 = 15 = 2^1 + 4^1 + 9^1$$
$$1^2 + 6^2 + 8^2 = 101 = 2^2 + 4^2 + 9^2$$
$$1^1 + 5^1 + 8^1 + 12^1 = 26 = 2^1 + 3^1 + 10^1 + 11^1$$
$$1^2 + 5^2 + 8^2 + 12^2 = 234 = 2^2 + 3^2 + 10^2 + 11^2$$
$$1^3 + 5^3 + 8^3 + 12^3 = 2{,}366 = 2^3 + 3^3 + 10^3 + 11^3$$
$$1^1 + 5^1 + 8^1 + 12^1 + 18^1 + 19^1 = 63 = 2^1 + 3^1 + 9^1 + 13^1 + 16^1 + 20^1$$
$$1^2 + 5^2 + 8^2 + 12^2 + 18^2 + 19^2 = 919 = 2^2 + 3^2 + 9^2 + 13^2 + 16^2 + 20^2$$
$$1^3 + 5^3 + 8^3 + 12^3 + 18^3 + 19^3 = 15{,}057 = 2^3 + 3^3 + 9^3 + 13^3 + 16^3 + 20^3$$
$$1^4 + 5^4 + 8^4 + 12^4 + 18^4 + 19^4 = 206{,}755 = 2^4 + 3^4 + 9^4 + 13^4 + 16^4 + 20^4$$

9.2. AMICABLE NUMBERS

What could possibly make two numbers amicable, or friendly? The friendliness of these two numbers will be shown in their "amicable" relationship to each other. This relationship is defined in terms of the proper divisors of these numbers. A *proper divisor* of a natural number n is any natural number that divides n, excluding n itself. For example, 12 has the proper divisors 1, 2, 3, 4, and 6 (but not 12). A number is called *perfect* if the sum of its proper divisors is equal to the number (see chapter 8, section 5). Two numbers are considered *amicable* if the sum of the proper divisors of one number equals the second number *and* the sum of the proper divisors of the second number equals the first number. Perhaps the best way to understand this is through an example. Let's look at the smallest pair of amicable numbers: 220 and 284.

- The divisors of **220** (other than 220 itself) are 1, 2, 4, 5, 10, 11, 20, 22, 44, 55, and 110.

- Their sum is $1 + 2 + 4 + 5 + 10 + 11 + 20 + 22 + 44 + 55 + 110$ $= \mathbf{284}$.
- The divisors of **284** (other than 284 itself) are 1, 2, 4, 71, and 142, and their sum is $1 + 2 + 4 + 71 + 142 = \mathbf{220}$.

This shows that the two numbers are amicable numbers.

This pair of amicable numbers was already known to Pythagoras by about 500 BCE.

A second pair of amicable numbers is 17,296 and 18,416. The discovery of this pair is usually attributed to the French mathematician Pierre de Fermat (1607–1665), although there is evidence that this discovery was anticipated by the Moroccan mathematician Ibn al-Banna al-Marrakushi al-Azdi (1256–ca. 1321).

The sum of the proper divisors of 17,296 is

$$1 + 2 + 4 + 8 + 16 + 23 + 46 + 47 + 92 + 94 + 184 + 188 + 368 + 376 +$$
$$752 + 1{,}081 + 2{,}162 + 4{,}324 + 8{,}648 = 18{,}416.$$

The sum of the proper divisors of 18,416 is

$$1 + 2 + 4 + 8 + 16 + 1{,}151 + 2{,}302 + 4{,}604 + 9{,}208 = 17{,}296.$$

Thus, they, too, are truly amicable numbers!

French mathematician René Descartes (1596–1650) discovered another pair of amicable numbers: 9,363,584, and 9,437,056. By 1747, Swiss mathematician Leonhard Euler (1707–1783) had discovered sixty pairs of amicable numbers, yet he seemed to have overlooked the second-smallest pair—1,184 and 1,210, which were discovered in 1866 by the sixteen-year-old B. Nicolò I. Paganini.

The sum of the divisors of 1,184 is

$$1 + 2 + 4 + 8 + 16 + 32 + 37 + 74 + 148 + 296 + 592 = 1{,}210.$$

And the sum of the divisors of 1,210 is

$$1 + 2 + 5 + 10 + 11 + 22 + 55 + 110 + 121 + 242 + 605 = 1,184.$$

To date we have identified over 363,000 pairs of amicable numbers, yet we do not know if there are an infinite number of such pairs. The table in the appendix, section 7, provides a list of the first 108 amicable numbers. An ambitious reader might want to verify the "friendliness" of each of these pairs. Going beyond this list, we will eventually stumble on an even larger pair of amicable numbers: 111,448,537,712 and 118,853,793,424.

Readers who wish to pursue a search for additional amicable numbers might want to use the following method for finding them: Consider the numbers

$$a = 3 \times 2^n - 1, b = 3 \times 2^{n-1} - 1, \text{ and } c = 3^2 \times 2^{2n-1} - 1,$$
where n is an integer ≥ 2.

If a, b, and c are prime numbers, then $2^n \times a \times b$ and $2^n \times c$ are amicable numbers. (For $n \leq 200$, only $n = 2, 4,$ and 7 would give us a, b, and c to be prime numbers.)

Inspecting the list of amicable numbers in the appendix, section 7, we notice that each pair is either a pair of odd numbers or a pair of even numbers. To date, we do not know if there is a pair of amicable numbers where one is odd and one is even. We also do not know if any pair of amicable numbers is relatively prime (that is, the numbers have no common factor other than 1). These open questions contribute to our continued fascination with amicable numbers.

9.3. OTHER TYPES OF AMICABILITY

There are other types of numbers that also have an amicable relationship, such as *imperfectly-amicable* numbers—two numbers the sums of
whose proper divisors are equal. For example, the numbers 20 and 38
are considered imperfectly-amicable numbers, since the proper divisors of 20 are 1, 2, 4, 5, 10, whose sum is 22, and the proper divisors
of 38 are 1, 2, 19, whose sum is also 22. Another pair of imperfectly-
amicable numbers are 69 and 133, since each has a sum of proper divisors equal to 27. You might want to verify that the numbers 45 and 87
are also imperfectly-amicable numbers.

We can always look for other nice relationships between numbers.
Some of them are truly mind-boggling! Take for example, the pair of
numbers 6,205 and 3,869, which we will call *structurally-amicable
numbers*, where the following relationship exists:

$$6{,}205 = 38^2 + 69^2,$$
$$62^2 + 05^2 = 3{,}869.$$

Notice the symmetry of the breakdown of these two given four-digit
numbers. The pair of numbers 5,965 and 7,706 has the same relationship:

$$5{,}965 = 77^2 + 06^2,$$
$$59^2 + 65^2 = 7{,}706.$$

There are also curious relationships between numbers that tie them
together in an amicable way, such as the pair of numbers 244 and 136,
which can be linked as follows:

$$244 = 1^3 + 3^3 + 6^3,$$
$$2^3 + 4^3 + 4^3 = 136.$$

Ambitious readers may seek other forms of number-pair friendliness!

9.4. PYTHAGOREAN TRIPLES AND THEIR PROPERTIES

Perhaps the most popular fact that adults remember from their high-school mathematics courses is the Pythagorean theorem, which they recall as $a^2 + b^2 = c^2$.

Some may even recall that the numbers a, b, and c can represent the lengths of the three sides of a right triangle. Some of the more frequently seen Pythagorean triples are (3, 4, 5), (5, 12, 13), and (7, 24, 25). We will define Pythagorean triples as follows:

- An ordered set of three natural numbers (a, b, c) that satisfy the Pythagorean relationship $a^2 + b^2 = c^2$ will be called a *Pythagorean triple*.

Pythagorean triples were known long before Pythagoras. The Babylonian cuneiform tablet in figure 9.1 (which you might try to decipher, using chapter 3) shows a large collection of Pythagorean triples. It was made more than a thousand years before the age of Pythagoras.

Figure 9.1: Plimpton 322, a clay tablet (13 cm × 9 cm), written in cuneiform, ca. 1820–1762 BCE, presently at Columbia University. (Plimpton Cuneiform 322 courtesy of the Rare Book & Manuscript Library, Columbia University in the City of New York.)

Pythagorean triples exhibit a very unique relationship among numbers. Typically, questions arise about the nature of these Pythagorean triples, such as the following: How many such triples are there? What are some properties of Pythagorean triples? Is there a general way in which one can find these triples without just trying various combinations of three numbers to see if they satisfy the relationship? As we continue our exploration of number relationships, we will explore the responses to some of these questions as well as to others that commonly arise about the Pythagorean triples.

When the only common factor of the three numbers of the Pythagorean triple is 1, then we call that a *primitive Pythagorean triple*. However, there are also multiples of those triples. For example, for the Pythagorean triple (3, 4, 5), multiples of this triple also satisfy the Pythagorean theorem, such as (6, 8, 10), and (15, 20, 25), since

$$6^2 + 8^2 = 36 + 64 = 100 = 10^2, \text{ and } 15^2 + 20^2 = 225 + 400 = 625 = 25^2.$$

We can justify this for the Pythagorean triple (3, 4, 5). We begin by representing a multiple of this triple as $(3n, 4n, 5n)$, where n is a positive integer. We need to show that these three numbers satisfy the Pythagorean equation $a^2 + b^2 = c^2$. Using $3^2 + 4^2 = 5^2$, we can do this as follows:

$$(3n)^2 + (4n)^2 = 3^2 n^2 + 4^2 n^2 = (3^2 + 4^2)n^2 = 5^2 n^2 = (5n)^2.$$

This verifies that $(3n, 4n, 5n)$ is also a Pythagorean triple. This allows us to conclude that there are an infinite number of Pythagorean triples that are multiples of (3, 4, 5).

Having established that there are infinitely many Pythagorean triples is not the whole picture, however, since the Pythagorean triples we have generated so far are all multiples of (3, 4, 5). Yet we know there are other Pythagorean triples that are not multiples of this triple, such as

(5, 12, 13), (8, 15, 17), and (7, 24, 25), to name just a few. Some of the corresponding right triangles are shown in figure 9.2. One is tempted to ask, how many such primitive Pythagorean triples exist? As you might expect, there are an infinite number of such primitive Pythagorean triples. Let's investigate this further by considering various ways to generate Pythagorean triples. (For more information about the Pythagorean theorem, see A. S. Posamentier, *The Pythagorean Theorem: The Story of Its Beauty and Power* [Amherst, NY: Prometheus Books, 2010].)

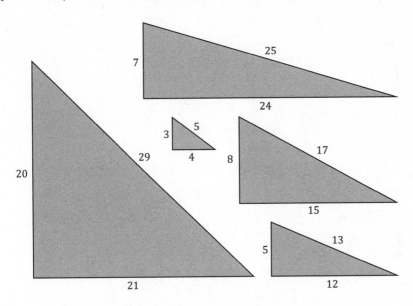

Figure 9.2: Some right triangles representing Pythagorean triples.

9.5. FIBONACCI'S METHOD FOR FINDING PYTHAGOREAN TRIPLES

Fibonacci was perhaps one of the most influential mathematicians of the thirteenth century (chapter 6, section 1). In 1225, he published a book titled *Liber quadratorum* (*Book of Squares*), in which he demonstrated another relationship of numbers and stated the following:

I thought about the origin of all square numbers and discovered that they arise out of the increasing sequence of odd numbers; for the unity is a square and from it is made the first square, namely 1; that to this unity is added 3, making a second square, namely 4, with root 2; if to the sum is added the third odd number, namely 5, the third square is created, namely 9, with root 3; and thus sums of consecutive odd numbers and a sequence of squares always arise together in order.[1]

Fibonacci was essentially describing the relationship that we discussed in chapter 4, section 4, that the sum of the first n odd numbers equals n^2:

$$1 + 3 + 5 + \ldots + (2n - 1) = n^2.$$

We visualize this statement again in figure 9.3. Notice how the squares beginning with the single square at the lower left increase in area by the consecutive odd numbers analogous to what we just established algebraically. This is a geometric analogue of this algebraic statement.

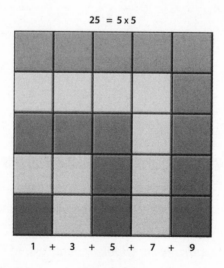

Figure 9.3

Fibonacci knew of the Pythagorean theorem and therefore was aware of Pythagorean triples—after all, he lived about 1,700 years after Pythagoras. He was able to generate these triples in the following way. Let's consider the sum of the first five odd natural numbers (as shown in figure 9.3), $1 + 3 + 5 + 7 + 9 = 5^2$, whose last term (9) is a square number. The sum in the parentheses: $(1 + 3 + 5 + 7) + 9 = 5^2$ is 16, which is also a square number because it is the sum of the first four odd numbers. So this equation can be rewritten as $16 + 9 = 25$, which, surprisingly, gives us the primitive Pythagorean triple $(3, 4, 5)$.

Let's consider another series of consecutive odd integers—one that ends with a square number—to convince ourselves that this scheme can really generate other primitive Pythagorean triples: Consider the sum of the first odd numbers up to the thirteenth, which happens to be a square number, $25 = 5^2$:

$$(1 + 3 + 5 + 7 + 9 + 11 + 13 + 15 + 17 + 19 + 21 + 23) + 25 = 169 = 13^2.$$

Using the same procedure as above, we add all the terms prior to the last one—within the parentheses. Because it is the sum of the first twelve odd numbers, it must be equal to the square number $12^2 = 144$. Therefore, we obtain $144 + 25 = 169$, or $12^2 + 5^2 = 13^2$. This gives us another primitive Pythagorean triple $(5, 12, 13)$.

In general terms, Fibonacci's construction of Pythagorean triples can be described as follows: Choose any odd number $a > 1$. Every odd number has an odd square, therefore a^2 is also an odd number, and we can write it in the form $a^2 = 2n + 1$ for some natural number n. For example, we obtain $a^2 = 9$ for $n = 4$, which is the fifth odd number, and $a^2 = 25$ (for $n = 12$) is the thirteenth odd number. In general, $2n + 1$ is the $(n + 1)$st odd number. Clearly, the sum of the first $n + 1$ odd numbers up to $a^2 = 2n + 1$ is equal to $(n + 1)^2$:

$$1 + 3 + 5 + \ldots + (a^2 - 2) + a^2 = (n + 1)^2.$$

Here the summand $(a^2 - 2)$ is the nth odd number, and the sum of the odd numbers up to $(a^2 - 2)$ is equal to n^2:

$$1 + 3 + 5 + \ldots + (a^2 - 2) = n^2.$$

From this we obtain the following:

$$n^2 + a^2 = (n + 1)^2.$$

To summarize, for any odd number a, we can write $a^2 = 2n + 1$, and the triple $(a, n, n + 1)$ is a Pythagorean triple. Curiously, it is even a primitive Pythagorean triple because the only common factor of n and $n + 1$ is 1. From $a^2 = 2n + 1$ we find $n = \frac{a^2 - 1}{2}$ and $n + 1 = \frac{a^2 + 1}{2}$. We see that Fibonacci's method determines, for every odd number $a = 3, 5, 7 \ldots$, the primitive Pythagorean triple $a, b = \frac{a^2 - 1}{2}, c = \frac{a^2 + 1}{2}$. We can, therefore, conclude that there are an infinite number of primitive Pythagorean triples because there are infinitely many odd numbers. A few of these triples are listed in table 9.1.

We can further evaluate the numbers b_n in table 9.1 as

$$b_n = \frac{a_n^2 - 1}{2} = \frac{(2n + 1)^2 - 1}{2} = \frac{4n^2 + 4n + 1 - 1}{2} = \frac{4n(n + 1)}{2} = 2n(n + 1).$$

Therefore, Fibonacci's result means that

$$(a_n = 2n + 1, \, b_n = 2n(n + 1), \, c_n = 2n(n + 1) + 1)$$

is a primitive Pythagorean triple for all natural numbers n.

We also note that $b_n = 4T_n$, where $T_n = \frac{n(n + 1)}{2}$ is the nth triangular number defined in chapter 4, section 5. The sequence of triangular numbers starts with $1, 3, 6, 10, 15, 21, 28, 36 \ldots$, and the b-values of table 9.1 are just four times these numbers. Therefore, Fibonacci's triples can also be written as $(2k + 1, 4T_k, 4T_k + 1)$.

k	a_k	$b_k = \dfrac{a_k^2 - 1}{2}$	$c_k = \dfrac{a_k^2 + 1}{2}$
1	3	4	5
2	5	12	13
3	7	24	25
4	9	40	41
5	11	60	61
6	13	84	85
7	15	112	113
8	17	144	145
9	19	180	181
10	21	220	221
11	23	264	265
12	25	312	313

Table 9.1: Primitive Pythagorean triples obtained using Fibonacci's method.

Figure 9.4 shows the right triangles corresponding to the first eight Pythagorean triples in table 9.1. Because the hypotenuse differs from the longer leg by 1, these triangles tend to become rather long and extended.

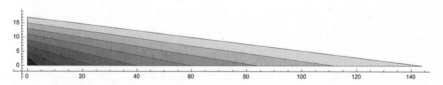

Figure 9.4: Triangles corresponding to the Pythagorean triples in table 9.1.

9.6. STIFEL'S METHOD FOR GENERATING PRIMITIVE PYTHAGOREAN TRIPLES

The following approach to generate Pythagorean triples is due to German mathematician Michael Stifel (1487–1567). He created a sequence of mixed numbers of the following form:

$$1 + \frac{1}{3},\ 2 + \frac{2}{5},\ 3 + \frac{3}{7},\ 4 + \frac{4}{9},\ 5 + \frac{5}{11},\ 6 + \frac{6}{13},\ 7 + \frac{7}{15},\ \dots.$$

This sequence is easy to remember: The whole number parts of the above mixed numbers are simply the natural numbers in order, the numerators of the fractions are the same number as the whole number, and the denominators of the fractions are consecutive odd numbers, beginning with 3.

Now we convert each of the mixed numbers in this sequence to a fraction. The fractions will then produce the first two members of a Pythagorean triple. For example, if we take the sixth term of this sequence, $6 + \frac{6}{13} = \frac{84}{13}$, we have the first two members of the Pythagorean triple $(13, 84, c)$. Then, to get the third member, we simply obtain c in the following manner: $c^2 = 13^2 + 84^2 = 169 + 7{,}056 = 7{,}225$, and then take the square root of $7{,}225$ to get 85. Thus, the complete Pythagorean triple is $(13, 84, 85)$.

What appears as a magic trick, is, in fact, easy to explain: We can write the nth number in Stifel's sequence as

$$n + \frac{n}{2n+1} = \frac{n(2n+1)+n}{2n+1} = \frac{2n^2+2n}{2n+1} = \frac{2n(n+1)}{2n+1}, \; n = 1, 2, 3, \ldots.$$

Compare this result with the general expressions for the Pythagorean triples obtained by Fibonacci's method toward the end of the last section: $a_n = 2n + 1$, $b_n = 2n(n + 1)$, and $c_n = b_n + 1$. You can see that each of Stifel's numbers is just the quotient of the first two members of this Pythagorean triple,

$$n + \frac{n}{2n+1} = \frac{b_n}{a_n}.$$

This shows that Stifel's method leads precisely to the same primitive Pythagorean triples as does Fibonacci's method.

9.7. EUCLID'S METHOD FOR FINDING PYTHAGOREAN TRIPLES

The question then arises, how can we more succinctly generate primitive Pythagorean triples? More importantly, how can we obtain all

Pythagorean triples? That is, is there a formula for achieving this goal? One such formula, attributed to the work of Euclid, for integers m and n, generates values of a, b, and c, where $a^2 + b^2 = c^2$, as follows:

$$a = m^2 - n^2, \, b = 2mn, \, c = m^2 + n^2 \text{ (assuming } m > n).$$

We can easily show that this formula will always yield a Pythagorean triple. First we will square each of the terms and then show that the sum of the first two squares is equal to the third square.

$$a^2 = (m^2 - n^2)^2, \, b^2 = (2mn)^2, \, c^2 = (m^2 + n^2)^2.$$

We will do this simple algebraic task by showing that the sum $a^2 + b^2$ is actually equal to c^2.

$$
\begin{aligned}
a^2 + b^2 &= (m^2 - n^2)^2 + (2mn)^2 \\
&= m^4 - 2m^2n^2 + n^4 + 4m^2n^2 \\
&= m^4 + 2m^2n^2 + n^4 = (m^2 + n^2)^2 = c^2.
\end{aligned}
$$

Therefore, $a^2 + b^2 = c^2$.

We can apply Euclid's formula to gain an insight into properties of Pythagorean triples.

When we insert some values of m and n, as in table 9.2, we should notice a pattern that would tell us when the triple will be primitive—which, you will recall, is when the largest common factor of the three numbers is 1—and also discover some other possible patterns.

An inspection of the triples in the list of table 9.2 would have us make the following conjectures—which, indeed, can be proved. For example, Euclid's formula $a = m^2 - n^2$, $b = 2mn$, $c = m^2 + n^2$ will yield *primitive* Pythagorean triples only when m and n are relatively prime—that is, when they have no common factor other than 1—and *exactly one* of these must be an even number, with $m > n$.

One can even show the fundamental result—that *all* primitive Pythagorean triples can be obtained with Euclid's formula:

- Every primitive Pythagorean triple can be written as

$$(m^2 - n^2,\ 2mn,\ m^2 + n^2)$$

with unique natural numbers m and n, which are relatively prime, $m > n$, and $m - n$ is odd.

m	n	$m^2 - n^2$	$2mn$	$m^2 + n^2$	Pythagorean Triple	Primitive
2	1	3	4	5	(3, 4, 5)	Yes
3	1	8	6	10	(6, 8, 10)	No
3	2	5	12	13	(5, 12, 13)	Yes
4	1	15	8	17	(8, 15, 17)	Yes
4	2	12	16	20	(12, 16, 20)	No
4	3	7	24	25	(7, 24, 25)	Yes
5	1	24	10	26	(10, 24, 26)	No
5	2	21	20	29	(20, 21, 29)	Yes
5	3	16	30	34	(16, 30, 34)	No
5	4	9	40	41	(9, 40, 41)	Yes
6	1	35	12	37	(12, 35, 37)	Yes
6	2	32	24	40	(24, 32, 40)	No
6	3	27	36	45	(27, 36, 45)	No
6	4	20	48	52	(20, 48, 52)	No
6	5	11	60	61	(11, 60, 61)	Yes
7	1	48	14	50	(14, 48, 50)	No
7	2	45	28	53	(28, 45, 53)	Yes
7	3	40	42	58	(40, 42, 58)	No
7	4	33	56	65	(33, 56, 65)	Yes
7	5	24	70	74	(24, 70, 74)	No
7	6	13	84	85	(13, 84, 85)	Yes
8	1	63	16	65	(16, 63, 65)	Yes
8	2	60	32	68	(32, 60, 68)	No
8	3	55	48	73	(48, 55, 73)	Yes
8	4	48	64	80	(48, 64, 80)	No
8	5	39	80	89	(39, 80, 89)	Yes
8	6	28	96	100	(28, 96, 100	No
8	7	15	112	113	(15, 112, 113)	Yes

Table 9.2: Using Euclid's Formula to Generate Pythagorean Triples

Euclid's formula has a nice geometric interpretation. This will enable us to provide a sketch of an elegant proof of Euclid's formula. Consider the Pythagorean relationship in the form $c^2 = b^2 + a^2$, and then we will divide this equation by c^2 to obtain

$$\frac{b^2}{c^2} + \frac{a^2}{c^2} = 1, \text{ or } x^2 + y^2 = 1 \text{ with } x = \frac{b}{c}, \ y = \frac{a}{c}.$$

Therefore (x,y) can be interpreted as the coordinates of a point P on the unit circle. Because a, b, and c are natural numbers, x and y are rational numbers (fractions). Figure 9.5 shows a triangle with vertex P on a circle with radius 1. The triangle has the same shape as the Pythagorean triangle with sides a, b, and c, but it is scaled to a size where the hypotenuse equals 1.

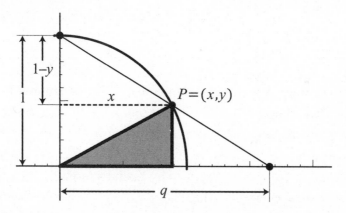

Figure 9.5: A scaled Pythagorean triangle.

Consider the construction in figure 9.5. It shows the unit circle and a point P with coordinates (x,y) satisfying $x^2 + y^2 = 1$. We draw a line from the point $(0,1)$ through the point P. This line intersects the horizontal axis at the point $(q,0)$, where q is some number greater than 1 (because it is outside the circle). It is clear that the number q in turn uniquely determines the point P on the circle. From figure 9.5, one can derive, with the help of some geometry and algebra (an ambitious reader might try to fill in the details), the following formulas, which allow us to determine q, if x and y are given, and conversely, to determine x and y, if q is given:

$$q = \frac{x}{1-y}, x = \frac{2q}{q^2+1}, y = \frac{q^2-1}{q^2+1}.$$

From these formulas, we may also conclude that x and y are rational numbers, whenever q is a rational number. Therefore, x and y are related to a Pythagorean triple whenever q is rational, that is, whenever $q = \frac{m}{n}$, with natural numbers m and n, where $m > n$. Inserting this term for q in the expressions for x and y, we obtain the following result

$$x = \frac{b}{c} = \frac{2\frac{m}{n}}{\left(\frac{m}{n}\right)^2+1} = \frac{2mn}{m^2+n^2}, y = \frac{a}{c} = \frac{\left(\frac{m}{n}\right)^2-1}{\left(\frac{m}{n}\right)^2+1} = \frac{m^2-n^2}{m^2+n^2}.$$

We conclude that any rational number $q = \frac{m}{n}$ determines a unique Pythagorean triple with $a = m^2 - n^2$, $b = 2mn$, $c = m^2 + n^2$, and vice versa: any Pythagorean triple determines a unique rational number by the construction in figure 9.5. This finally leads to the conclusion that every Pythagorean triple can be described by Euclid's formula.

Figure 9.6 shows the shapes of the right triangles with sides $a = m^2 - n^2$, $b = 2mn$, and $c = m^2 + n^2$, and with vertices of the right angles situated at the points (m,n). As it was the case in table 9.2, the triangles are all scaled to a smaller size (with the hypotenuse equal to 1). The primitive triangles are black.

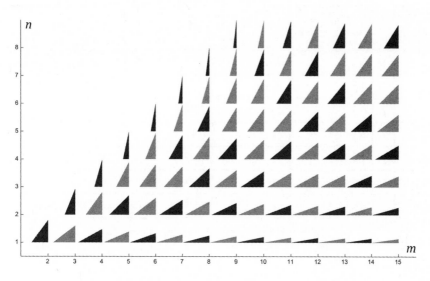

Figure 9.6: Shapes (proportions) of Pythagorean triangles.

9.8. EXPLORING PYTHAGOREAN TRIPLES

Euclid's formula will allow us to discover many other relationships that exist among these Pythagorean triples. For example, when we inspect the values of m and n that determine a primitive Pythagorean triple and, in addition, where $n = 1$ (corresponding to the row of triangles at the bottom of figure 9.6), we notice that in table 9.3, the hypotenuse c (that is, the third member of the triple) will differ from one of the legs by 2.

Algebraically, this is easily demonstrated. Once again, consider the formula for generating all Pythagorean triples: $a = m^2 - n^2$, $b = 2mn$, and $c = m^2 + n^2$. When $n = 1$, we get $a = m^2 - 1$, $b = 2m$, $c = m^2 + 1$. Therefore, the difference between c and a is $c - a = (m^2 + 1) - (m^2 - 1) = 2$. Table 9.3 shows a few cases of primitive Pythagorean triples where $n = 1$.

m	n	$a = m^2 - n^2$	$b = 2mn$	$c = m^2 + n^2$	Pythagorean Triple
2	1	3	4	5	(3, 4, 5)
4	1	15	8	17	(8, 15, 17)
6	1	35	12	37	(12, 35, 37)
8	1	63	16	65	(16, 63, 65)
14	1	195	28	197	(28, 195, 197)
18	1	323	36	325	(36, 323, 325)
22	1	483	44	485	(44, 483, 485)

Table 9.3: Some primitive triples with $n = 1$.

9.9. CONSECUTIVE MEMBERS OF A PYTHAGOREAN TRIPLE

The special Pythagorean triples, found by Fibonacci's method in table 9.1, all have the property that $c = b + 1$. For which values of m and n will this be the case? Inspection of the various values listed in table 9.2 reveals that when $m - n = 1$, then $c - b = 1$. By now it should be relatively easy to justify this finding algebraically. Indeed, we have

$$c - b = m^2 + n^2 - 2mn = (m - n)^2.$$

You can see that $c - b = 1$ in exactly the cases where $(m - n)^2 = 1$. This leads us to $m - n = 1$, or $m - n = -1$. However, $m - n$ cannot be negative, since we always assume $m > n$. Therefore, the condition $c - b = 1$ is equivalent with $m - n = 1$, or $m = n + 1$. This verifies our conjecture about the difference of $c - b = 1$. All Pythagorean triples (a, b, c) with $c - b = 1$ have $m - n = 1$.

We can now ask whether there are some Pythagorean triples with the property $c - b = 1$ that are not among those obtained by Fibonacci's method. In order to answer this question, we apply Euclid's formula, which is known to give all Pythagorean triples. Because we know that any triple with $c - b = 1$ must have $m - n = 1$, we consider Euclid's formula with consecutive natural numbers n and $m = n + 1$, and find

$$a = m^2 - n^2 = (n + 1)^2 - n^2 = 2n + 1,$$
$$b = 2mn = 2(n + 1)n,$$
$$c = m^2 + n^2 = (n + 1)^2 + n^2 = 2n^2 + 2n + 1 = b + 1.$$

But this is precisely the formula ($a = 2n + 1$, $b = 2n(n + 1)$, $c = b + 1$) for the Pythagorean triples constructed by Fibonacci's method that we obtained in section 5. Fibonacci's method already provided us with all triples with the property $c - b = 1$.

When we further inspect the list of Pythagorean triples in table 9.1, we notice not only that $c = b + 1$, but also that $a^2 = b + c$, a truly remarkable pattern among Pythagorean triples. For example, for the Pythagorean triple (7, 24, 25) we have $25 = 24 + 1$, and at the same time we also have $7^2 = 24 + 25 = 49$. This result follows easily from the formulas above:

$$a^2 = (2n + 1)^2 = 4n^2 + 4n + 1 = 4n(n + 1) + 1 =$$
$$2n(n + 1) + (2n(n + 1) + 1) = b + c.$$

While we are considering number relationships, we ought to consider a peculiarity among the b terms (i.e., $2mn$) in this pattern-rich list shown in table 9.1, namely, 4, 12, 24, 40, 60, 84 . . . . They just happen to fold nicely into the following pattern:

$$3^2 + [4^2] = 5^2$$
$$10^2 + 11^2 + [12^2] = 13^2 + 14^2$$
$$21^2 + 22^2 + 23^2 + [24^2] = 25^2 + 26^2 + 27^2$$
$$36^2 + 37^2 + 38^2 + 39^2 + [40^2] = 41^2 + 42^2 + 43^2 + 44^2$$
$$55^2 + 56^2 + 57^2 + 58^2 + 59^2 + [60^2] = 61^2 + 62^2 + 63^2 + 64^2 + 65^2$$

Searching for patterns among the Pythagorean triples can be a lot of fun. For example, perhaps the most obvious pattern to consider is that in which the three numbers of the triple are consecutive numbers, such as (3, 4, 5). To search for such triples, we can let the three members of the Pythagorean triple be represented by $a = b - 1$, b, $c = b + 1$. Now to secure them into the Pythagorean relationship we get:

$$(b - 1)^2 + b^2 = (b + 1)^2,$$
$$b^2 - 2b + 1 + b^2 = b^2 + 2b + 1,$$
$$b^2 = 4b, \text{ therefore } b = 4.$$

The resulting Pythagorean triple is then (3, 4, 5), where the three members are in consecutive order (i.e., each of the members differ by 1). This indicates to us that the Pythagorean triple (3, 4, 5) is the only Pythagorean triple where all three members are consecutive numbers.

But there are some Pythagorean triples that have two consecutive numbers as the first two numbers of the triple. Our most familiar Pythagorean triple, (3, 4, 5), already meets this criterion, since 3 and 4 are consecutive numbers. Table 9.4 lists some others.

n	a_n	$b_n = a_n + 1$	c_n
1	3	4	5
2	20	21	29
3	119	120	169
4	696	697	985
5	4,059	4,060	5,741
6	23,660	23,661	33,461
7	137,903	137,904	195,025
8	803,760	803,761	1,113,689
9	4,684,659	4,684,660	6,625,109
10	27,304,196	27,304,197	38,613,965

Table 9.4

Again if you search for it, a pattern can be found—as is so often the case in mathematics. However, this pattern is a bit different from earlier ones. The Pythagorean triple where the first two numbers are consecutive can be constructed as follows: Starting with $a_1 = 3$ and $a_2 = 20$, we obtain the third triple in table 9.4 with the formula $a_3 = 6\,a_2 - a_1 + 2 = 120 - 3 + 2 = 119$. The general formula for the a-values in table 9.4 is

$$a_n = 6a_{n-1} - a_{n-2} + 2, \text{ for } n = 3, 4, 5 \ldots$$

Once we know a_n, we get the other consecutive number b_n of the Pythagorean triple just by adding 1. The third member, c_n, can then be found by applying the Pythagorean relation, that is, by taking the square root of $c_n^2 = a_n^2 + b_n^2$.

9.10. SOME OTHER PYTHAGOREAN CURIOSITIES

The following list of curiosities will further illuminate the practically boundless relationships that exist among the three members of a Pythagorean triple—once again demonstrating number relationships that further enhance one's appreciation for the beauty of mathematics.

Pythagorean Curiosity 1

We begin with any primitive Pythagorean triple—say (a, b, c). We will substitute these values of a, b, and c into the following three sets of formulas (table 9.5). Curiously, each will generate a new primitive Pythagorean triple (x, y, z):

	x	y	z
Formula 1	$a - 2b + 2c$	$2a - b + 2c$	$2a - 2b + 3c$
Formula 2	$a + 2b + 2c$	$2a + b + 2c$	$2a + 2b + 3c$
Formula 3	$-a + 2b + 2c$	$-2a + b + 2c$	$-2a + 2b + 3c$

Table 9.5

To see how this works, we will apply the three formulas to the primitive Pythagorean triple (5, 12, 13), which gives the three primitive Pythagorean triples (7, 24, 25), (55, 48, 73), and (45, 28, 53) (see table 9.6).

	x	y	z
Formula 1	$5 - 2 \times 12 + 2 \times 13 = 7$	$2 \times 5 - 12 + 2 \times 13 = 24$	$2 \times 5 - 2 \times 12 + 3 \times 13 = 25$
Formula 2	$5 + 2 \times 12 + 2 \times 13 = 55$	$2 \times 5 + 12 + 2 \times 13 = 48$	$2 \times 5 + 2 \times 12 + 3 \times 13 = 73$
Formula 3	$-5 + 2 \times 12 + 2 \times 13 = 45$	$-2 \times 5 + 12 + 2 \times 13 = 28$	$-2 \times 5 + 2 \times 12 + 3 \times 13 = 53$

Table 9.6

Essentially, we can use any primitive Pythagorean triple to generate three others with these three formulas. In fact, all primitive Pythagorean triples can be generated in this way from the triple (3, 4, 5), and every triple is obtained exactly once. For example, repeatedly applying formula 2 of table 9.5 to the triple (3,4,5) would generate all the triples of table 9.4.

Pythagorean Curiosity 2

Recall the sequence of Fibonacci numbers F_n, starting with 1, 1, 2, 3, 5, 8, 13 . . . (see chapter 6, section 1). Starting with any four consecutive Fibonacci numbers F_{k-1}, F_k, F_{k+1}, F_{k+2}, from the table in the appendix, section 1, one obtains the first number of a Pythagorean triple by multiplying the outer two numbers ($F_{k-1} \times F_{k+2}$), the second number by multiplying the middle two numbers and doubling the result ($2 \times F_k \times F_{k+1}$), and the third number by adding the squares of the middle two numbers (which is again a Fibonacci number F_{2k+1}).

We can express this as a compact formula as follows:

$(F_{k-1} \times F_{k+2}, 2 \times F_k \times F_{k+1}, F_{2k+1})$ is a Pythagorean triple for any natural number $k > 1$.

For example, with $k = 2$, we obtain the triple $(3, 4, 5)$, because

$$F_1 \times F_4 = 1 \times 3 = 3, \ 2 \times F_2 \times F_3 = 2 \times 1 \times 2 = 4, \ \text{and} \ F_5 = 5.$$

And for $k = 10$, we obtain (consulting the table in the appendix, section 1):

$$F_9 \times F_{12} = 34 \times 144 = 4896, \ 2 \times F_{10} \times F_{11} = 2 \times 55 \times 89 = 9790, \ F_{21} = 10946.$$

Indeed, one can verify that $4896^2 + 9790^2 = 10946^2$.

This observation is a consequence of the following formulas, which hold for all Fibonacci numbers:

$$(F_{k+1})^2 + (F_k)^2 = F_{2k+1}, \ \text{and} \ (F_{k+1})^2 - (F_k)^2 = F_{k-1} \times F_{k+2}.$$

We can use Euclid's formula for generating primitive Pythagorean triples and insert for n and m two consecutive Fibonacci numbers, F_k and F_{k+1}. Then we get the Pythagorean triple

$$(m^2 - n^2, 2mn, m^2 + n^2) = ((F_{k+1})^2 - (F_k)^2, \ 2 \times F_k \times F_{k+1}, (F_{k+1})^2 + (F_k)^2).$$

From the formulas above, we get the Pythagorean triple

$$(F_{k-1} \times F_{k+2}, 2 \times F_k \times F_{k+1}, F_{2k+1}).$$

Pythagorean Curiosity 3

Inspection of the list of Pythagorean triples will convince you that the product of the first two members of a Pythagorean triple is always a multiple of 12. Therefore, we find the following result:

- The product $a \times b$ of the two smaller members of a Pythagorean triple is always a multiple of 12.

One member of a Pythagorean triple is always a multiple of 5. From this we obtain:

- The product $a \times b \times c$ of all three numbers of a Pythagorean triple is always a multiple of 60.

A further still-unanswered question is whether there are two Pythagorean triples (primitive or nonprimitive) with the same product of its members.

The area of the right triangle with the legs a and b is $\frac{a \times b}{2}$. If a and b are members of a Pythagorean triple, the area is always a multiple of 6. A curious Pythagorean triple is (693, 1924, 2045), which just happens to have an area of 666,666. Readers involved in numerology will recognize this as a sort of double 666, which is often referred to as the "Number of the Beast" as described in the Book of Revelation (13:17–18) in the New Testament of the Christian Bible. Obviously, some Pythagorean curiosities are just that: nothing but an unusual number that raises eyebrows.

Pythagorean Curiosity 4

Pierre de Fermat posed a problem in 1643 and finally found an answer himself. He sought a Pythagorean triple where the sum of the two smaller numbers is a square integer and the larger number is also a square integer. Symbolically, he sought to find a Pythagorean triple where $a + b = p^2$ and $c = q^2$, where p and q are integers. He found one such Pythagorean triple to be (4,565,486,027,761; 1,061,652,293,520; 4,687,298,610,289), where $a + b$ = 4,565,486,027,761 + 1,061,652, 293,520 = 5,627,138,321,281 = $2,372,159^2$. The third number is also a square number: c = 4,687,298,610,289 = $2,165,017^2$. Besides discovering this Pythagorean triple, Fermat also proved it was the *smallest* Pythagorean triple having this property! It is hard to imagine the next larger such Pythagorean triple.

Pythagorean Curiosity 5

We can generate a "family" of rather unusual Pythagorean triples by using the formula:

$$a = 2n + 1, b = 2n(n + 1), c = 2n(n + 1) + 1,$$

as shown in table 9.7.

n	$a = 2n + 1$	$b = 2n(n + 1)$	$c = 2n(n + 1) + 1$
10	21	220	221
10^2	201	20,200	20,201
10^3	2,001	2,002,000	2,002,001
10^4	20,001	200,020,000	200,020,001
10^5	200,001	20,000,200,000	20,000,200,001
10^6	2,000,001	2,000,002,000,000	2,000,002,000,001

Table 9.7

You will be pleasantly surprised when you generate a similar list with powers of 20, 40, and so on in place of the powers of 10 we used in table 9.7. For $n = 20$, you will get $41^2 + 840^2 = 841^2$, and for 20^2 you will get $401^2 + 80,400^2 = 80,401^2$. See what other patterns of this kind you can discover.

Pythagorean Curiosity 6

There are an infinite number of primitive Pythagorean triples where the third member is the square of a natural number. This demonstration is rather simple:

Start with two natural numbers x and y, which are relatively prime, satisfy $x > y$, and are of different parity (that is, one odd and the other even). Using Euclid's formula, we define a primitive Pythagorean triple by

$$m = x^2 - y^2, n = 2xy, h = x^2 + y^2.$$

(In fact, any Pythagorean triple can be obtained in this way). We can also obtain a second primitive Pythagorean triple (a, b, c) by using the Euclidean formula with m and n. In general, we could either have $m > n$ or $n > m$, but in both cases,

$$c = m^2 + n^2 = (2xy)^2 + (x^2 - y^2)^2 = 4x^2y^2 + x^4 - 2x^2y^2 + y^4 = x^4 + 2x^2y^2 + y^4 = (x^2 + y^2)^2.$$

Thus we have shown that the third member of a Pythagorean triple will be a square number whenever we use the first two members of another Pythagorean triple in the Euclidean formula to generate this second Pythagorean triple. Then this second Pythagorean triple has its third member as a square of a natural number.

For example, from the primitive Pythagorean triple (8, 15, 17) we can generate the primitive Pythagorean triple (161, 240, 289) where the third member, 289, is a square number (17^2).

Pythagorean Curiosity 7

In a similar fashion, we can also show that there are infinitely many primitive Pythagorean triples where one of the first two members is a square number. An example where the odd member is a square is the triple (9, 40, 41), and where the even member is a square we have (16, 63, 65). There are infinitely many of these. Pierre de Fermat proved that there are no Pythagorean triples where *both* first two members are square numbers. Table 9.8 shows a few examples of Pythagorean triples that have as their smallest member a square number, and some where the smallest member is a perfect cube.

n	Primitive Pythagorean Triples with the Smallest Member a Square Number, n^2	Primitive Pythagorean Triples with the Smallest Member a Cube, n^3
3	(9; 40; 41)	(27; 364; 365)
4	(16; 63; 65)	(64; 1,023; 1,025)
5	(25; 312; 313)	(125; 7,812; 7,813)
6	(36; 77; 85)	(216; 713; 745)
6		(216; 11,663; 11,665)
7	(49; 1,200; 1,201)	(343; 58,824; 58,825)
8	(64; 1,023; 1,025)	(512; 65,535; 65,537)
9	(81; 3,280; 3,281)	(729; 265,720; 265,721)
10	(100; 621; 629)	(1,000; 15,609; 15,641)
10		(1,000; 249,999; 250,001)
11	(121; 7,320; 7,321)	(1,331; 885,780; 885,781)

Table 9.8

Pythagorean Curiosity 8

Another property of Pythagorean triples (a, b, c) is that they all have the following relationship:

$$\frac{(c-a)(c-b)}{2}$$ is always a square number.

For example, for the Pythagorean triple (7, 24, 25) we find $\frac{(25-7)(25-24)}{2}=\frac{18\times1}{2}=9$, which is a square number. The converse of this statement is not true—for example, the same relationship also holds for the triple (6, 12, 18), but it is *not* a Pythagorean triple.

The statement is easy to demonstrate using Euclid's formula, $a = m^2 - n^2$, $b = 2mn$, and $c = m^2 + n^2$:

$$\frac{(c-a)(c-b)}{2}=\frac{\left(m^2+n^2-\left(m^2-n^2\right)\right)\left(m^2+n^2-2mn\right)}{2}=\frac{\left(2n^2\right)(m-n)^2}{2},$$

which is equal to the square number $(n(m-n))^2$.

Pythagorean Curiosity 9

Another curiosity embedded among the many Pythagorean triples is one that relates to the Pythagorean triple (5, 12, 13). If we place the digit 1 before each member of the triple, then we get (15, 112, 113), which, curiously, is also a Pythagorean triple. This is conjectured to be the only time a single digit can be placed to the left of each member of a Pythagorean triple to generate another Pythagorean triple.

Pythagorean Curiosity 10

Some symmetric Pythagorean triples are also worth highlighting. One is where the second and third members are reverses of one another, and the first member is a palindromic number (see chapter 7, section 7). Here are two such examples: (33, 56, 65) and (3,333; 5,656; 6,565). Can you find other such "symmetric" pairs of Pythagorean triples?

There are also Pythagorean triples where the first two members are reverses of one another, such as (88,209; 90,288; 126,225). Are there more such triples?

Naturally, we can create palindromic Pythagorean triples by multiplying each of the members of the triple (3, 4, 5) by 11, 111, 111, 1111 . . . , or by 101, 1001, 10001 . . . , and so on. We would get Pythagorean triples that will look like (33, 44, 55), (333, 444, 555) . . . , or like (303, 404, 505), (3003, 4004, 5005). . . .

On the other hand, there are some Pythagorean triples that contain a few palindromic numbers. Some of these are: (20, 99, 101), (252, 275, 373), and (363, 484, 605). There are some where the first two members are palindromes, such as (3,993; 6,776; 7,865), (34,743; 42,824; 55,145), or (48,984; 886,688; 888,040). A more comprehensive list of Pythagorean triples with a pair of palindromic numbers is shown in the appendix, section 8.

Pythagorean Curiosity 11

For any pair of Pythagorean triples (a, b, c) and (p, q, r) the expression

$$(c + r)^2 - (a + p)^2 - (b + q)^2$$

is a square number. Take, for example, (7, 24, 25) and (15, 8, 17). Applying his relationship we get: $(25 + 17)^2 - (24 + 8)^2 - (7 + 15)^2 = 42^2 - 32^2 - 22^2 = 1,764 - 1,024 - 484 = 256 = 16^2$.

You might want to try this for other pairs of Pythagorean triples. Again, this relationship can be proved with the help of Euclid's formula.

Pythagorean Curiosity 12

For those readers who remember complex numbers from high-school algebra, we present an unexpected connection between complex numbers and Pythagorean triples. A *complex number z* is composed of a *real part a* = Re z and an *imaginary part b* = Im z, and appears in the form $z = a + ib$. Here $i = \sqrt{-1}$ is the *imaginary unit*, which is characterized by the property that $i^2 = -1$. Using this property, we can easily compute the square of any complex number:

$$z^2 = (a + ib)^2 = (a + ib)(a + ib) = a^2 + 2iab + i^2b^2 = (a^2 - b^2) + i(2ab).$$

Assuming $a = m$ and $b = n$ are natural numbers with $m > n$, the square of the complex number $m + in$ is a number with real part $m^2 - n^2$ and imaginary part $2mn$.

$$z^2 = (m + in)^2 = (m^2 - n^2) + i(2mn),$$
$$\text{therefore Re}(z^2) = m^2 - n^2, \text{Im}(z^2) = 2mn.$$

These are not only natural numbers, but, according to Euclid's formula, the first two members of a Pythagorean triple. The third member is $m^2 + n^2 = (\text{Re } z)^2 + (\text{Im } z)^2$, which is the square of the absolute value $|z|$ of z.

Hence for two natural numbers m and n, with $m > n$, the complex number $z = m + in$ defines a Pythagorean triple by $(\text{Re}(z^2), \text{Im}(z^2), |z|^2)$.

9.11. DIVISIBILITY OF NUMBERS

Number relationships can also manifest themselves in ways that assist us in making arithmetic judgments. In the base-10 number system, we are able to determine by inspection (and sometimes with a bit of simple arithmetic) when a given number is divisible by other numbers. For example, we know that when the last digit of a number is an even number, then the number is divisible by 2, such as with the numbers 30, 32, 34, 36, and 38. Of course, if the last digit is not divisible by 2, then we know that we cannot divide the number exactly by 2.

Divisibility by Powers of 2

Just as we look at the terminal digit on the number to determine if it is divisible by 2, so can we extend this to determine when a number is divisible by 4. In this case, when a number's last *two* digits (considered as a number) is divisible by 4, then, and only then, is the entire number also divisible by 4. For example, the underlined portion of each of the following numbers 1<u>24</u>, 1<u>28</u>, 3<u>56</u>, and 7<u>68</u> is each divisible by 4, therefore, each of these numbers is also divisible by 4. On the other hand, the last two digits of the number 322, namely 22, is not divisible by 4, and therefore, the number 322 is not divisible by 4.

Furthermore, we can conclude that when, and only when, the last *three* digits of a number (considered as a number) is divisible by 8, the entire number also divisible by 8. A clever person would then extend this rule to a number whose last *four* digits form a number that is divisible by 16 to conclude that only then is the entire number divisible by 16, and so on for succeeding powers of 2.

Divisibility by Powers of 5

An analogous rule to that for powers of 2 can be used for divisibility by 5. We know that only when the last digit is either a 5 or a 0 is the number divisible by 5. Again, only when the last two digits (considered as a number) is divisible by 25 is the number divisible by 25. Some such examples of where the last two digits considered as number is divisible by 25 are 325, 450, 675, and 800, and each of these numbers is therefore divisible by 25 as well. This rule continues for powers of 5 (i.e., 5, 25, 125, 625, etc.) just as it did for powers of 2 earlier.

Divisibility by 3 and 9

A different rule is used to determine if a number is divisible by 3. Here we inspect the sum of the digits of the number. Only when the sum of the digits of the number being considered is divisible by 3 will the entire number be divisible by 3. For example, to determine if the number 345,678 is divisible by 3, we simply check to see if the sum of the digits $3 + 4 + 5 + 6 + 7 + 8 = 33$ is divisible by 3. In this case, it is; therefore, the number 345,678 is divisible by 3.

A similar rule can be used to determine divisibility by 9. If the sum of the digits of a given number is divisible by 9, then the number is divisible by 9. An illustration of this is the number 25,371, where the sum of the digits is $2 + 5 + 3 + 7 + 1 = 18$, which is divisible by 9. Therefore, the number 25,371 is divisible by 9.

It is interesting to see why these rules work. Consider the number 25,371 and break it down as follows:

$$25,371 = 2 \times (9,999 + 1) + 5 \times (999 + 1)$$
$$+ 3 \times (99 + 1) + 7 \times (9 + 1) + 1.$$

Doing the indicated arithmetic operations, and some rearrangement, we get

$$= (2 \times 9,999 + 5 \times 999 + 3 \times 99 + 7 \times 9)$$
$$+ (2 + 5 + 3 + 7 + 1)$$

We can see that the term $(2 \times 9,999 + 5 \times 999 + 3 \times 99 + 7 \times 9)$ is a multiple of 9 (and a multiple of 3, as well). Therefore, we only need to have the remainder of the number, $(2 + 5 + 3 + 7 + 1)$, to be a multiple of 9 (or 3)—which just happens to be the sum of the digits of the original number 825,372—in order for the entire number to be divisible by 9 (or 3). In this case, $2 + 5 + 3 + 7 + 1 = 18$, which is divisible by 9 and 3. Hence the number 25,371 is divisible by 9 and 3.

For example, the number 789 is not divisible by 9, because 7 + 8 + 9 = 24, and 24 is not divisible by 9. Yet the number 789 is divisible by 3, since 24 is divisible by 3.

Divisibility by Composite Numbers

With the exception of 6 and 7, we have established divisibility rules to test for the numbers up to 10. Before we consider a test for divisibility by 7, we ought to make a statement about divisibility testing of composite (nonprime) numbers. To test divisibility by a composite number, we employ the divisibility tests for its "relatively-prime factors"—that is, numbers whose only common factor is 1. For example, the test for divisibility by the composite number 12 would require applying the divisibility test for 3 and 4, which are its relatively-prime factors (not 2 and 6, which are not relatively prime). The divisibility test for 18 requires applying the test for divisibility by 2 and for 9—which are relatively prime—and not the rules for divisibility by 3 and 6, whose product is also 18, but which are not relatively-prime factors, since they have a common factor of 3.

We can summarize the divisibility by composite numbers by inspecting the table below (table 9.9), which shows the first few composite numbers and their relatively-prime factors.

To Be Divisible By	6	10	12	14	15	18	20	21	24	26
The Number Must Be Divisible By	2, 3	2, 5	3, 4	2, 7	3, 5	2, 9	4, 5	3, 7	3, 8	2, 13

Table 9.9

The inclusion of some prime numbers in this chart now leads us to consider the divisibility rules for other prime numbers. We find the rules may be a bit cumbersome and not realistic for use in everyday-life situations—especially since the calculator is so pervasive. We will, therefore, present these divisibility rules largely for entertainment purposes, rather than as a useful tool.

The Rule for Divisibility by 7

Delete the last digit from the given number, and then subtract *twice* this deleted digit from the remaining number. If and only if the result is divisible by 7 will the original number be divisible by 7. This process may be repeated if the result is still too large for a simple visual inspection for divisibility by 7.

To better understand this divisibility test, we will apply it to determine if the number 876,547 is divisible by 7—without actually doing the division.

We begin with 876,547 and delete its units digit, 7, and then subtract its double, 14, from the remaining number: $87,654 - 14 = 87,640$. Since we cannot yet visually determine if 87,640 is divisible by 7, we shall continue the process.

We take this resulting number 87,640 and delete its units digit, 0, and subtract its double, still 0, from the remaining number; we get $8,764 - 0 = 8,764$.

This did not help us much in this case, so we shall continue the process. We delete its units digit, 4, from this resulting number, 8,764, and subtract its double, 8, from the remaining number to get $876 - 8 = 868$. Since we still cannot visually inspect the resulting number, 868, for divisibility by 7, we will continue the process.

This time we delete its units digit, 8, from the resulting number 868 and subtract its double, 16, from the remaining number to get: $86 - 16 = 70$, which we can easily determine is divisible by 7. Therefore, the original number, 876,547, is divisible by 7.

Now for the beauty of mathematics! That is, showing why this engaging procedure actually does what we say it does—test for divisibility by 7. Being able to show in a rather simple way why this procedure works contributes to what we call the wonders of mathematics.

Each step of the procedure actually amounts to a subtraction. For example, 8,764 will be reduced by subtracting 4 from the units and 2 × 4 = 8 from the tens. This reduces the original number by 84, which is a multiple of 7. The number 8,764 will become 8,764 – 84 = 8,680 = 868 × 10. This is divisible by 7 if and only if 868 is divisible by 7 (because 10 is not divisible by 7). Therefore, we can just ignore the 0 at the end, which explains why we can just "drop" the last digit. In each step, the process actually takes away "bundles of 7" from the original number. Whenever the remaining part is divisible by 7, then the original number is divisible by 7.

Terminal Digit	Number Subtracted from Original
0	0 = 0 × 7
1	20 + 1 = 21 = 3 × 7
2	40 + 2 = 42 = 6 × 7
3	60 + 3 = 63 = 9 × 7
4	80 + 4 = 84 = 12 × 7
5	100 + 5 = 105 = 15 × 7
6	120 + 6 = 126 = 18 × 7
7	140 + 7 = 147 = 21 × 7
8	160 + 8 = 168 = 24 × 7
9	180 + 9 = 189 = 27 × 7

Table 9.10

To justify the technique of determining divisibility by 7, consider the various possible terminal digits (that you are "dropping") and the corresponding subtraction that is actually being done in that step of the procedure. In table 9.10 you will see how the terminal digit together with its double in the tens place results in a number that is a multiple of 7. In all cases, the original number gets reduced by that multiple of 7, thereby changing the last digit of the original number to 0, which may then be dropped. If the remaining number is divisible by 7, then so is the original number divisible by 7.

The Rule for Divisibility by 11

Checking divisibility by 11 could be done in a similar manner as in the case of 7. But, since 11 is one more than the base (10), we have an even simpler test:

- Find the sums of the alternate digits and then take the difference of these two sums. If and only if that difference is divisible by 11 is the original number divisible by 11.

To better grasp this technique, consider, as an example, the number 246,863,727. First, we find the sums of the alternate digits: $2 + 6 + 6 + 7 + 7 = 28$, and $4 + 8 + 3 + 2 = 17$. The difference of these two sums is: $28 - 17 = 11$, which is clearly divisible by 11. Therefore, the original number is divisible by 11.

This rule rests on the observation that each of the following numbers are divisible by 11:

$$11 = 10^1 + 1, \quad 1001 = 10^3 + 1, \quad 100001 = 10^5 + 1, \quad \ldots$$
$$99 = 10^2 - 1, \quad 9999 = 10^4 - 1, \quad 999999 = 10^6 - 1, \quad \ldots$$

To see how this rule for divisibility by 11 works, we shall break down the number 25,817 in the following way:

$$
\begin{aligned}
25{,}817 &= 2 \times 10^4 + 5 \times 10^3 + 8 \times 10^2 + 1 \times 10^1 + 7 \times 10^0 \\
&= 2 \times (10^4 + 1 - 1) + 5 \times (10^3 + 1 - 1) + 8 \times (10^2 + 1 - 1) + 1 \times (10^1 + 1 - 1) + 7 \times 1 \\
&= 2 \times (10^4 - 1) + 2 \times 1 + 5 \times (10^3 + 1) - 5 \times 1 + 8 \times (10^2 - 1) + 8 \times 1 + 1 \times (10^1 + 1) - 1 \times 1 + 7 \times 1 \\
&= 2 \times \mathbf{(10^4 - 1)} + 5 \times \mathbf{(10^3 + 1)} + 8 \times \mathbf{(10^2 - 1)} + 1 \times \mathbf{(10^1 + 1)} + 2 - 5 + 8 - 1 + 7
\end{aligned}
$$

Each of the bold terms are divisible by 11. Therefore, we need to just ensure that the sum of the rest of the terms is also divisible by 11. They are $2 - 5 + 8 - 1 + 7 = 11$, which clearly is divisible by 11. Notice that this is the difference of the sums of the alternate digits, $2 - 5 + 8 - 1 + 7 = (2 + 8 + 7) - (5 + 1)$, of the number 25,817.

The Rule for Divisibility by 13

Delete the last digit from the given number, and then subtract *nine times* this deleted digit from the remaining number. If and only if the result is divisible by 13 will the original number be divisible by 13. Repeat this process if the result is too large for simple inspection of the divisibility by 13.

This is similar to the rule for testing divisibility by 7, except that the 7 is replaced by 13 and instead of subtracting twice the deleted digit, we subtract nine times the deleted digit each time. Let's check the number 5,616 for divisibility by 13. Begin with 5,616 and delete its units digit, 6, and subtract its multiple of 9, namely 54, from the remaining number: $561 - 54 = 507$.

Since we still cannot visually inspect the resulting number for divisibility by 13, we continue the process.

Continue with the resulting number 507, and delete its units digit and subtract nine times this digit from the remaining number: $50 - 63 = -13$, which is divisible by 13. Therefore, the original number is divisible by 13.

To determine the "multiplier," 9, we sought the smallest multiple of 13 that ends in a 1. That was 91, where the tens digit is 9 times the units digit. Once again, consider the various possible terminal digits and the corresponding subtractions in the following table (table 9.11).

Terminal Digit	Number Subtracted from Original
1	$90 + 1 = 91 = 7 \times 13$
2	$180 + 2 = 182 = 14 \times 13$
3	$270 + 3 = 273 = 21 \times 13$
4	$360 + 4 = 364 = 28 \times 13$
5	$450 + 5 = 455 = 35 \times 13$
6	$540 + 6 = 546 = 42 \times 13$
7	$630 + 7 = 637 = 49 \times 13$
8	$720 + 8 = 728 = 56 \times 13$
9	$810 + 9 = 819 = 63 \times 13$

Table 9.11

In each case, a multiple of 13 is being subtracted one or more times from the original number. Hence, only if the remaining number is divisible by 13 will the original number be divisible by 13.

As we proceed to the divisibility test for the next prime, 17, we shall once again use this technique. We seek the multiple of 17 with a units digit of 1, that is, 51. This gives us the "multiplier" we need to establish the following rule.

The Rule for Divisibility by 17

Delete the units digit, and subtract *five times* the deleted digit each time from the remaining number until you reach a number small enough to determine if it is divisible by 17.

We can justify the rule for divisibility by 17 as we did the rules for 7 and 13. Each step of the procedure subtracts a "bunch of 17s" from the original number until we reduce the number to a manageable size and can make a visual inspection of divisibility by 17.

The patterns developed in the preceding three divisibility rules (for 7, 13, and 17) should lead you to develop similar rules for testing divisibility by larger primes. The following chart (table 9.12) presents the "multipliers" of the deleted digits for various primes.

To Test Divisibility By	7	11	13	17	19	23	29	31	37	41	43	47
Multiplier	2	1	9	5	17	16	26	3	11	4	30	14

Table 9.12

You may want to extend this chart; it can be fun, and challenging. In addition to extending the rules for divisibility by prime numbers, you may also want to extend your knowledge of divisibility rules to include composite (i.e., nonprime) numbers. Remember that to test divisibility for composite numbers we need to consider the rules for the number's relatively-prime factors—this guarantees that we will be using independent divisibility rules. These rules for divisibility should enhance your appreciation for the relationship of numbers in a mathematical context.

At this point we have presented a rather exhaustive illustration of how numbers can relate to one another. Many of these illustrations are unexpected and, therefore, that much more appreciated. We hope to have motivated readers to seek out other illustrations of number relationships.

CHAPTER 10
NUMBERS AND PROPORTIONS

10.1. COMPARING QUANTITIES

When the mathematicians of ancient Greece spoke about "numbers," they meant "natural numbers." Although they knew about fractions from Babylonian and Egyptian science, and could use them in a practical sense, they did not accept fractions as numbers. To them, a fraction was not a number but a certain relationship between two quantities, and it would have been called a proportion rather than a number.

Today, the word *proportion* occurs mainly in geometry, when we speak of the proportions, or similarity, of geometrical shapes. Indeed, the proportion of the sides of a rectangle is very familiar from everyday life and is often called the "aspect ratio" (or the format) of the rectangle. For example, a modern television screen would have the aspect ratio 16:9 (see figure 10.1), and there are two common proportions of images produced by digital cameras, 3:2 and 4:3.

A rectangle that is 16 inches long and 9 inches wide would be an example of the proportion 16:9, as would be a rectangle 32 inches long and 18 inches wide. The lengths of the sides are not important, but their relation is what matters because this defines the shape of the rectangle. Rectangles with the same proportion have the same shape, even if they are of a different size.

Figure 10.1: The proportion 16:9.

As we said, ancient Greek mathematicians did not describe a proportion *a:b*, which we today see as a fraction, as a number. They considered only the counting numbers, or natural numbers. Consequently, a proportion of quantities could not be represented by a single number; it had to be considered as a relationship between quantities. The founder of a general theory of proportions was Eudoxus of Cnidus (ca. 395–340 BCE), one of the greatest Greek mathematicians. We know about his work only through the reports of others, and Eudoxus's theory of proportions is contained in Euclid's *Elements*.

In some cases, a proportion can be expressed through two natural numbers, like the proportion 16:9. In the Pythagorean era, the only theory of proportions that was available was the one between natural numbers. Pythagoreans used this theory to explain the universe. Musical scales may serve as an example, because, according to Archimedes, "they presumed the whole heaven to be a musical scale and a number."[1]

The Pythagorean scale is built on simple proportions that can be demonstrated on a monochord, which is a musical instrument that has a single string stretched over a resonance box, as shown in figure 10.2. The string oscillates with a particular frequency, producing the fundamental tone of the instrument. The length of the oscillating part of the string can be adjusted by moving a slider to vary the sound. The Pythagoreans discovered that one would obtain especially pleasing (consonant) musical intervals if the ratio between the whole string and the oscillating part of the string could be expressed in terms of small integers. If the length of the string was halved (by positioning the slider at 6), it would produce, when plucked, a pitch an octave higher, and the frequency would be twice the frequency of the original, realizing a proportion of 2:1. If you divide the string at position 8, as shown in figure 10.2, you would create a proportion of 3:2. This would produce a pitch that is a fifth higher than the fundamental tone. And if you place the slider at 9, the pitch would be a fourth higher than the fundamental tone, corresponding to the proportion 4:3. The whole Pythagorean musical scale is built around these simple proportions of string lengths.

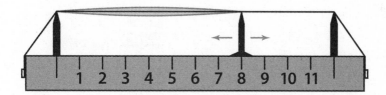

Figure 10.2: The monochord.

Impressed by the explanatory force of proportions, the Pythagoreans believed that one should be able to express any proportion of any two quantities in terms of natural numbers n and m, and they regarded proportions involving small natural numbers as particularly pleasing.

10.2. PROPORTIONS OF LENGTHS

Let us represent two magnitudes geometrically by lengths a and b. Consider, for example, the two line segments shown at the top of figure 10.3. How can we learn something about the relation between the two line segments—that is, about their proportion? If possible, we would like to find the natural numbers n and m, such that the proportion $a:b$ could be expressed as $n:m$.

In order to achieve this, the Pythagoreans devised the following method, which we illustrate in figure 10.3.

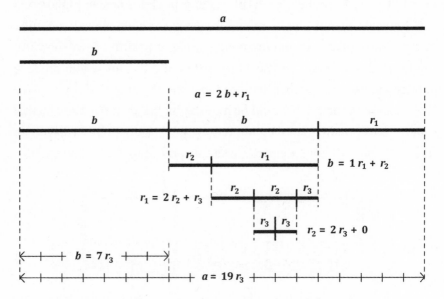

Figure 10.3: Determining the proportion of two line segments.

One starts by examining how often the shorter line segment b would fit into the longer one. Obviously, b fits into a twice, and then a short part r_1 of the segment a would be left over. Hence, we write $a = 2 \times b + r_1$, where r_1 is shorter than b. The next question would be, how often would r_1 fit into b? Figure 10.3 shows that, obviously, $b = 1 \times r_1 + r_2$, with $r_2 < r_1$. The next step shows that the remainder r_2 would fit into r_1 twice, with an even smaller remainder r_3. Finally, r_3 fits exactly twice into r_2. That is, there is no remainder r_4, or $r_4 = 0$.

What have we achieved now? Obviously, we found a length r_3 (the last nonvanishing remainder) that fits a whole number of times into all previous lengths, and, therefore, $a = 19r_3$, and b = $7r_3$. Both line segments a and b can be expressed with the help of the small line segment r_3; they are both integer multiples of r_3. The line segment r_3 is a "unit" that allows us to measure a and b simultaneously and is called the "greatest common measure" of a and b. If r_3 would be taken as the unit of length, then we would have $a = 19$ and $b = 7$. We then say that a is to b in the same relation as 19 to 7. One says the proportion a:b equals 19:7. Today, this proportion would be considered a fraction with the numerical value of 19 divided by 7. In decimal notation this would be

$$\frac{19}{7} = 2.7\overline{142857}14285$$

(the bar over the six digits after the point indicates that they are continuously repeated).

Greek philosophers in the fifth century BCE seemed to have held the belief that this method of finding an integer proportion would actually always work for any two quantities and would come to an end after a finite number of steps. Philosophers Leucippus and Democritus claimed that any extended continuous quantity cannot be divided infinitely. It was the birth of the theory of atomism—that is, that any division of an extended quantity would finally terminate in atoms, which cannot be further divided. Likewise, the method shown in figure 10.3 would terminate, in the worst case, when the remainder was the size of an atom and hence indivisible.

10.3. EUCLID'S ALGORITHM AND CONTINUED FRACTIONS

In number theory, the method described in figure 10.3 is known as Euclid's algorithm for finding the greatest common divisor of two natural numbers a and b. It allows us to express the proportions between two numbers in the simplest possible way. To illustrate this, we will try to find the greatest common divisor of $a = 1{,}215$ and $b = 360$. We begin by writing

$1{,}215 = 3 \times 360 + 135$, in general: $a = k_1 \times b + r_1$,
$360 = 2 \times 135 + 90$, $b = k_2 \times r_1 + r_2$,
$90 = 1 \times 90 + 45$, $r_1 = k_3 \times r_2 + r_3$,
$90 = 2 \times 45 + 0$, $r_2 = k_4 \times r_3 + r_4$.

The last nonvanishing remainder is 45. This is the common unit of 1,215 and 360, which is the greatest common divisor. Obviously, 1,215 is 27×45 and $360 = 8 \times 45$, and, therefore,

$$\frac{1215}{360} = \frac{27 \times 45}{8 \times 45} = \frac{27}{8}.$$

The algorithm also leads to another nice representation of the quotient of a and b:

$$a = k_1 b + r_1 \text{ implies } \frac{a}{b} = k_1 + \frac{r_1}{b}.$$

The next step of the algorithm gives $b = k_2 \times r_1 + r_2$. Inserting this in the expression above gives us

$$\frac{a}{b} = k_1 + \frac{r_1}{k_2 r_1 + r_2} = k_1 + \cfrac{1}{k_2 + \cfrac{r_2}{r_1}}.$$

Here, the last expression was obtained by dividing the numerator and denominator by r_1. We continue by inserting $r_1 = k_3 \times r_2 + r_3$, and so on:

$$\frac{a}{b} = k_1 + \cfrac{1}{k_2 + \cfrac{r_2}{k_3 r_2 + r_3}} = k_1 + \cfrac{1}{k_2 + \cfrac{1}{k_3 + \cfrac{r_3}{r_2}}} = k_1 + \cfrac{1}{k_2 + \cfrac{1}{k_3 + \cfrac{1}{k_4 + \cfrac{r_4}{r_3}}}} \ldots$$

This process can be continued until one of the remainders is zero. In the case of $a = 1{,}215$ and $b = 360$, the remainder r_4 turned out to be 0, while $k_1 = 3$, $k_2 = 2$, $k_3 = 1$, and $k_4 = 2$. Thus, we obtain the continued-fraction representation

$$\frac{1215}{360} = \frac{27}{8} = 3 + \cfrac{1}{2 + \cfrac{1}{1 + \cfrac{1}{2}}} .$$

The coefficients in the continued-fraction expansion describe precisely how often the shorter segment fits into the longer segment in each step of the method described in figure 10.3. In that example, the coefficients of the continued fraction expansion are $k_1 = 2$, $k_2 = 1$, $k_3 = 2$, and $k_4 = 2$. We then get

$$\frac{a}{b} = \frac{19}{7} = 2 + \cfrac{1}{1 + \cfrac{1}{2 + \cfrac{1}{2}}} .$$

There is another geometric interpretation of this method. Assume that we have a rectangle with length a and width b. Let us try to divide the rectangle into a grid of squares. For example, we start with a rectangle with sides $a = 19$ and $b = 7$, as shown in figure 10.4. The coefficients determined from Euclid's algorithm would describe how often the various squares would fit into the rectangle. There are k_1 big squares (with side b), k_2 of the next smaller size, and so on.

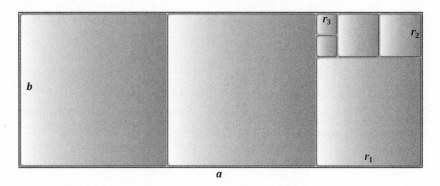

Figure 10.4: Tiling of a rectangle.

10.4. CREATING RECTANGLES FROM SQUARES

Consider figure 10.5, which shows how to create a rectangular domain from square tiles whose side lengths are a multiple of a common unit. We start with a square with side length 1 and add a second tile of the same size. On the longer side of the resulting rectangle we can place a square with side length 2 to obtain a new, larger rectangle. On its longer side we could fit a square with side length 3. Figure 10.5 should give you an idea of how to proceed. We continue to create larger and larger rectangles by attaching, in each step, a square to the longer side of the rectangle obtained in the previous step.

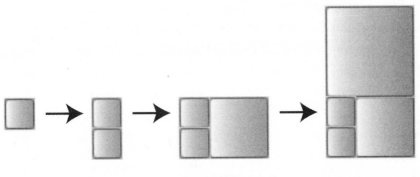

Figure 10.5

After a few steps, for example, we obtain the rectangle of figure 10.6. It has the proportion 55:34.

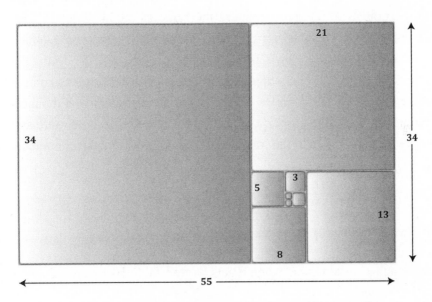

Figure 10.6: The Fibonacci numbers forming a golden rectangle.

Looking closely, you may, perhaps, recognize the numbers representing the successive side lengths of the squares. They form the sequence 1, 1, 2, 3, 5, 8, 13, 21, 34 . . . . We encountered this sequence before, in chapters 5 and 6. These are the Fibonacci numbers F_n, where each number is the sum of the two preceding numbers.

By the way, the area of the rectangle is

$$1^2 + 1^2 + 2^2 + 3^2 + 5^2 + 8^2 + 13^2 + 21^2 + 34^2 = 34 \times 55.$$

Indeed, this explains the following general formula for the sum of the squares of the Fibonacci numbers and holds for an arbitrary number n of members of the Fibonacci sequence.

The rectangles created successively in figure 10.6 get bigger and bigger, but they seem to have very similar proportions. Looking closely, we see that the smaller rectangle with sides 21 and 13 has a similar shape as the larger rectangle with sides 55 and 34. That is, the proportion 21:13 is approximately the same as 55:34. Indeed, if we evaluate the corresponding quotients numerically, we would obtain $\frac{21}{13} \approx 1.6154$, $\frac{55}{34} \approx 1.6177$, which is quite close. So we ask the following question: If we would continue the construction of rectangles to obtain larger and larger rectangles whose sides are consecutive Fibonacci numbers, would the proportions of these rectangles become more and more the same? Would the fractions of consecutive Fibonacci numbers

$$a_n = \frac{F_{n+1}}{F_n}$$

approach a certain value when n gets larger? Table 10.1 shows the first fifteen values.

	Proportion	(Approx.) Value
a_1	1:1	1.000000
a_2	2:1	2.000000
a_3	3:2	1.500000
a_4	5:3	1.666667
a_5	8:5	1.600000
a_6	13:8	1.625000
a_7	21:13	1.615385
a_8	34:21	1.619048
a_9	55:34	1.617647
a_{10}	89:55	1.618182
a_{11}	144:89	1.617978
a_{12}	233:144	1.618056
a_{13}	377:233	1.618026
a_{14}	610:377	1.618037
a_{15}	987:610	1.618033

Table 10.1: Ratios of consecutive Fibonacci numbers.

From the numbers in the last column in table 10.1 it appears that the sequence of the a_n values indeed approaches a certain value for large n. This value is close to 1.618. We are going to denote this limiting value ϕ (the Greek letter *phi*). From the property of the Fibonacci sequence, namely $F_{n+1} = F_n + F_{n-1}$, we can learn more about this number ϕ.

$$a_n = \frac{F_{n+1}}{F_n} = \frac{F_n + F_{n-1}}{F_n} = 1 + \frac{F_{n-1}}{F_n}\ .$$

Next we will write the last summand, using the definition of the a_n, as

$$\frac{F_{n-1}}{F_n} = \frac{1}{\dfrac{F_n}{F_{n-1}}} = \frac{1}{a_{n-1}} \ .$$

Inserting this into the formula above, we obtain the following formula for a_n:

$$a_n = 1 + \frac{1}{a_{n-1}}$$

and when n is so large that both a_n and a_{n-1} are approximately equal to their limit ϕ, we can infer that the number ϕ must satisfy the relation

$$\phi = 1 + \frac{1}{\phi} \text{ or } \phi^2 = \phi + 1.$$

The mathematically inclined reader will probably know how to solve this equation using the formula for the solution of the general quadratic equation

$$ax^2 + bx + c = 0, \ x = \frac{-b \pm \sqrt{b^2 - 4ac}}{2a}$$

The positive solution is

$$\phi = \frac{1 + \sqrt{5}}{2} = 1.61803398874989484820458683436563811772\ldots$$

The sequence of numbers behind the decimal point would not terminate, hence the number is usually rounded off as

$$\phi \approx 1.618.$$

The number ϕ is one of the most famous numbers in mathematics; it is known as the *golden ratio.*

10.5. THE GOLDEN RATIO

Although the proportion was already known in ancient Greece, the name "golden ratio" was coined in the nineteenth century by the German mathematician Martin Ohm (1792–1872). During the Renaissance, the Italian mathematician and Franciscan friar Luca Pacioli (1445–1517) called it the "divine proportion." He wrote a book, *De Divina Proportione*, which contains illustrations by his friend Leonardo da Vinci (1452–1519). Moreover, Pacioli investigated proportions in nature, art, and architecture, and he explored the design principles behind the letters of the alphabet. The logo of the Metropolitan Museum of Art in New York City, showing the letter M, is based on one his designs (see figure 10.7).

Figure 10.7: Study of the letter M by Luca Pacioli, 1509.

Leonardo da Vinci is said to have incorporated the golden ratio ϕ, into some of his drawings and paintings—for example, his *Vitruvian Man*, a study of human proportions according to the Roman architect Vitruvius (see figure 10.8). Here, the ratio of the radius of the circle and the side of the square is approximately the golden ratio, ϕ.

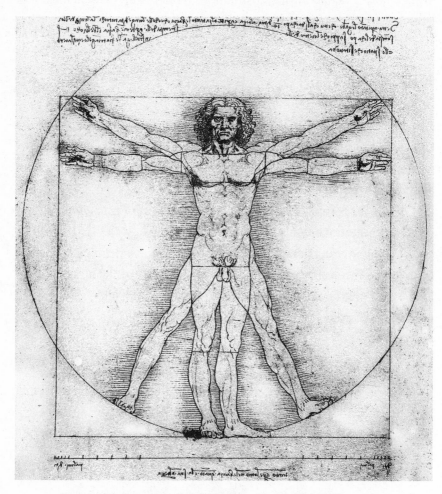

Figure 10.8: Leonardo da Vinci's *Vitruvian Man*, ca. 1490.
(Original located at Campo della Carità, Dorsoduro 1050, Venice, Italy.)

A rectangle where the length a and width b are in the proportion $a{:}b = \phi{:}1$, is called a *golden rectangle,* as shown in figure 10.9. The proportions of a golden rectangle can be approximated by the "Fibonacci rectangles" in figure 10.6 because $\phi{:}1$ approximates $F_{n+1}{:}F_n$ for large natural numbers n.

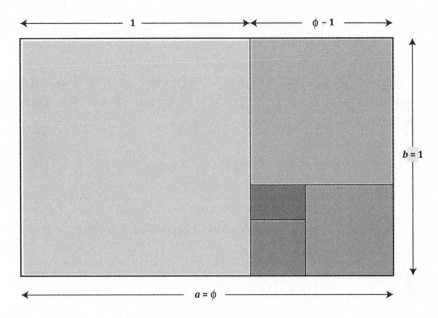

Figure 10.9: The golden rectangle.

The golden rectangle has the strange property that if we cut off a square, the remaining figure will again be a golden rectangle. This can be seen as follows. The remaining figure, as shown in figure 10.9, has the proportion $1{:}(\phi - 1)$. Consider the following computation, which makes use of the "defining equation" $\phi^2 = \phi + 1$, or $\phi^2 - \phi = 1$:

$$\frac{1}{\phi - 1} = \frac{\phi}{\phi^2 - \phi} = \frac{\phi}{1}.$$

The proportion of the sides of the small rectangle is again ϕ:1. That makes the small rectangle again a golden rectangle.

The relation $\phi^2 = \phi + 1$, which defines the golden ratio, also implies (upon dividing both sides by ϕ) that

$$\frac{\phi + 1}{\phi} = \phi.$$

Consider a line segment divided into two parts, a and b, such that their proportion is a:$b = \phi$:1, which is why the golden ratio is often called the *golden section*. (See figure 10.10.) The just-derived proportions,

$$1{:}(\phi - 1) = \phi{:}1 = (\phi + 1){:}\phi,$$

are the same if the two line segments are not of length ϕ and 1 but of length a and b, where a:$b = \phi$:1. The equality of proportions above would then be equivalent to

$$b{:}(a - b) = a{:}b = (a + b){:}a.$$

Expressed as a proportion of two line segments, these relations are shown in figure 10.10, illustrating the following facts: If you cut a straight line into two segments a and b, in the proportion of the golden ratio, then

(a) the shorter line segment b divides the longer line segment a in the proportion of the golden ratio, that is, a:$b = \phi$;

(b) the longer line segment a divides the sum $a + b$ in the proportion of the golden ratio, that is, $(a + b)$:$a = \phi$; and

(c) the difference $a - b$ divides the shorter line segment in the proportion of the golden ratio, that is, b:$(a - b) = \phi$.

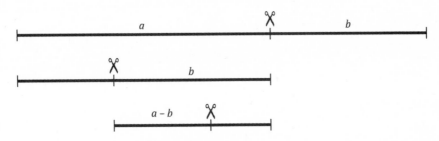

Figure 10.10: The golden section.

10.6. INCOMMENSURABILITY

The simple observation presented in figure 10.10 seems to destroy the Pythagorean belief that all proportions can be expressed as proportions of natural numbers.

In section 10.2, we described a method of mutually marking off a shorter length on a longer line segment in order to express the proportion of two lengths a and b as a proportion of natural numbers. Let us now consider any two lengths that are in the proportion of the golden ratio—$a{:}b = \phi{:}1$. Figure 10.10 clearly implies that the shorter segment b would fit just once into the longer segment a. The remainder is $a - b$. But b is in the same proportion to $a - b$ as a is to b. All these segments are in the golden ratio $\phi{:}1$. Unfortunately, this step has not brought us closer to finding the common unit of a and b. We have the same situation as in the beginning, namely two line segments in the proportion of the golden ratio. We could repeat that procedure indefinitely, and whenever you compare the shorter segment with the longer, it would just fit once, dividing the longer segment in the proportion of the golden ratio.

Hence the algorithm for finding a common unit for a and b would not terminate. There exists no common unit for two lengths that are in the proportion $a{:}b = \phi{:}1$. We cannot find a line segment r, such that $a = nr$ and $b = mr$ with some natural numbers n and m. If expressed as a continued fraction, the proportion $a{:}b$ would have the representation

$$\frac{a}{b} = \phi = 1 + \cfrac{1}{1 + \cfrac{1}{1 + \cfrac{1}{1 + \cfrac{1}{1 + \cfrac{1}{1 + \cfrac{1}{1 + \cfrac{1}{1 + \cfrac{1}{1 + \cdots}}}}}}}}$$

and this could be continued indefinitely. As we explained earlier, this just expresses the fact that b fits into a once, with a remainder that fits into b once, and so on. We could have seen this earlier, from the equation

$$\phi = 1 + \frac{1}{\phi}.$$

Inserting this very expression for ϕ into the denominator of the quotient on the right, we would obtain

$$\phi = 1 + \cfrac{1}{1 + \cfrac{1}{\phi}}$$

and we could repeat this process indefinitely, which would lead to the same infinite continued fraction as shown above:

$$\phi \; = \; 1+\cfrac{1}{1+\cfrac{1}{\phi}} = 1+\cfrac{1}{1+\cfrac{1}{1+\cfrac{1}{\phi}}} = 1+\cfrac{1}{1+\cfrac{1}{1+\cfrac{1}{1+\cfrac{1}{\phi}}}} + \; \ldots$$

Any number that cannot be expressed as the ratio of two natural numbers is called *irrational*. From a more geometrical point of view, two lengths whose proportion *a:b* cannot be expressed as a ratio of integers are called *incommensurable* lengths, which means that there exists no common measure, that is, no common unit *e* such that $a = ne$ and $b = me$. We mentioned this possibility in chapter 1, and now we have shown that it indeed occurs. If you try to measure the length ϕ, it is not possible to obtain it as a unit 1 plus a multiple of a fraction of that unit.

Among the real numbers, the rational numbers are the exception. There are so many irrational numbers that if you pick a random number on the number line, it would almost certainly be irrational. Likewise, two randomly chosen line segments in the plane would almost certainly be incommensurable.

A particularly famous example is the square root of two. This is the proportion between the side of a square and its diagonal. In Euclid's tenth book of his *Elements*, we find a number-theoretic proof of their incommensurability. The result was probably obtained much earlier, likely in the fifth century BCE. It is usually attributed to the Pythagorean philosopher Hippasus of Metapontum, who was a member of the Pythagorean order, for which the city Metapont was one of the centers in southern Italy (see chapter 4). One of the symbols of the Pythagorean order, the pentagram, was investigated around that time. The regular pentagram is formed by the diagonals of a regular pentagon. It turns out that the side of the pentagon and its diagonal are in the proportion of the golden ratio (see figure 10.11). Here, the indicated line segments are all in the proportion of the golden ratio: $a{:}b = a'{:}b' = a''{:}b'' = \phi{:}1$. It is ironic that Pythagoreans' symbol, which is an apparent violation of their fundamental belief that everything can be expressed through proportions of natural numbers had always been right in their eyes.

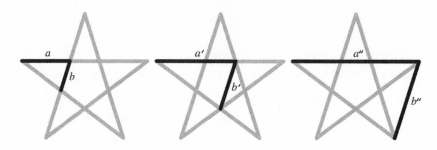

Figure 10.11: Golden proportions in the pentagram.

The golden ratio manifests itself in art and architecture as well as in nature. For a complete discussion of the golden ratio and its manifestations in geometry, its relationship to other famous numbers, and its many physical representations, we recommend the book *The Glorious Golden Ratio*, by Alfred S. Posamentier and Ingmar Lehmann (Amherst, NY: Prometheus Books, 2012).

10.7. THE FAMOUS NUMBER π

In mathematics some numbers have taken on a special status. This can be the result of either their frequent appearance, or perhaps because enough observers have been enchanted by their special properties. The number that we will now consider seems to be most prevalent in the recollection of one's school mathematics. We are referring to the number that is typically represented by the Greek letter π. In mathematics, this letter is usually associated with either of the two formulas involving the circle—namely, the formula for the circumference of a circle ($C = 2\pi r$) and that for the area of a circle ($A = \pi r^2$). The value that is usually associated with this letter π is 3.14. For some people, π is nothing more than a touch of the button on their pocket calculator, where, then, a particular number appears on the readout; for others, this number holds an unimaginable fascination. Depending on size of the calculator's display, the number shown might be

3.1415927, or
3.14159265358979323846264338332795, or even longer.

This push of a button still doesn't tell us what π actually is. We merely have a slick way of getting the decimal value of π. Actually, the number π represents a proportion: the ratio of the circumference of a circle to its diameter (see figure 10.12). As a proportion, it does not depend on the size of the circle.

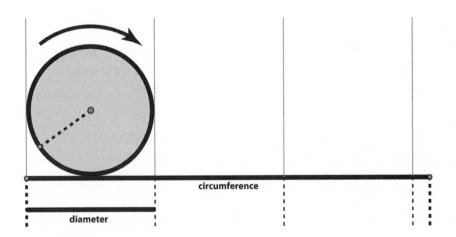

Figure 10.12: The proportion circumference:diameter is called π.

The number π is also the proportion between the area of the circle and the area of the square over the radius (see figure 10.13).

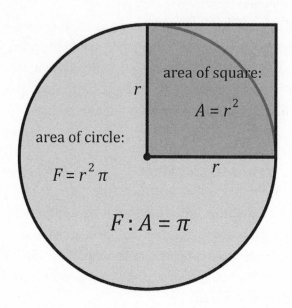

Figure 10.13: Proportion between areas of a circle and a square.

German mathematician Johann Heinrich Lambert (1728–1777) was the first to rigorously prove that π is irrational—that is, that π cannot be precisely represented as a fraction with integers in the numerator and denominator. His method of proof was to use a continued fraction expansion of the tangent function to show that if tan(x) is rational, then x cannot be rational. But if $\tan\left(\frac{\pi}{4}\right) = 1$ is a rational number, then $\frac{\pi}{4}$, or π cannot be rational. In 1770 Lambert produced the following continued fraction for π.

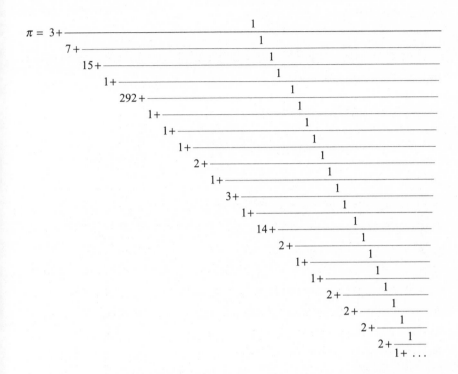

Because π is an irrational number, the continued fraction would expand indefinitely. Inspecting this lengthy expression, we find that good approximations can be obtained if we end the expansion at some place with a large denominator. For example, we could replace the fraction with 292 in the denominator just by zero. This leads to the very good approximation

$$\pi \approx 3 + \cfrac{1}{7 + \cfrac{1}{15 + \cfrac{1}{1+0}}} = \frac{355}{113} \approx 3.1415929,$$

which differs from the actual value of π only in the seventh decimal place. This approximation is also easy to remember because the denominator and the numerator combine into the easily memorable pattern 113355.

For everyday purposes, we could end the expansion even earlier and use the approximation

$$\pi \approx 3 + \frac{1}{7+0} = \frac{22}{7} \approx 3.1429.$$

Computing the circumference of a circle with a diameter of 1 meter with this approximate value 3.1429 would create an error of about 1 mm.

It is interesting that no obvious pattern has been discovered in the sequence of denominators in the continued fraction expansion. It can be proved that these numbers are not repeated periodically. This is related to the fact that π is a transcendental number—one that is *not* the solution of a nonzero polynomial equation with rational coefficients (unlike the golden ratio, ϕ, which is the solution of the polynomial equation $\phi^2 = \phi + 1$). On the other hand, there are different types of continued fraction expansions, which exhibit a striking regularity. The following result was obtained by English mathematician William Brouncker (1620–1684):

$$\pi = \cfrac{4}{1 + \cfrac{1^2}{2 + \cfrac{3^2}{2 + \cfrac{5^2}{2 + \cfrac{7^2}{2 + \cfrac{9^2}{2 + \cdots}}}}}}$$

The following continued fraction was discovered in 1869 by James

Joseph Sylvester (1814–1897), who is also known for his role in founding the *American Journal of Mathematics*:

$$\pi = 2 + \cfrac{2}{1 + \cfrac{1 \times 2}{1 + \cfrac{2 \times 3}{1 + \cfrac{3 \times 4}{1 + \cfrac{4 \times 5}{1 + \cfrac{5 \times 6}{1 + \cdots}}}}}}$$

In the decimal representation, the value of π can be expanded to an indefinite number of places. Mathematicians are always in search of patterns among these digits. So far, they have not found any pattern. It appears that the sequence of digits behaves like an arbitrary random sequence. Therefore, if you look far enough, you can find any given sequence of numbers of finite length among the decimal places of π. However, strange coincidences do occur. British mathematician John Conway (1937–) has indicated that if you separate the decimal expansion of π into groups of ten places, the probability of each of the ten digits appearing in any of these blocks is about 1 in 40,000. Yet he shows that it does occur in the seventh such group of ten places, as you can see from the grouping below:

π = 3.1415926535 8979323846 2643383279 5028841971 6939937510 5820974944 **5923078164** 0628620899 8628034825 3421170679 8214808651 3282306647 0938446095 5058223172 5359408128. . . .

10.8. THE AMAZING HISTORY OF π

You may wonder how this famous ratio came to be represented by the Greek letter π. In 1706, English mathematician William Jones (1675–1749), in his book *Synopsis Palmariorum Matheseos: or, A New Introduction to the Mathematics*, used the symbol π for the first time to actually represent the ratio of the circumference of a circle to its diameter. However, the true popularity of the symbol π to represent this ratio came in 1748, when one of mathematics' most prolific contributors, Swiss mathematician Leonhard Euler (1707–1783), used the symbol π in his book *Introductio in Analysin Infinitorum* to represent the ratio of the circumference of a circle to its diameter. A brilliant mathematician with an uncanny memory and ability to do complex calculations, Euler developed numerous methods for calculating the value of π, some of which approached the true value of π more quickly (that is, in fewer steps) than procedures developed by his predecessors. He calculated π to 126-place accuracy. The series below is particularly interesting, since it involves the reciprocals of the squares of all natural numbers:

$$\frac{\pi^2}{6} = 1 + \frac{1}{2^2} + \frac{1}{3^2} + \frac{1}{4^2} + \frac{1}{5^2} + \cdots$$

Multiplication by 6 and taking the square root would give you the value of π.

Many curiosities evolve from that number. For example, the quest to expand the decimal approximation of π to the largest number of places has been a fascinating challenge for centuries. You may ask, why do we need such accuracy for the value of π? We actually don't. If you want to compute the circumference of the whole observable universe from its radius (which is about 10^{27} m, the largest observable distance), you would need about sixty-two digits of π in order achieve a precision of a Planck length (10^{-35} m, the shortest observable distance). So the hunt for more digits has no practical purpose. The methods of calculation are simply used to check the accuracy and speed of the computer and the sophistication of the calculating procedure (sometimes referred to as an algorithm). That is, to determine how accurate and efficient the computer and software being tested is.

The current record for the most number of decimal places for the value of π is held by Alexander Yee and Shigeru Kondo, who used the software y-cruncher by Alexander Yee to compute 13.3 trillion digits in 2014. The number of digits of π will surely continue to increase.

It might be worthwhile to consider the magnitude of 13.3 trillion. How old do you think a person who has lived 13.3 trillion seconds might be? The question may seem irksome, since it requires having to consider a very small unit a very large number of times. However, we know how long a second is. But how big is one trillion? A trillion is 1,000,000,000,000, or one thousand billion. In one year there are 365 × 24 × 69 × 60 seconds. Therefore, 13.3 trillion seconds is equal to

$$13.3 \times \frac{1,000,000,000,000}{365 \times 24 \times 69 \times 60} \approx 421,740 \text{ years.}$$

One would have to be in his 421,740th year of life to have lived 13.3 trillion seconds!

From these accurate approximations of π we might want to look back at some of the earliest estimates of π, which for many years was thought to be 3. One always relishes the notion that hidden codes can reveal long-lost secrets. Such is the case with the common interpretation of the value of π as 3 in the Bible. Let us look at one of the more amazing modern interpretations of ancient knowledge. There are two places in the Bible where the same sentence appears, identical in every way, except for one word, which is spelled differently in the two citations. The description of a pool or fountain in King Solomon's temple is referred to in the passages that may be found in 1 Kings 7:23 and 2 Chronicles 4:2, and it reads as follows:

> And he made the molten sea of ten cubits from brim to brim, round in compass, and the height thereof was five cubits; and *a line* of thirty cubits did compass it round about.

The circular structure described here is said to have a circumference of 30 cubits and a diameter of 10 cubits. From this we notice that the Bible has $\pi = \frac{30}{10} = 3$. This is obviously a very primitive approximation of π. A late eighteenth-century rabbi, Elijah of Vilna (1720–1797), was one of the great modern biblical scholars, who earned the title "Gaon of Vilna" (meaning "brilliance of Vilna"). He came up with a remarkable discovery that, although the common interpretation of the value of π in the Bible was 3, brought the value of π in the Bible to much greater accuracy. Elijah of Vilna noticed that the Hebrew word for "line measure" was written differently in each of the two biblical passages mentioned above.

In 1 Kings 7:23 it was written as קוה, whereas in 2 Chronicles 4:2 it was written as קו. Elijah applied the ancient biblical analysis technique called Gematria, which is still used by Talmudic scholars today. This technique involves having the Hebrew letters take on their appropriate numerical values according to their sequence in the Hebrew alphabet. The letter values are: ק = 100, ו = 6, and ה = 5. Therefore, the spelling for "line measure" in 1 Kings 7:23 is קוה = 5 + 6 + 100 = 111, while in 2 Chronicles 4:2 the spelling קו = 6 + 100 = 106. Using the process of Gematria, he then took the ratio of these two values: $\frac{111}{106} = 1.0472$ (to four decimal places), which he considered the necessary "correction factor." By multiplying the Bible's apparent value (3) of π by this "correction factor," one gets 3.1416, which is π correct to four decimal places! "Wow!" is a common reaction. Such accuracy is quite astonishing for ancient times. If ten people were to take a piece of string and with it measure the circumference and diameter of some circular object and take their quotient, and then we take the average of these ten quotients, we would be hard-pressed to get the usual two-place accuracy, namely, $\pi = 3.14$. Now imagine getting π accurate to four decimal places—it might be nearly impossible with typical string measurements. Try it, if you need convincing.

On the other hand, the occurrence of the "correction factor" in the Bible could be pure coincidence. The two spellings of "line measure" also occur in other places in the Bible, in contexts where the correction factor seems to have no significance. Claims derived from Gematria are therefore not regarded as scientific. Whether or not you believe it also depends on the importance you attribute to the words in the Bible. Actually, it is rather improbable that a more accurate version of π was available at the time the biblical text was written (probably before 300 BCE). The greatest scholar of antiquity, Archimedes (ca. 267–212 BCE), found what was then the most accurate approximation of π, placing its value at

$$3.1408450704 \approx 3 + \frac{10}{71} < \pi < 3 + \frac{10}{70} = 3.\overline{142857}.$$

Archimedes arrived at these estimates by comparing the area of an inscribed polygon with the area of a circumscribed polygon, as shown in figure 10.14. In order to achieve this precision, he had to compute the circumference of the inscribed and circumscribed ninety-six-sided regular polygon.

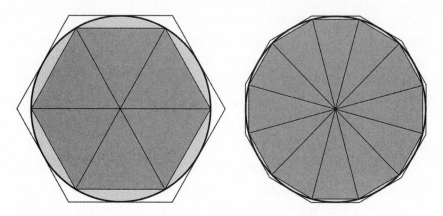

Figure 10.14: Approximation of circles by polygons.

The constant π is sometimes called *Archimedes's constant* because this algorithm was the best one to determine the value of π until modern times. Methods based on polygons were also used in ancient China, where, for example, Zu Chongzhi (428–500 CE) obtained a precision of seven digits by considering a 24,576-sided polygon. This was a record that held for over nine hundred years. In 1424, Persian astronomer and mathematician Jamshīd al-Kāshī (ca. 1380–1429) obtained an accuracy of nine sexagesimal digits (corresponding to an accuracy of sixteen digits in the decimal system) by computing the perimeter of a regular polygon with more than 800 million sides.

At about the same time, Indian astronomer and mathematician Madhava of Sangamagrame (ca. 1340–1425) invented a different method, which was based on the infinite sum

$$\pi = 4\left(1 - \frac{1}{3} + \frac{1}{5} - \frac{1}{7} + \frac{1}{9} - \frac{1}{11} \pm \right) \cdots$$

This beautiful representation of π was later rediscovered in Europe by Gottfried Leibniz (1646–1716) and is now called the *Madhava-Leibniz formula.* Madhava even found representations that were more useful in the actual computation of π and managed to determine π with a precision of eleven digits.

The next record was set by Ludolph van Ceulen (1540–1610) in about 1600. Using Archimedes's method, he approximated the circle using a regular polygon with 2^{62} sides. This task occupied him for most of his life. In his honor, π is sometimes referred to as the *Ludolphian number*. His value of π is engraved onto his tombstone in St. Pieter's Kerk in Leiden, Holland.

Euler introduced more efficient methods to compute π, based on series expansions of the inverse tangent function. At the beginning of the twentieth century, Srinivasa Ramanujan (1887–1920) developed new expressions for π in the form of rapidly converging infinite sums, which later helped to design efficient algorithms for computers to compute the value of π. With the invention of the computer, the hunt for more digits continued and accelerated. In 1962, Daniel Shanks and his team published the first one hundred thousand digits, and the first billion digits were computed by David and Gregory Volfovich, known as the Chudnovsky brothers. Around 2000, Yasumasa Kanada and his team broke the 1 trillion digit threshold, which then led to the current record holders, Alexander Yee and Shigeru Kondo.

For a plethora of further information about the fascinating number π, we recommend the book *Pi: A Biography of the World's Most Mysterious Number*, by Alfred S. Posamentier and Ingmar Lehmann (Amherst, NY: Prometheus Books, 2004).

10.9. FAMOUS NUMBERS IN THE GREAT PYRAMID

Khufu, second pharaoh of the fourth dynasty, better known under his Hellenized name Cheops, ruled the Old Kingdom in ancient Egypt in about 2600 BCE. His tomb is the oldest and largest of the monumental pyramids and is built on the plateau of Giza, near Cairo. How could people who had just escaped the Stone Age have built such a gigantic monument in a perfect geometrical shape? The base of the pyramid is an almost-perfect square. The average length of its sides is $s = 230.36$ m, with a maximal deviation of 3.2 cm. Indeed, this precision is remarkable because the Egyptians had, at best, rather primitive tools at their disposal, which, however, were accurately manufactured and handled with extreme care. Today, it is not an easy task to measure the dimensions of the pyramid, because natural erosion and stone robberies left it severely damaged. It was, however, possible to measure the side-length of the base precisely because some of the casing stones of the base were still preserved in their original position. In particular, the location of the corner stones could be precisely determined.

It is believed that the side length of the square at the base was exactly 440 royal cubits. A royal cubit is the ancient Egyptian unit of length, divided into 7 palms or 28 fingers. Using 230.36 m for the side length of the pyramid, we find that a royal cubit must be 52.35 cm, which agrees quite well with other sources.

Much more difficult than a measurement at the base is the determination of the height of the pyramid. Today, the top of the pyramid is gone. It is still 138.75 m high, but originally it was almost 8 m higher. A rather plausible estimate for the original height is 146.5 m. From this, we can compute the slant angle of the faces of the pyramid to be $51°49'30''$, which corresponds well to the measured values (which are not expected to be very accurate).

Throughout the history of pyramid research, people have often speculated that the dimensions of the pyramid contain some hidden message. Pseudoscientists, pyramidologists, and numerologists have looked for curious relationships in the numbers characterizing the proportions of the pyramid. In the nineteenth century, Scottish astronomer Charles Piazzi Smyth (1819–1900) conducted several expeditions in order to measure all dimensions of the Great Pyramid. He combined the measured values in all possible ways and evaluated all possible ratios in order to find some hidden treasures in the numbers provided by the pyramid. We will consider these findings and compute a few of the typical proportions of the great pyramid.

We shall start by computing the height of the triangular face of the pyramid, the slant height d, as shown in figure 10.15. We can evaluate d with the help of the known length s of the side and the known pyramid height h. The shaded triangle in figure 10.15 is a right triangle, hence we apply Pythagorean theorem as follows:

$$d^2 = h^2 + \left(\frac{s}{2}\right)^2 \text{, or } d = \sqrt{h^2 + \left(\frac{s}{2}\right)^2}.$$

Inserting $h = 146.5$ m and $s = 230.36$ m, we obtain $d = 186.356$ m. Let us now compute the ratio between d and $\frac{s}{2}$—that is, the proportion of the two bold lines in figure 10.15.

$$d : \frac{s}{2} = \frac{186.356}{115.18} = 1.61795.$$

This number is surprisingly close to the golden ratio $\phi \approx 1.61803$. Could it be that the architects of the pyramid indeed chose to encode the golden ratio in the proportions of this pyramid?

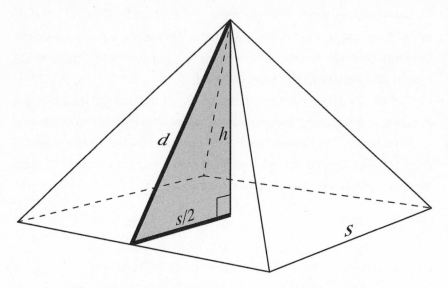

Figure 10.15: Slant height:half-side length = the golden ratio.

We can determine the height h of the pyramid that would be required for the above proportion to be *exactly* the golden ratio. The computation gives h = 146.511 m. This value deviates only 11 mm from the original estimate (146.5 m). It is clear that such minimal differences could not make any difference in a structure of that magnitude. Moreover, the inaccuracy of determining the height is much larger given the present-day condition of the pyramid. Therefore, the measured dimensions of the pyramid would support the following hypothesis:

Hypothesis 1: *The Great Pyramid of Cheops is designed in such a way that the slant height and the half-side of its base are in the proportion of the golden ratio:*

$$d : \frac{s}{2} = \phi.$$

Now the big question is whether it was planned that way or whether everything is mere coincidence. It is safe to assume that the Egyptians of that time did not know the golden ratio and that to them this proportion could not have the mathematical importance that was assigned to it more than two thousand years later by Greek mathematicians. So why should they encode the golden ratio in the dimensions of the pyramid?

An often-quoted explanation has its origin in the writings of American pyramidologist John Taylor (1781–1864), who referred to the Greek historian Herodotus (ca. 484–425 BCE). He said that the Egyptians did not intend to encode the golden ratio in the pyramid, instead they applied the following idea:

Hypothesis 2: *The Great Pyramid of Cheops is designed in such a way that every face of the pyramid is of the same area as the square of the height.* (See figure 10.16.)

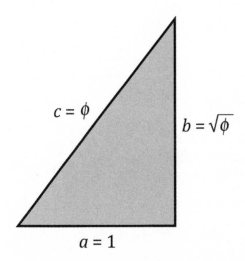

$$c = \phi$$

$$b = \sqrt{\phi}$$

$$a = 1$$

Figure 10.16: The two shaded areas are approximately equal in size.

From this hypothesis, it is a matter of elementary algebra to derive the above-mentioned proportion involving the golden ratio. The two hypotheses are consequently mathematically equivalent. But is Taylor's interpretation correct? Actually, the corresponding quote of Herodotus, the only one considering the dimensions of the pyramid, reads as follows: "The pyramid itself was twenty years in the making. Its base is square, each side eight plethra long, and its height is the same; the whole is of stone polished and most exactly fitted; there is no block of less than thirty feet."[2] Here one plethron is 100 Greek feet; its exact measure is unknown. Canadian mathematician Roger Herz-Fischler, in his book *The Shape of the Great Pyramid*, points out that this source is not sufficient to prove Taylor's claim. The height seems to be just a rough estimate and not accurate. Taylor seems to have undertaken a radical reinterpretation of this text when he claimed that the words "and its height is the same" should not be understood as an equality in lengths, but as a quadratic equality, according to which the area of the face is the same as the area of the square with side h. Today, Taylor's interpretation appears rather dubious. Let us look for other possible explanations.

10.10. THE "PI-RAMID"

Using again $s = 230.36$ m and $h = 146.5$ m, we compute half the circumference $2s$ of the base and its ratio with the height h of the pyramid. We obtain

$$2s : h = 3.14485.$$

This number is suspiciously close to $\pi \approx 3.14159$. If the originally assumed height was about 15 cm higher, then this proportion would be exact. The deviation is larger than it was with the golden ratio, but

15 cm is still within the range of uncertainty when determining the original height of an eroded monument. Therefore we can be sure that the following hypothesis is fully compatible with the empirically measured dimensions of the Cheops pyramid:

Hypothesis 3: *The Great Pyramid of Cheops is designed in such a way that*

$$2s : h = \pi, \text{ or } 4s = 2h\pi,$$

that is, the circumference of the base equals the circumference of a circle whose radius is the height of the pyramid.

Again, this hypothesis has the problem that π was not known in ancient Egypt. We know about Egyptian mathematical knowledge, in particular, from the Rhind papyrus (about 1650 BCE). This is a collection of exercises concerning different mathematical problems that were important at that time. It also contains problems about the area of circles that were approximately computed by dividing its area into a number of square regions. The method effectively amounts to using an approximate value of 3.16 for π (while the value in the pyramid has a much higher precision). It is also clear from this source that the Egyptians had no idea of the proportionality between the area of a circle and the square of the radius. Obviously, the concept of π was unknown to the Egyptians of that era. How then is it possible that the number π is present in the pyramid with such "unreasonable" accuracy?

Moreover, how can it be that the golden ratio appears together with π in the dimensions of the pyramid? This is indeed a marvelous coincidence. In order to appreciate this, we should look again at the right triangle in figure 10.15, which is formed by the half-side of the base, the height of the pyramid, and its slant height. This triangle is a very good approximation of the proportions of the so-called Kepler triangle, shown in figure 10.17. The Kepler triangle is defined as a right triangle with hypotenuse

$c = \phi$ and sides $a = 1$ and $b = \sqrt{\phi}$. In this case, the Pythagorean theorem, $a^2 + b^2 = c^2$, is equivalent to the defining equation for the golden ratio: $1 + \phi = \phi^2$. This triangle demonstrates exactly the golden ratio according to Hypothesis 1. But it also has the number π hidden in its proportions. The reason is an amazing, but nevertheless purely accidental, similarity of the following two numerical values:

$$\frac{1}{\sqrt{\phi}} = 0.786151... \approx \frac{\pi}{4} = 0.785398....$$

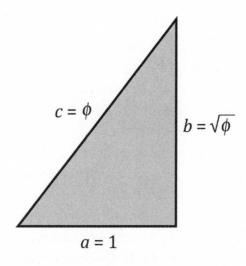

Figure 10.17: The Kepler triangle.

Because of this coincidence, the proportion of the legs of the Kepler triangle is related to π in the following way:

$$1 : \sqrt{\phi} \approx \pi : 4.$$

In the pyramid, this corresponds approximately to the proportion $\frac{s}{2}:h$. For this reason, a pyramid that has ϕ in its proportions, has, in a very good approximation, also π in its proportions, and vice versa. Therefore, an explanation of Hypothesis 3 concerning the occurrence of π would at the same time explain Hypothesis 1 on the appearance of ϕ in the pyramid.

A very plausible explanation has been offered British physicist Kurt Mendelssohn (1906–1980) in his book *The Riddle of the Pyramids*. He assumes that horizontal distances were measured by the number of revolutions of a surveyor's wheel. In contrast, vertical distances would have to be measured with the diameter of the wheel as the unit of length, which we pictured in figure 10.18. Thus, the side of the base of the pyramid would correspond to a certain multiple of the circumference of the surveyor's wheel, while the height would correspond to a certain multiple of its diameter. In that way, the proportion π between circumference and diameter of the wheel would be included in the proportions of the pyramid without the Egyptians having been aware of it.

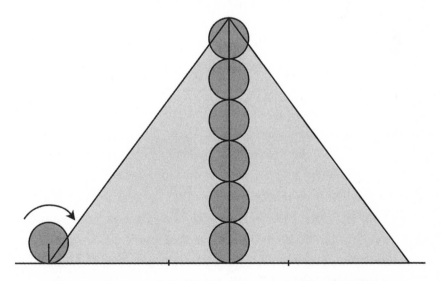

Figure 10.18: Surveyor's wheel for measuring lengths and heights.

Unfortunately, this explanation has not been supported scientifi-cally. While there are ancient pictures showing Egyptians handling instruments, there is no hint that they could have used wheels for mea-suring lengths. The assumption has no other justification than to explain the occurrence of the number π in the Great Pyramid, as we have done here. If one could come up with a simpler explanation, it would be preferred over any assumption of a forgotten knowledge about higher geometry. Indeed, such an explanation exists.

10.11. THE HISTORICAL EXPLANATION

The Rhind papyrus mentioned earlier contains some mathematical prob-lems dealing with the slope of pyramids. In this connection we learned about the *seked*. The seked measures the slope of the face of a pyramid. The seked is the horizontal distance (in fingers) needed for a rise of one royal cubit (1 royal cubit = 28 fingers). Therefore, a steeper pyramid would have a smaller seked. The Great Pyramid of Cheops seems to have been planned with a seked of 22 fingers, which we have shown in figure 10.19.

Assuming a base length of 440 royal cubits and a seked of 22, we can again determine the height of the pyramid (actually one of the problems in Rhind papyrus). We would obtain precisely 280 cubits = 146.59 m, which is 9 cm more than the original estimate and in between the heights obtained from Hypotheses 1 and 3, respectively. Again, this result is completely in accordance with the empirically measured height of the pyramid.

Many pyramids in Egypt were built with a certain seked, and the sekeds 22 and 21 were used more than once. In order to achieve a good impression, they tried to build the pyramid as steep as possible, which led to technical problems concerning the stability of the structure. It can be assumed that a seked of 22 was the technical optimum at the time of Cheops. When this seked is chosen, the triangle in figure 10.15 becomes virtually indistinguishable from the Kepler triangle. Figure 10.20 shows three triangles: the left-most exhibits the golden ratio (Hypothesis 1 and 2), the one in the middle assumes that the horizontal side is measured by n revolutions of a surveyor's wheel, and the vertical side is given by $4n$ diameters of the same wheel. Finally, the right-most triangle realizes a seked of 22 (vertical gain of 28 fingers over a horizontal distance of 22 fingers). Even if they were to have been drawn on a larger sheet of paper, the three triangles would be identical to within the thickness of the line of a pencil.

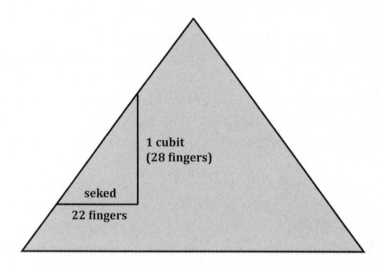

1 cubit
(28 fingers)

seked

22 fingers

Figure 10.19: A seked of 22—the slope of the pyramid.

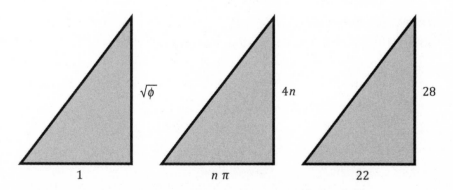

$\sqrt{\phi}$ $4n$ 28

1        $n\,\pi$ 22

Figure 10.20: Three triangles with very similar proportions.

The explanation is simple. Just remember that a good approxima-
tion of the number π is given by the fraction $\frac{22}{7}$. Hence the fractions

$$\frac{22}{28} = \frac{1}{4} \times \frac{22}{7} \approx \frac{\pi}{4} \approx \frac{1}{\sqrt{\phi}}$$

are all approximately the same. A seked of 22 automatically, and coincidentally, creates the proportion of the golden ratio together with the
proportion of π in the dimensions of the pyramid. Actually, there is
nothing mysterious about it, and in part it is just numerical coincidence. By the way, a seked of 21 would, instead of the triangle in figure
10.20, lead to a triangle with sides 21, 28, and 35, which is a Pythagorean triple (with proportions 3:4:5). This would lead to a pyramid with
simple integer proportions, which certainly would have been preferred
by the Pythagoreans.

We learn from this that not every occurrence of the golden ratio or
of π, be it in architecture or in art or in nature, has a "hidden" meaning
or was implemented intentionally. Very often, apparently meaningful
number relations might pop up unexpectedly, seemingly hinting at a
deeper reason. This gives rise to a long tradition of mystic speculations
about the occurrence of certain numbers. One has to be careful about
making unjustifiable generalizations. Very often, the scientific explanation, which, admittedly, is less romantic than number mysticism,
reveals that there is nothing behind these speculations.

CHAPTER 11

NUMBERS AND PHILOSOPHY

11.1. NUMBERS—INVENTED OR DISCOVERED?

For several thousand years, numbers have been involved in a wide spectrum of research and have been the focus of research as well. Mathematicians have developed and refined our understanding of numbers and accumulated a vast amount of knowledge about them and their application. They have developed sophisticated procedures using numbers for a wide variety of purposes in many different contexts. Apart from natural numbers, mathematicians have introduced new types of numbers—for example, negative numbers, rational numbers, real numbers, and complex numbers. And, of course, they have kept thinking about the nature of numbers, that is, about "what numbers really are" and why they play such a formidable role in the universe.

We have already seen in chapter 1 that the number concept reflects some basic properties of our world; in particular, the possibility to group objects into sets of distinguishable elements. Evolution has provided us (and some animal species) with a rudimentary number sense, which is exact for small numbers and approximate for large numbers. Counting arbitrary sets requires a synthesis of these aspects and thus requires mental abilities specific to *Homo sapiens*. Numbers were invented in early human societies as people started to become settled, and therefore numbers belong to the first cultural achievements of humankind. Numbers seem to be a human creation—a tool for the human mind to

329

create an adequate and useful mental representation of certain aspects of our world. And the process of simplification and information reduction leading to an abstract number concept appears to be even more of a mental construction, a function of the human brain that helps to organize thought processes in an economic manner.

Mathematicians, however, often think differently about the nature of numbers or other mathematical objects. When mathematicians get deeply immersed in research, they have the impression that they are dealing with entities that are not just a human creation but exist in a more objective sense. They believe that numbers have been discovered, not invented, and that their laws and properties can be explored in the same sense as a physicist would explore the properties of elementary particles. The only difference seems to be that numbers are objects that exist in a nonphysical, and also nonpsychological, manner, while elementary particles exist in the physical universe. But, like elementary particles, numbers seem to exist independently of the human mind. And where a physicist would use experiments and measuring devices, mathematicians use their intuition, logical thinking, and abstract reasoning to discover the beauty and truth in a previously unexplored terrain. The world where mathematicians conduct their research is an abstract world populated by mathematical objects and ideas. When they find unexpected relationships, patterns, and structures, then a new range of mathematical knowledge, a new region of that abstract world, becomes accessible. The mathematician would then feel like an explorer of a past century who had discovered a new, previously unknown region of the earth.

This view cannot be discarded easily. For example, when we played with square numbers in chapter 4, we "discovered" the amazing result that the sum of the first n odd numbers equals $n \times n$. We found that this result must be true from the obvious, and geometrically intuitive, way in which the next square number is constructed from a given square number. This sense of truth is further confirmed by algebraic methods, which do not rely on geometric visualization at all. And therefore it is the general consent among mathematicians that this statement is indeed true for all natural numbers, n. Once convinced of its truth, one has the feeling that this statement expresses more than just a psychological conviction or a social convention. Indeed, the result is an inevitable conclusion of logical reasoning, and hence is independent of human belief or attitude.

This gives the impression that the result represents an objective truth that existed and was true even before it was formulated and proved. It evokes the idea that there is a metaphysical realm of numbers that exists independently of the physical universe. In other words, if the whole universe disappeared tomorrow, the eternal world of numbers would still exist.

We have just described two contradicting philosophical positions concerning numbers: One position holds that numbers have a mind-independent existence in a metaphysical world "out there." The other position is that numbers exist "in here" as creations of the human mind, and are designed to help us in various tasks, like classifying and ordering sets of objects.

11.2. THE PLATONIC POINT OF VIEW

The philosophical position that mathematical objects (such as numbers, triangles, equations, etc.) exist by themselves in some "realm of mathematics," which is outside the world of physical objects, and also outside our mind, is called *Platonism*, named after the famous Greek philosopher Plato (428/427–348/347 BCE). In his "theory of forms," Plato claimed that ideas possess a more fundamental kind of reality than material objects. Ideas (also called "forms") are nonmaterial and abstract, and they exist in a metaphysical world of ideas. Material objects that are perceived through our senses are just "shadows" or "instances" of their ideal forms—their true essence. Human beings are like a caveman who sits with his back toward the entrance of the cave and can observe only the shadows of the outside reality on the wall in front of him. Consequently, real insight can only be gained through the study of ideas that are not directly accessible through our senses but are accessible through reason.

Until the twentieth century, this was indeed the common belief concerning the nature of numbers. Mathematicians considered numbers to be "real" objects in an immaterial realm of abstract ideas existing independently of human beings. While modern mathematicians usually do not go so far as to declare the material world as unreal, many of them would still uphold the Platonic view of the reality of mathematical objects. For example, as French mathematician Charles Hermite (1822–1901) stated: "I believe that numbers and functions of analysis are not the arbitrary result of our minds; I think that they exist outside of us, with the same character of necessity as the things of objective reality, and we meet them or discover them, and study them, as do the physicists, the chemists and the zoologists."[1]

Elsewhere, he wrote, "There exists, if I am not mistaken, an entire world, which is the totality of mathematical truths, to which we have access only with our mind, just as a world of physical reality exists, the one like the other independent of ourselves, both of divine creation."[2]

In the book *A Mathematician's Apology* (1940), the well-known British mathematician Godfrey Harold Hardy (1877–1947) expressed his belief as follows: "I believe that mathematical reality lies outside us, that our function is to discover or observe it, and that the theorems which we prove, and which we describe grandiloquently as our 'creations,' are simply our notes of our observations."[3]

11.3. AN ONGOING DISCUSSION

In 2007, an article titled "Let Platonism Die," by British mathematician E. Brian Davies (1944–), revived the discussion. Davies points out that the belief in the independent existence of an abstract mathematical world makes implicit assumptions about the working of our brain. Platonists seem to believe that the brain can make a connection to the Platonic realm and thus reach beyond the confines of space and time into an abstract cosmos. For Davies, this view "has more in common with mystical religion than with modern science."[4] He points out that scientific studies about how the brain creates mathematics indicate that mathematical thought processes have a purely physiological basis, and that these studies "owe nothing to Platonism, whose main function is to contribute a feeling of security in those who are believers. Its other function has been to provide employment for hundreds of philosophers, vainly trying to reconcile it with everything we know about the world. It is about time that we recognized that mathematics is not different in type from all our other, equally remarkable, mental skills and ditched the last remnant of this ancient religion."[5]

In 2008, two follow-up articles by American mathematicians Reuben Hersh (1927–) and Barry Mazur (1937–) took the discussion further. The question of the reality of mathematical objects has nothing to do with the fact that mathematics is a human and culturally dependent pursuit. Thus, numbers could well have an independent existence, even if a basic understanding of numbers was provided by evolution, and even if the mental images of numbers created in our mind depend on sociological factors. Dr. Mazur gives the following example: If we were not interested in numbers, but in "*writing a description of the Grand Canyon*, and if a Navajo, an Irishman, and a Zoroastrian were each to set about writing their descriptions, you can bet that these descriptions will be culturally dependent, and even dependent upon the moods and education and the language of the three describers."[6] But this does not "undermine our firm faith in the existence of the Grand Canyon."

According to Reuben Hersh, Platonism "expresses a correct recognition that there are mathematical facts and entities, that these are not subject to the will or whim of the individual mathematician but are forced on him as objective facts and entities."[7] But in his opinion, the "fallacy of Platonism is in the misinterpretation of this objective reality, putting it outside of human culture and consciousness. Like many other cultural realities, it is external, objective, *from the viewpoint of any individual*, but internal, historical, socially conditioned, *from the viewpoint of the society or the culture as a whole*."

11.4. PHILOSOPHY OF MATHEMATICS

At the beginning of the twentieth century, philosophers, logicians, and mathematicians tried to formulate the proper foundations of mathematics. This resulted in the so-called foundational crisis of mathematics, from which several schools emerged, fiercely opposing each other, each with a radically different view of the right approach. During the first half of the twentieth century, the three most influential schools were known as logicism, formalism, and intuitionism. As numbers form an essential element of mathematics, the various philosophical schools also had different approaches to the concept of number.

Logicism, for example, whose most famous members were German mathematician Gottlob Frege (1848–1925) and British mathematician Bertrand Russell (1872–1970), tried to base all of mathematics on pure logic. In particular, they believed numbers should be identified with basic entities from set theory and their arithmetic should be derived from first logical principles. This was an important goal because all traditional pure mathematics can in fact be derived from the properties of natural numbers together with the propositions of pure logic. This idea already appeared in the work of German mathematician Richard Dedekind (1831–1916), who, in 1889, wrote "I consider the number-concept entirely independent of the notions or intuitions of space and time. . . . I rather consider it an immediate product of the pure laws of thought."[8] In 1903, Russell wrote that the object of logicism was "the *proof* that all pure mathematics deals exclusively with concepts definable in terms of a very small number of fundamental logical concepts, and that all its propositions are deducible from a very small number of fundamental logical principles."[9] The program of logicism was to reduce the notion of number to elementary ideas founded in pure logic, to "establish the whole theory of cardinal integers as a special branch of logic."[10] In this way, Russell hoped to give the notion of numbers a definite meaning.

Formalism, on the other hand, did not attempt to give meaning to mathematical objects. In the formalist approach—whose main proponent was German mathematician David Hilbert (1862–1943)—the goal was to define a mathematical theory in terms of a small set of axioms, mathematical propositions that are simply *assumed* to be true. From these axioms, mathematical theorems were derived by the rules of logical inference. In formalism, one is not interested in the nature of numbers, or in the question of whether numbers have a meaning. Rather one is only interested in the formal properties of numbers; that is, in the rules that govern their relations. Any set of objects that follows these rules could then serve as numbers. The formalist's view is best expressed in a famous statement usually attributed to David Hilbert: "Mathematics is a game played according to certain simple rules with meaningless marks on paper."[11]

Intuitionism, originating in the work of L. E. J. Brouwer (1881–1966), is non-Platonic because its philosophy is based on the idea that mathematics is a creation of the human mind. Because mathematical statements are mental constructions, the validity of a statement is ultimately a subjective claim asserted by the intuition of the mathematician. The mathematical formalism is just a means of communication. By restricting the allowed methods of logical reasoning (denying the validity of the principle of the excluded middle—that any proposition must either be true or its negation must be true), intuitionism strongly deviates from classical mathematics and the other philosophical schools, in particular in the accepted methods of proof. For an intuitionist, a mathematical object (e.g., the solution of an equation) would exist if it could be constructed explicitly. This is in contrast to classical mathematics, where the existence of an object can be proved indirectly by deriving a contradiction from the assumption of its nonexistence. The main differences occur, however, in how intuitionism deals with infinity, but that subject is beyond the scope of this book. Statements concerning the arithmetic of finite numbers generally remain true, and, in this context, intuitionism and classical mathematics have a lot in common.

Logicism, formalism, and intuitionism all made valuable contributions to the foundations of mathematics, but they all ran into unexpected difficulties of a rather technical nature. These difficulties finally prevented any of these programs from being fully realized.

11.5. THE LOGICIST DEFINITION OF A CARDINAL NUMBER

In his book *Introduction to Mathematical Philosophy*, Bertrand Russell used the concept of a set and the bijection principle (see chapter 1) to define the abstract concept of a cardinal number. This abstract definition makes clear that a number does not refer to any particular group of objects but to the whole class of sets with the same number of objects in it. For this, it is important to remember that in order to determine whether two sets have the same number of elements, it is not necessary to count them; one only has to find a one-to-one correspondence (a bijection) between the elements of the two sets.

One first defines that two sets A and B are equivalent (for the purpose of counting), whenever there is a bijection between these two sets. Thus, when two sets are equivalent, then the elements of one set can be paired with the elements of the other, with no leftover elements in either set. The set of fingers on the right hand is equivalent to the set of fingers on the left hand, the pairing being established by putting the fingertips together. Equivalent sets cannot be distinguished by counting. Hence the definition of "number" must refer to the whole bunch of equivalent sets.

Russell, therefore, defines the "cardinal number of a set A" simply as the collection of all sets that are equivalent to A:

- The class of all sets that are equivalent (for the purpose of counting) is called a *cardinal number*.

Here we could have said "the set of all sets that are equivalent." The word "class" is used because in set theory this describes a particular type of (infinite) collection that avoids certain logical problems encountered with arbitrary "sets of sets"—logical problems that were discovered in 1901 by Russell.

Numbers thus become "equivalence classes" of sets. This just means that all sets containing a certain number of elements contribute to the definition of that number. Conceptually, it is the whole collection of equivalent sets that describes best their common property—and this property is the number. According to Bertrand Russell, this is the same process of abstraction that happens in everyday life, where, for example, the best description of what is meant by the abstract concept "table" is the whole collection of all objects that are called "table." Only in that way could the abstract notion "table" encompass everything that might be called "table."

Russell describes his idea of reducing the concept of number to set theory with the following words: "We naturally think that the class of couples (for example) is something different from the number 2. But there is no doubt about the class of couples: it is indubitable and not difficult to define, whereas the number 2, in any other sense, is a metaphysical entity about which we can never feel sure that it exists or that we have tracked it down. It is therefore more prudent to content ourselves with the class of couples, which we are sure of, than to hunt for a problematical number 2 which must always remain elusive."[12]

Thus, according to Russell, the number 2 is the collection of *all* pairs—it consists of *all* sets that contain precisely two elements because all these sets are indistinguishable by counting and are, therefore, considered equivalent. So, we consider the whole collection of sets that are equivalent to a pair of shoes and call it "number 2." Any particular set of two objects is then just an example of the number 2, a representative, very much in the same sense as any particular dining table is just a representative of the abstract notion "table."

At first sight, this definition might seem circular, because how could we define the collection of all two-element sets unless we already know what "two" is. But actually, one can define a two-element set without mentioning the number 2, as follows: We say that a set A contains two elements, whenever the following conditions are fulfilled:

(a) A contains an element x and an element y, such that x is not equal to y,
(b) for all elements z belonging to A we have either $z = x$ or $z = y$.

These conditions express in purely logical terms (as an equality), what we mean by a set of two elements. In an analogous way, one can define a set of three elements, four elements, and so on. Thus, the goal seems to have been reached to define the finite numbers on the basis of pure logic.

Hungarian-American mathematician John von Neumann (1903–1957) described a purely set-theoretic method to construct the natural numbers. In the axioms of mathematical set theory, there is a unique set that contains no elements at all. It is called the "empty set" or "null set" and denoted by $\emptyset$ or occasionally by $\{\ \}$. Neumann took the empty set $\emptyset$ to represent the number 0. Next, we can form the set that contains the empty set as its only element; this is the set $\{\emptyset\}$. We take it to represent the number 1 because it obviously contains precisely one element. Next, we can form the set with the elements $\emptyset$ and $\{\emptyset\}$, that is $\{\emptyset,\{\emptyset\}\}$. Remember that the collection of all sets equivalent to this set would be number 2. For the number 3, we combine all previously obtained objects, namely $\emptyset$, $\{\emptyset\}$, and $\{\emptyset,\{\emptyset\}\}$, into a new set and proceed in an analogous way. In this way, we construct a sequence of certain prototypical sets using only the basic notions of set theory, nothing else, and the numbers are then the collections of all sets that are equivalent to the prototypical sets.

0 . . . represented by . . . $\emptyset$ (empty set)
1 . . . represented by . . . $\{\emptyset\}$
2 . . . represented by . . . $\{\emptyset, \{\emptyset\}\}$
3 . . . represented by . . . $\{\emptyset, \{\emptyset\}, \{\emptyset, \{\emptyset\}\}\}$
4 . . . represented by . . . $\{\emptyset, \{\emptyset\}, \{\emptyset, \{\emptyset\}\}, \{\emptyset, \{\emptyset\}, \{\emptyset, \{\emptyset\}\}\}\}$

and so on.

If x is a set constructed in this way, then the successor of x is always defined as the union of x and $\{x\}$. On the nth stage, x is a set containing n elements, and the set $\{x\}$ contains just one element, namely x. The union of these two sets will, thus, contain $n + 1$ elements. This will serve as a new prototypical set with $n + 1$ elements representing the number $n + 1$. All other sets that are in one-to-one correspondence with this prototypical set would together be the "number $n + 1$." In that way, natural numbers are created, one after another, out of the empty set $\emptyset$, that is, "out of nothing." We also note that the cardinal numbers thus obtained are naturally ordered by size.

From here, it is still a long way to a complete mathematical exposition of numbers and their arithmetic, or even to a precise mathematical definition of the set of all natural numbers. This approach actually needs a lot of experience in abstract logical reasoning, and we shall not pursue it any further here. We should note that in most situations not even a mathematician would think of the number 4 as the "class of all sets equivalent with $\{\emptyset, \{\emptyset\}, \{\emptyset, \{\emptyset\}\}, \{\emptyset, \{\emptyset\}, \{\emptyset, \{\emptyset\}\}\}\}$." This construction, nevertheless, serves to show that the abstract notion of a number can be defined on a strictly logical basis, using only very elementary definitions from set theory. The mathematical definition alluded to here formalizes the somewhat vague statement, "The number 4 describes what 4 apples and 4 people have in common."

The concepts described here have paved the way to a logically rigorous analysis of mathematically advanced ideas about infinity, leading to definitions of infinite cardinal numbers and infinite ordinal numbers that have opened up a huge field of research for the mathematicians of the twentieth century.

11.6. A FORMALIST'S DEFINITION OF NUMBER

Unlike a logicist, a formalist would not be concerned about the nature or meaning of number. He would accept any kind of objects, whether they exist in nature or in one's imagination, to play the role of numbers, as long as these objects satisfy certain properties qualifying them for that role.

The properties required for natural numbers were first described by Italian mathematician Giuseppe Peano (1858–1932). The Peano axioms are typically given in the form of five statements, which we will describe below. They formulate the properties of an otherwise-not-specified set, whose elements are called *natural numbers*. This description focuses on the intuitive idea that every natural number n has a unique "next number," which is here called the "successor" $S(n)$. Of course, what we have in mind is that the successor of n is simply $n + 1$, but at this stage addition has not yet been defined. The first Peano axiom singles out one element of the set as the first of the natural numbers and gives it the name "0."

1. 0 is a natural number.
2. Every natural number n has a unique successor $S(n)$, which is also a natural number.
3. There is no natural number that has 0 as successor.
4. Different natural numbers have different successors.
5. Any property that
   a. holds for 0, and
   b. holds for $S(n)$ whenever it holds for n actually holds for all natural numbers.

The last axiom is the most difficult to understand. It is called the *induction axiom* and is an essential tool in proving properties of natural numbers.

As a set fulfilling these axioms, we could take, for example, the sequence of bullet points shown in figure 11.1:

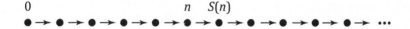

Figure 11.1: A sequence of dots as a model for the Peano axioms.

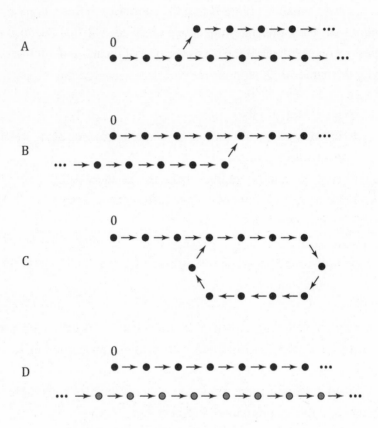

Figure 11.2: Counter examples to Peano's axioms.

In figure 11.1, it is assumed that the sequence of bullets can be continued indefinitely toward the right. In this set of points (they all look
the same, but are distinguished by their position), the successor of any
point is the next point to the right, as indicated by the arrow. We call
the first and leftmost point "0." This element is not a successor (axiom
3), but every other point is the unique successor of the point to its left
(axiom 2). The set of points cannot end, because in that case there
would be some point that has no successor, thus violating axiom 2.
The other Peano axioms exclude, for example, the possibility of loops,

bifurcations, or parallel chains of points. They also ensure that 0 is the only element without a predecessor. Hence, this simple chain of points starting at some point and extending indefinitely is, in fact, the only possibility. If you want to do an exercise in logical reasoning, you may look at the examples of point sets with a successor relation in figure 11.2. Each one of the sets A, B, C, and D, violates a particular Peano axiom. Can you figure out which Peano axiom is violated by which example? The solution will be given at the end of this section.

Starting with the element "0," we obtain step-by-step all elements of the natural numbers by going to the successor: The successor of 0 will be called 1, the successor of 1 will be called 2, and so on:

$$1 = S(0),$$
$$2 = S(1) = S(S(0)),$$

and so on, and this sequence will never end. If we want to define, say, the addition of two numbers, we define first $m + 0 = m$ and then $m + S(n) = S(m + n)$.

From this definition, we can now derive, for example, that the successor of m is $m + 1$ because

$$m + 1 = m + S(0) = S(m + 0) = S(m).$$

And from here, it is not so difficult for a trained mathematician to develop the whole arithmetic of natural numbers.

It is very interesting that in all these considerations it was not necessary to assume anything about the nature of the elements in the set of natural numbers. This set could consist of counting words, of dots, or of the sets $\{\emptyset, \{\emptyset\}, \{\emptyset, \{\emptyset\}\}, \ldots\}$ that were described in the previous section. The only thing that matters is that one can define a "successor function" for this set, which has the properties required by the Peano axioms. Once this can be done, the elements of this set can be just treated as natural numbers, and all further properties of natural numbers can be derived by logical reasoning from the Peano axioms. For the mathematical formalist, the nature of the objects that fulfill the Peano axioms is indeed irrelevant. All that counts is the mathematical structure that governs the behavior of these objects. And if these objects behave like numbers in every respect, a formalist would simply take these objects for numbers and go on to more important business.

It does not come as a surprise that Bertrand Russell was not at all happy with the formalist's point of view. He complains that

> any progression may be taken as the basis of pure mathematics: we may give the name '0' to its first term, the name 'number' to the whole set of its terms, and the name 'successor' to the next in the progression. The progression need not be composed of numbers: it may be composed of points in space, or moments of time, or any other terms of which there is an infinite supply. . . . It is assumed that we know what is meant by '0,' and that we shall not suppose that this symbol means 100 or Cleopatra's Needle or any of the other things that it might mean. . . . We want our numbers not merely to verify mathematical formulae, but to apply in the right way to common objects. We want to have ten fingers and two eyes and one nose.[13]

Solution to the exercise in logical thinking:

A: This set violates axiom 2, since there is a point whose successor is not unique, because it has two successors.

B: Violates axiom 4 because two different points have the same successor.

C: Again violates axiom 4 because two different points have the same successor.

D: Violates axiom 5. The set D contains two sequences of dots, which are not connected by the successor property. For clarity, they are distinguished by color. 0 is black and for every black point, also its successor is black. Hence, by axiom 5, all points should be black, which is obviously not the case. (The color is not really necessary for this argument. Instead of the property "black," we could use in the same way, for example, the property "being a point in the upper row.")

11.7. THE STRUCTURALIST'S POINT OF VIEW

Structuralism is a philosophy of mathematics that evolved in the second half of the twentieth century. It holds that mathematical objects, like numbers, are only meaningful as parts of a larger structure. In order to understand this point of view, we need to go back to chapter 1, where we dealt with numbers in the context of counting. We learned that counting is done according to certain principles, and we saw how these principles imposed a certain structure on the set of number words. According to structuralism, number words do not refer to "abstract numbers." They have no meaning in their own right; they get their meaning exclusively from the structure of the whole set of number words.

The number words "one," "two," "three," . . . are distinguished by their strict and invariable ordering. This strict order of that sequence implies that any particular number word, for example, "eight," defines in a unique way a part of the sequence of the number words. This is the initial sequence of all number words from "one" to "eight":

"eight" $\Rightarrow$ ("one", "two", "three", "four", "five", "six", "seven", "eight").

When we count a set and find that it contains eight elements, this simple statement actually means that this set contains exactly as many elements as there are number words in the initial sequence defined by "eight." This means that there is a one-to-one correspondence between the given set and the initial section, "one" through "eight," of the sequence of number words (see figure 11.3).

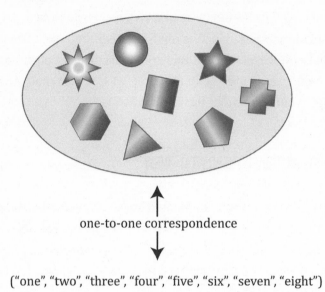

one-to-one correspondence

("one", "two", "three", "four", "five", "six", "seven", "eight")

Figure 11.3: "This set has eight elements."

As explained in chapter 1, the statement "This set has eight elements" is actually an abbreviation for "There is a one-to-one correspondence between the set and the initial section of the sequence of number words up to 'eight.'" This latter sentence describes precisely what we do when we count a collection of objects. We point with a finger to each of the objects in turn, tagging each with a number word and using the number words in a strict order. In that way, we pair each object with a number word, thereby establishing the one-to-one correspondence between the objects and the initial section of the number-word sequence. The last number word in that initial section has been called the *cardinal number* of the set.

The whole procedure is reported in an abbreviated form—"There are eight objects." The short form does not mention the initial section of the number-word sequence and its relation with the set of objects. This gives the impression that "eight" is a property of the collection of objects, while the statement "there are eight objects" actually tells us something about the relation between a set of objects and a certain part of the number-word sequence.

In the discussion above, the English number words just served as an example. Another sequence of counting tags—for example, the German number words (like "eins," "zwei," "drei," . . .)—would serve the same purpose. Any set of words or symbols that can be arranged in a linearly ordered sequence could serve as a sequence of counting tags. In a more universal fashion, we could just use the ordered sequence of symbols: (1, 2, 3, 4, 5, 6, 7, 8, 9, 10, 11, 12, 13, 14, 15, . . .) in the same way as we use number words, for tagging the counted objects. The sentence "a set has 8 elements" is a short form of the statement that there is a one-to-one correspondence between the set and the initial section (1, 2, 3, 4, 5, 6, 7, 8) of the symbol sequence.

If taken in this sense, the symbol "8" (or the word "eight," or any other symbol or word that we might use in counting) does not refer to any particular mathematical object at all. It is the ordered structure of the whole sequence that makes its members suitable for the purpose of counting, not a property of the individual members of the sequence. The individual number is meaningless; the meaning lies in the structure of the number sequence. This point of view was expressed by German mathematician Hermann Weyl (1885–1955) in the article "Mathematics and the Laws of Nature" as follows: "But numbers have neither substance, nor meaning, nor qualities. They are nothing but marks, and all that is in them we have put into them by the simple rule of straight succession."[14] American mathematician Paul Benacerraf (1931–), who is one of the main proponents of structuralism in mathematical philosophy, writes in his article "What Numbers Could Not Be": "What is important is not the individuality of each element, but the structure which they jointly exhibit."[15] He argues that it is utterly pointless to ask whether any particular set-theoretic object, like the set {∅, {∅}}, could replace the number 2 because "'objects' do not do the job of numbers singly; the whole system performs the job or nothing does."

Moreover, this type of identification of a "number" with a mathematical object could never be done in a unique way. Therefore Benacerraf comes to the conclusion "that numbers could not be objects at all; for there is no more reason to identify any individual number with any one particular object than with any other."[16] For arithmetical purposes, all that matters is that the collection of numbers has the structure of a linear progression. Additional individual properties of numbers would not matter at all; they are of no consequence to arithmetic. He continues, "but it would be only these properties that would single out a number as this object or that." Hence the question of whether a number is any particular sort of abstract object is completely irrelevant. This question misses the point of what arithmetic is all about. The arithmetic of natural numbers is the science of describing the structure of a linear progression. This is the structure of all ordered sequences where we have a first element and where each element has a successor, as described in the Peano axioms (see section 6). Arithmetic is not the search for which particular objects the numbers really represent—"there is no unique set of objects that are the numbers. Number theory is the elaboration of the properties of *all* structures of the order type of the numbers. The number words do not have single referents."

Mathematical structuralism emphasizes the description of structural properties as the real goal of mathematics. Therefore, it has much in common with formalism, but to a structuralist the purely formalist point of view that mathematics is just a game played according to certain rules with meaningless symbols goes too far. The statement "this set has 8 elements" does have a definite meaning. But the meaning of "8" (or "eight" or "acht") can only be explained if we know the position of that element within the whole sequential structure to which it belongs. It is not a property of the individual member of the sequence, but its relation to the other members of the sequence that gives a meaning to the number word or symbol.

The structuralist point of view says that "8" does not refer to an abstract object. This is in accordance with a modern linguist's point of view. After a linguistic analysis, German scholar Heike Wiese (1966–), in her book *Numbers, Language, and the Human Mind*, also comes to the conclusion that number words are nonreferential—a number word does not refer to any real or abstract object; it just works as an element of the number-word sequence. Number words are special because "unlike other words, they do not have any meaning, they do not refer to anything in the outside world. This is because they are not names for numbers, they *are* numbers. Counting words are tools that we use in number assignments, and for this job they do not need any referentiality."[17]

11.8. THE UNREASONABLE EFFECTIVENESS OF MATHEMATICS

The great German philosopher Immanuel Kant (1724–1804) was less concerned about the reality of abstract objects than he was in statements and propositions about these objects. Consider the following statements concerning the role of mathematics in the physical world:

A: Mathematics is rooted in observations and knowledge about the physical world, and its results tell us something about empirical reality.

B: Mathematics is a system of propositions, each of which is true in itself and needs no empirical verification or confirmation.

Kant distinguished between a priori knowledge and a posteriori knowledge. Knowledge is a priori if it is independent of any experience about the physical world (such as "all triangles have three sides"). Knowledge is a posteriori if it depends on empirical evidence (such as "there are six items in that box").

A statement is called *analytic* if it is true in itself—for example, "all husbands are married." This statement can be seen to be true because the word *husband* only refers to a married person. Understanding the meaning of the words is sufficient to judge the truth of the sentence. A statement where the meanings of the words alone do not imply whether the statement is true or false is called *synthetic*. The statement "all husbands are happy" (whether or not it is true) is synthetic because the word *husband* alone does not imply happiness. It appears that analytic statements are not very interesting, because they can be made in advance (that is, "a priori"), without referring to anything that is not already contained in the definition of the words in that statement. Synthetic statements seem to be more interesting because they make a claim that is not self-evident and that does not already follow from the meaning of the words in the statement. The correctness of a synthetic statement cannot be inferred just by analyzing its content. From this we see that synthetic statements tend to be a posteriori. One typically has to refer to experience and observation for determining the truth of a synthetic statement. Kant's big question now is whether any synthetic knowledge exists a priori and whether a mathematical statement such as $5 + 3 = 8$ would be synthetic and a priori.

As a mathematical result about numbers, $5 + 3 = 8$ is a statement that follows logically from the structure of the number sequence $(1, 2, 3, 4 \ldots)$. As such, it is true because of the way mathematicians draw conclusions from the axioms and because of the way they use the rules of logic to determine the truth of statements. Assuming that the sequence of symbols $(1, 2, 3, 4 \ldots)$ has the properties required by the Peano axioms, the truth of arithmetic statements follows inevitably from elementary rules of logic. The statement thus represents a priori knowledge and it is analytic because it just expresses a formal property of the ordered sequence to which the symbols 3, 5, and 8 belong.

As a statement about cardinal numbers, "$5 + 3 = 8$" means that if we combine two sets with cardinalities 5 and 3 (that is, if we form the "union" of these sets), we will obtain a set with cardinality 8. The truth of this result can also be verified in reality, where we can check by counting that 5 objects combined with 3 objects indeed gives a set with 8 objects. As a statement about empirical reality, it seems to be a synthetic statement.

The statement "5 + 3 = 8" is therefore obtained a priori, just by logical derivation from the axioms defining the properties of the number sequence, and yet it appears to be a synthetic statement that tells something about the physical universe. In view of our considerations on the psychology of numbers, it may indeed be doubted whether anything about this statement is a priori. The concept of number is clearly rooted in elementary knowledge about properties of the universe—knowledge that is, in part, acquired through evolution and innate, and in part culturally acquired. The same could be said about the logical rules, which are by no means totally self-evident (as we notice in the discussion about the rule of the excluded middle between intuitionists and classical mathematicians). But the logical rules are probably also, in part, rooted in core-knowledge systems, ingrained in our brain by evolutionary processes, and thus reflect some elementary properties of the (causal) mechanisms in the world that surrounds us.

The philosophical position that mathematics is not a priori but that all its objects have their origin in empirical knowledge is called *empiricism*. According to this view, mathematics is, after all, not so different from other natural sciences. American philosopher Willard van Quine (1908–2000), an important proponent of empiricism, said that mathematical entities, like numbers, exist as the best explanation for experience. Thus, mathematical results, like 5 + 3 = 8, are not completely certain, because they refer to observations that, at least in principle, could be wrong. Fortunately, mathematics is very central to all of science, and a large web of trusted knowledge depends on it, and thus it would be extremely difficult to change mathematics. This gives the impression that the results of mathematics are completely certain and not likely to be revised.

Indeed, mathematical considerations of much higher complexity than mere additions are routinely and successfully applied to predict phenomena in the physical world. People have often wondered how it is possible that abstract mathematics is so successful in describing reality.

Albert Einstein (1879–1955), in an address given in 1921 at the Prussian Academy of Sciences in Berlin, formulated this problem as follows: "How can it be that mathematics, being, after all, a product of human thought, which is independent of experience, is so admirably appropriate to the objects of reality?"[18]

Later, in 1960, Hungarian physicist Eugene P. Wigner (1902–1995) coined the expression of "the unreasonable effectiveness of mathematics in the natural sciences."[19] It is obviously a question that has remained a topic of discussion among mathematicians, natural scientists, and philosophers. In the introduction to the 2007 edition of *The Oxford Handbook of Philosophy of Mathematics and Logic*, American philosopher Stewart Shapiro (1951–) says, "Mathematics seems necessary and a priori, and yet it has something to do with the physical world. How is this possible? How can we learn something important about the physical world by a priori reflection in our comfortable armchairs?"[20]

Einstein, in his 1921 address, attempts to give an answer: "As far as the laws of mathematics refer to reality, they are not certain; and as far as they are certain, they do not refer to reality."[21] This expresses the point of view of an empiricist and applied mathematician who is reluctant to accept that mathematics has any a priori relevance for the physical world. To the applied mathematician, any application of mathematics to reality can be understood as a process of mathematical modeling.

11.9. MATHEMATICAL MODELS

A mathematical model is a representation of a typical real-world situation, expressed in the language of mathematics. It is usually created to solve a problem, or to answer some question. Depending on the scope of the model, the mathematical representation could be an entire field of mathematics (for example, the axioms, definitions, and theorems of Euclidean geometry) or just a mathematical equation. The process of translating the real-world situation into the mathematical model is fittingly called *mathematization* (see figure 11.4).

As an example of a real-world situation, consider the weather in some region of the world. We could ask, for example, "What is the weather going to be tomorrow?" Translating this question into mathematical language is by no means simple. The mathematical model needed for a weather forecast can be derived from the physical laws that govern the temporal change, and the mutual influence of physical variables related to weather phenomena, like wind velocity, air pressure, temperature, and humidity. The physical laws typically lead to a mathematically rather complicated system of differential equations. The question about tomorrow's weather would then translate into a question about the solutions of these equations. One would then try to solve these equations by a numerical computation, which uses as input a collection of initial data describing today's weather (temperature, wind velocity, etc., measured at various locations of the country) and which takes into account boundary conditions describing the local geography (mountains, coastline, etc.).

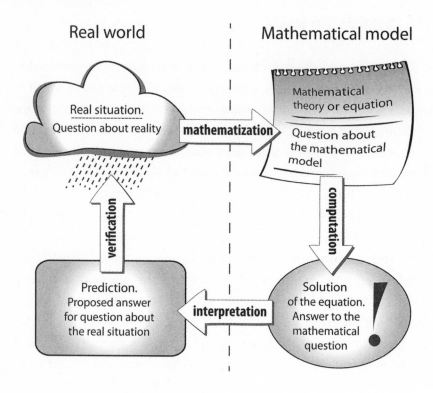

Real world | Mathematical model

Real situation. Question about reality

mathematization

Mathematical theory or equation

Question about the mathematical model

computation

verification

Prediction. Proposed answer for question about the real situation

interpretation

Solution of the equation. Answer to the mathematical question

Figure 11.4: Applications of mathematics—the modeling cycle.

Once we have obtained a mathematical answer within the model, how can we know that the mathematical result is true? The answer is we can never be sure. In a complicated model, there are many sources of possible errors. In order to make weather forecasts feasible within a reasonable time allowed for the computation, the model usually contains simplifications and approximations, and, moreover, the initial state that defines the starting point of the computation is only known with limited accuracy.

One, therefore, has to test the model and examine its validity by comparing the outcome of the mathematical computation with observations in reality. In order to do this, one first has to interpret the result of the mathematical computation in terms of the real-world situation at hand. The mathematical expressions (the solutions of the model equations) have to be translated into tomorrow's values of the physical variables related to weather phenomena. From these values, a meteorologist can conclude tomorrow's weather conditions. The interpreted result is, thus, a prediction about an observation concerning physical reality. The comparison of the prediction with the actual situation (called *validation*, or *verification*) will either confirm the model or reveal a flaw. Whenever the prediction disagrees with the observation, we will have to adjust and improve the model. This can be done, for example, by taking into account more details, or by correcting any error that might have occurred in the whole process.

Natural numbers and their arithmetic can likewise be interpreted as a mathematical model of certain aspects of reality. We are just not accustomed to think that way, because natural numbers and arithmetic are so fundamental and are applied as a matter of course. And yet, the process of modeling certain phenomena with numbers and arithmetic has the same structure as in the case of weather forecasting.

In the example depicted in figure 11.5, the real-world situation is about combining sets of discrete objects. If we put first five and then three apples into a basket, how many apples would be in the basket? The answer is so obvious that we are usually not aware that we in fact used a mathematical model—natural numbers and their addition. The mathematical representation of this situation involves just two natural numbers, 5 and 3, and the question about the real-world situation can be translated into the mathematical question, "What is the result of $5 + 3$?"

As always, the mathematical model may lack some details—in this case, the model completely disregards the concrete nature of the objects and just describes their numbers. A model need not have more details than necessary to answer the question at hand. So the mathematical question is just about numbers, not about the geometrical shape or the arrangement of the apples. The next step is computation, and, unlike the computations involved in a weather forecast step, it does not require a computer to produce the result "8." Next, the result has to be interpreted in view of the real situation. We remember that we wanted to know the number of apples in a basket and made the prediction that it would contain eight objects. The final step would be a reality check (validation, verification) in order to compare the outcome of the mental operation with the real situation. Here, this is simply done by counting the number of apples in the basket.

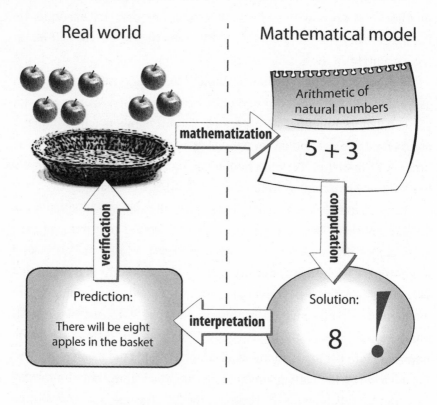

Figure 11.5: Arithmetic of numbers as a model.

If a model makes valid predictions in many concrete cases, if it already has been applied and tested successfully in many situations, we have some right to trust in that model. By now, we believe in the model "natural numbers and their arithmetic" and in its predictions without having to check it every time. We do not expect that the result might be wrong; hence the verification step is not needed any longer for validating the model. If the model had a flaw, it would have been eliminated already in the past. We trust the model so much that if the basket contained only seven apples, we would not look for an error in the model but instead would look for the thief who stole the apple.

But we also have to know that every model has its limitations. The model of natural numbers and their sums is very successful to determine the number of objects in the union of two different groups of well-distinguished objects. But as a mathematical model, the arithmetic of numbers is not generally true but only validated and confirmed for certain well-controlled situations.

In some situations, the arithmetic of numbers does not give the appropriate answer. For example, if you put together 5 cm^3 of water and 3 cm^3 of salt, you would not obtain 8 cm^3. The mixture would have a smaller volume than 8 cm^3 because a large part of the salt would dissolve in the water and the solution is denser than pure water, thus occupying less space.

Assume you walk forward at the rate of 3 miles per hour on a ship that moves with a speed of 5 miles per hour with respect to the coast. How fast would you be moving with respect to the coast? Addition of natural numbers is also a good model for this situation, and the answer would be that you would be moving at the rate of 8 miles per hour with respect to the shore. But this answer is only approximately true. For the addition of velocities, it would be more accurate to use the relativistic addition of velocities. According to this model, your velocity, as seen from the coast, would be less than 8 miles per hour. It would be 7.99999999999999973 . . . miles per hour.

The error is so small that it was unknown before 1905, when Albert Einstein set up the special theory of relativity. According to this theory, velocities cannot be added in the usual way, as you would add apples.

Instead one has to use the following formula to determine the sum of a velocity u and a velocity v: If the velocity of the ship with respect to earth is v and you move in the same direction with velocity u with respect to the ship, then your velocity with respect to earth would be not simply $u + v$, but rather

$$\frac{u + v}{1 + \dfrac{u \times v}{c^2}}$$

Here c represents the velocity of light, which is 670,616,629 miles per hour (its exact value is 299,792,458 meters per second). Because the speed of light is so large, the fraction $\frac{u \times v}{c^2}$ is extremely small, and so the denominator $1 + \frac{u \times v}{c^2}$ is very close to 1 in all situations of everyday life, and one can usually neglect it. But the denominator inhibits any relative velocities greater than or equal to c. For example, if you added the velocities $\frac{c}{2}$ and $\frac{c}{2}$, you would not get c, but only $\frac{4}{5}c$.

11.10. LIMITS OF THE MODEL OF NATURAL NUMBERS

Numbers do not always have a clear and unique meaning when applied to natural phenomena. Within quantum mechanics, the state of the physical system with a discrete energy spectrum is usually described in terms of a few "quantum numbers." For example, one of the simplest quantum systems is the so-called harmonic oscillator. This is the quantum analog of a mass point attached to a spring—a particle bound by an attractive force that increases with the distance from the center of force. In quantum mechanics, the energy of the harmonic oscillator is quantized—that is, only integer multiples of an energy unit (a quantum of energy) can occur as the result of an energy measurement. When the harmonic oscillator is in a state with quantum number 5, this means that it has a total energy of 5 energy quanta (ignoring, for simplicity, the ground-state energy). Thus, measuring the energy effectively means to count the number of energy quanta belonging to a certain state of the harmonic oscillator. But the states with definite energy quantum numbers are, in fact, the exception. There are infinitely many others, which are *superpositions* of states with different quantum numbers. One could combine a state with quantum number 3 and a state with quantum number 4 into a new state, whose quantum number remains undetermined prior to an energy measurement (which would sometimes give the result 3 and sometimes the result 4). In the mathematical formalism of quantum mechanics, the energy is therefore not represented by a number (as in classical mechanics), but by a more complicated mathematical object, which is associated with all possible quantum numbers simultaneously. (In mathematical terms: the physical quantity is represented by a "linear operator" and its possible values, the quantum numbers, are represented by the so-called *eigenvalues* of the linear operator. In quantum mechanics, therefore, a physical quantity need not have a definite value but may have many different values at the same time.) In general, the number of energy quanta in a particular state is undetermined, and then, according to the standard interpretation of quantum mechanics, the precise value of the energy is not just unknown to us, but it does not exist as a precise value, only as a probability distribution of the possible values.

Strangely enough, in quantum mechanics of many particle systems, the property "number of particles" is also represented by a linear operator with many different possible values. Hence, the quantum system has states where the number of particles is not determined; a system could have 2 or 3 or 4 particles with equal probability. Creation and annihilation processes occurring with certain probabilities would soon turn a state with n particles into a new state, which is a superposition of states with different particle numbers. In such a state, the precise number of particles of the quantum system is undetermined and the physical system then simply does not have the property of containing a definite number of particles. At the level of elementary particles, the notion of number, thus, seems to lose much of its clarity. It is not even clear what it means to talk of a set of particles if the particles have no individuality and if their number is unsharp and undetermined.

11.11. THE PROBLEM WITH REALLY HUGE NUMBERS

Mathematicians firmly believe that the statement

$$8{,}864{,}759{,}012 + 7{,}938{,}474{,}326 = 16{,}803{,}223{,}338$$

is correct in the same sense as $5 + 3 = 8$, although it appears to be fairly impossible to verify that claim simply by counting.

On the other hand, numbers as large as these appear in economics, and we even have a name for the result: sixteen billion eight hundred three million two hundred twenty-three thousand three hundred thirty-eight. The Gross Domestic Product of the United States was worth 16.8 trillion US dollars in 2013, and this number is still a thousand times larger than the result above. Obviously, numbers like these are handled without any problems in our culture.

The sequence of number words is built according to a system that at least in principle, has no end. It does have practical limits, however, because we tend to run out of names and symbols for extremely large numbers. According to the common system that is used in the United States, we have names for

a billion . . . 1,000,000,000
a trillion . . . 1.000,000,000,000
a quadrillion . . . 1,000,000,000,000,000
a quintillion . . . 1,000,000,000,000,000,000

and so on, with a new name for every three zeros added to the number. The prefixes *bi-, tri-, quadri-*, and so on are derived from Latin number words. With this system, we will eventually arrive at a centillion, which would be a 1 followed by 303 zeros. But these numbers play no role in our life and are typically not named at all. Besides, any naming scheme is only a temporal solution, and one can easily construct an example that is outside the range of names, and then the problem of finding appropriate number words would arise again. How would one name a number like a 1 followed by a quintillion zeros?

The scientific notation uses powers of 10 to describe large numbers. For example, a billion would be 10^9, a trillion 10^{12}, and so on. Here the exponent gives the number of zeros following the leading 1 and describes how often 10 is multiplied by itself in order to produce that number. Hence, for example,

$$1 \text{ billion} = 10^9 = 10 \times 10 \times 10 \times 10 \times 10 \times 10 \times 10 \times 10 \times 10 = 1,000,000,000,$$

and a quintillion would be 10^{18}.

This notation is by far more effective than the attempt to give all numbers an English name. In fact, we can easily write down a number like $10^{10^{18}} = 10^{\left(10^{18}\right)}$, which would be a 1 followed by 10^{18} (a quintillion) zeros. But the numbers in scientific notation are also only approximate. For most practical purposes, the following number would simply be written as 3×10^{15}:

$$3,000,000,000,219,325 = 3.000000000219325 \times 10^{15}.$$

We see from this that for most numbers, in order to describe them exactly, the scientific notation would not really provide a notational abbreviation. In particular, one could generally not describe a number with a quintillion arbitrary digits with scientific notation, except in an approximate sense.

Among the very large numbers, a few have gained special popularity. In their book *Mathematics and the Imagination*, Edward Kasner and James Newman describe extremely large numbers and introduce the name "googol" for the number 10^{100}. It is said that this name was invented by Kasner's nine-year-old nephew in about 1920. Later, the name was used in a slightly changed form to name the Internet search engine Google, thus indicating the huge amount of data on the World Wide Web. A googol is indeed unimaginably large, and the total number of particles in the observable universe is usually estimated to be much lower (putting aside the fact that the number of particles is a rather ill-defined quantity). Yet one can easily define, using mathematical notation, even much larger numbers, such as

$$10^{\text{google}} = 10^{\left(10^{10^{100}}\right)} \quad \text{(which is sometimes called a } googolplex)$$

or even googolplex^{googolplex}, which would be a googolplex multiplied by itself a googolplex times.

What about the meaning of huge numbers like googolplex? They have absolutely no use in counting, because nobody can count that far and there are no collections in the observable physical universe that have nearly that many elements. Obviously, we have a precise algorithmic description for some of these huge numbers, like googol^{googol}, which is 10^{100} taken to the power of 10^{100}. But the numbers having a relatively short description are the exception. A "typically huge" number, which in the usual decimal notation would have, say, about a googolplex digits, is not just a 1 followed by the corresponding number of zeros. Rather, the digits 0, 1, 2 . . . 9 would follow each other in a fairly random manner, and in general there is no rule or notation or "compression algorithm" that could describe all these typical numbers in a shorter way.

Can we say that such a huge number, for which we have no means to properly write it, exists in any reasonable sense? What would we mean by the word *existence*, when there is not even a symbolic representation? There is no collection of concrete objects that this number would represent. If there is a number with about googolplex digits, but if the number of digits is unknown, it would be forever beyond the reach of humanity to determine the *exact* number of its digits. Hence, you could not even distinguish this number from a number that is about a million times as large, because how would you distinguish a number with about googolplex digits from a number with six more digits? You could not do the simplest arithmetic with this number, and of course you could not write it down, because there is not enough matter, space, and time in the universe for that task. Considering the impossibility of realizing such a number, or even describing it exactly, do these large numbers have any meaning? How could we claim that every huge number has a unique successor? Of course, we could just call any huge number by the variable name n, and write "$n + 1$" for its successor. But (except in a few cases) we do not have exact expressions to substitute for the variable name. Hence one could not describe exactly to which concrete number the letter n would refer. Exact huge numbers, in general, are not represented in this universe, not even in symbolic notation. Therefore, nothing in reality or imagination would correspond to the successor of that number n, because nothing in reality or imagination corresponds exactly to n.

The branch of philosophy of mathematics that would *not* accept objects or expressions that nobody can construct in any practical sense is called *ultrafinitism*. According to this view, not even the concept of natural numbers would be accepted without restrictions, and, of course, an ultrafinitist would refuse to talk about infinity. To most mathematicians, this view would be too extreme. Reducing mathematics to finite and not-too-large objects would restrict mathematics and its usefulness in an intolerable way. (And, by the way, how would one define "not too large"?)

Most mathematicians are not particularly worried by the fact that there are natural numbers so huge that they cannot be conceptualized exactly. Typically, when applying numbers to reality, approximate quantities are sufficient, and extremely large numbers would rarely be needed. In theory, the natural numbers are just a sequence whose structure is axiomatically described by the Peano axioms. As a mathematician, one typically does not care about the practical realizability of particular numbers. That every number has a unique successor is simply true by assumption; it needs no practical verification. Mathematicians usually think not in terms of concrete realizations but in terms of rules that are given axiomatically. Mathematics is the art of arguing with some chosen logic and some chosen axioms. As such, it is simply one of the oldest games with symbols and words.

And, moreover, the usefulness of mathematics is by no means limited to finite objects or to those that can be represented with a computer. Mathematical concepts depending on the idea of infinity, like real numbers and differential calculus, are useful models for certain aspects of physical reality.

11.12. SHUT UP AND CALCULATE!

Philosophical questions rarely find generally accepted answers. Also, the question of whether numbers were discovered or invented has never found a definite answer accepted by a clear majority of mathematicians. Probably, there will always be a certain plurality of ideas and approaches, as described in the previous sections.

But for most mathematicians the questions mentioned above have little influence on the actual mathematical practice. Some even have a very skeptical position toward the usefulness of philosophy. In his 1994 book *Dreams of a Final Theory*, American physicist Steven Weinberg (1933–) writes in a chapter called "Against Philosophy" that we should not expect philosophy "to provide today's scientists with any useful guidance about how to go about their work or about what they are likely to find."[22] Indeed, any strict philosophical position could inhibit free and unprejudiced thought and thus make progress more difficult. For example, if you strictly adhered to the ultrafinitist position, you would exclude yourself from most of mathematics—in particular from many branches that are very useful for practical applications.

Many mathematicians therefore consider thoughts about the foundations of their science as a "waste of time." In 2013, in notes for a paper titled "Does Mathematics Need a Philosophy?" British philosopher Thomas Forster (1948–) writes, "Unfortunately most of what passes for Philosophy of Mathematics does not arise from the praxis of mathematics. In fact I even believe that the entirety of the activity of 'Philosophy of Mathematics' as practiced in philosophy departments is—to a first approximation—a waste of time, at least from the point of view of the working mathematician."[23]

The success of mathematics, when applied to the solution of concrete problems, fortunately does not depend on philosophical positions. Even if two mathematicians disagree about the foundations of their science, they would usually agree about the result of a concrete calculation. Whether or not you believe in the independent existence of numbers, a statement like "5 + 3 = 8" remains valid and useful in many concrete situations. All that is important is that the existing framework of mathematics allows us to solve real problems. It is quite a common position that as long as the application of mathematical models is successful, we need no philosophical interpretations. This is called the "shut up and calculate" position. The expression was coined by American physicist David Mermin (1935–), who used it to describe a common attitude of physicists toward philosophical problems with the interpretation of quantum mechanics.

According to Reuben Hersh, most mathematicians seem to oscillate between Platonism and a formalist's point of view. As these two positions are rather incompatible, one can see that philosophy is not a typical mathematician's primary concern. On the other hand, a few sentences could hardly ever be "a full and honest expression of some flesh-and-blood mathematician's view of things," as stated by Barry Mazur, who describes the meandering between philosophical positions and motivations for doing mathematics as follows:

> When I'm working I sometimes have the sense—possibly the illusion—of gazing on the bare platonic beauty of structure or of mathematical objects, and at other times I'm a happy Kantian, marveling at the generative power of the intuitions for setting what an Aristotelian might call the *formal conditions of an object*. And sometimes I seem to straddle these camps (and this represents no contradiction to me). I feel that the intensity of this experience, the vertiginous imaginings, the leaps of intuition, the breathlessness that results from "seeing" but where the sights are of entities abiding in some realm of ideas, and the passion of it all, is what makes mathematics so supremely important for me. Of course, the realm might be illusion. But the experience?[24]

APPENDIX
TABLES

A.1. TABLE OF FIBONACCI NUMBERS

No.	Fibonacci	No.	Fibonacci	No.	Fibonacci	No.	Fibonacci
1	1	21	10946	41	165580141	61	2504730781961
2	1	22	17711	42	267914296	62	4052739537881
3	2	23	28657	43	433494437	63	6557470319842
4	3	24	46368	44	701408733	64	10610209857723
5	5	25	75025	45	1134903170	65	17167680177565
6	8	26	121393	46	1836311903	66	27777890035288
7	13	27	196418	47	2971215073	67	44945570212853
8	21	28	317811	48	4807526976	68	72723460248141
9	34	29	514229	49	7778742049	69	117669030460994
10	55	30	832040	50	12586269025	70	190392490709135
11	89	31	1346269	51	20365011074	71	308061521170129
12	144	32	2178309	52	32951280099	72	498454011879264
13	233	33	3524578	53	53316291173	73	806515533049393
14	377	34	5702887	54	86267571272	74	1304969544928657
15	610	35	9227465	55	139583862445	75	2111485077978050
16	987	36	14930352	56	225851433717	76	3416454622906707
17	1597	37	24157817	57	365435296162	77	5527939700884757
18	2584	38	39088169	58	591286729879	78	8944394323791464
19	4181	39	63245986	59	956722026041	79	14472334024676221
20	6765	40	102334155	60	1548008755920	80	23416728348467685

No.	Fibonacci	No.	Fibonacci
81	37889062373143906	91	4660046610375530309
82	61305790721611591	92	7540113804746346429
83	99194853094755497	93	12200160415121876738
84	160500643816367088	94	19740274219868223167
85	259695496911122585	95	31940434634990099905
86	420196140727489673	96	51680708854858323072
87	679891637638612258	97	83621143489848422977
88	1100087778366101931	98	135301852344706746049
89	1779979416004714189	99	218922995834555169026
90	2880067194370816120	100	354224848179261915075

A.2. TABLE OF THE FIRST PRIME NUMBERS UNDER 10,000

2	3	5	7	11	13	17	19	23	29	31	37
41	43	47	53	59	61	67	71	73	79	83	89
97	101	103	107	109	113	127	131	137	139	149	151
157	163	167	173	179	181	191	193	197	199	211	223
227	229	233	239	241	251	257	263	269	271	277	281
283	293	307	311	313	317	331	337	347	349	353	359
367	373	379	383	389	397	401	409	419	421	431	433
439	443	449	457	461	463	467	479	487	491	499	503
509	521	523	541	547	557	563	569	571	577	587	593
599	601	607	613	617	619	631	641	643	647	653	659
661	673	677	683	691	701	709	719	727	733	739	743
751	757	761	769	773	787	797	809	811	821	823	827
829	839	853	857	859	863	877	881	883	887	907	911
919	929	937	941	947	953	967	971	977	983	991	997
1009	1013	1019	1021	1031	1033	1039	1049	1051	1061	1063	1069
1087	1091	1093	1097	1103	1109	1117	1123	1129	1151	1153	1163
1171	1181	1187	1193	1201	1213	1217	1223	1229	1231	1237	1249
1259	1277	1279	1283	1289	1291	1297	1301	1303	1307	1319	1321
1327	1361	1367	1373	1381	1399	1409	1423	1427	1429	1433	1439

1447	1451	1453	1459	1471	1481	1483	1487	1489	1493	1499	1511
1523	1531	1543	1549	1553	1559	1567	1571	1579	1583	1597	1601
1607	1609	1613	1619	1621	1627	1637	1657	1663	1667	1669	1693
1697	1699	1709	1721	1723	1733	1741	1747	1753	1759	1777	1783
1787	1789	1801	1811	1823	1831	1847	1861	1867	1871	1873	1877
1879	1889	1901	1907	1913	1931	1933	1949	1951	1973	1979	1987
1993	1997	1999	2003	2011	2017	2027	2029	2039	2053	2063	2069
2081	2083	2087	2089	2099	2111	2113	2129	2131	2137	2141	2143
2153	2161	2179	2203	2207	2213	2221	2237	2239	2243	2251	2267
2269	2273	2281	2287	2293	2297	2309	2311	2333	2339	2341	2347
2351	2357	2371	2377	2381	2383	2389	2393	2399	2411	2417	2423
2437	2441	2447	2459	2467	2473	2477	2503	2521	2531	2539	2543
2549	2551	2557	2579	2591	2593	2609	2617	2621	2633	2647	2657
2659	2663	2671	2677	2683	2687	2689	2693	2699	2707	2711	2713
2719	2729	2731	2741	2749	2753	2767	2777	2789	2791	2797	2801
2803	2819	2833	2837	2843	2851	2857	2861	2879	2887	2897	2903
2909	2917	2927	2939	2953	2957	2963	2969	2971	2999	3001	3011
3019	3023	3037	3041	3049	3061	3067	3079	3083	3089	3109	3119
3121	3137	3163	3167	3169	3181	3187	3191	3203	3209	3217	3221
3229	3251	3253	3257	3259	3271	3299	3301	3307	3313	3319	3323
3329	3331	3343	3347	3359	3361	3371	3373	3389	3391	3407	3413
3433	3449	3457	3461	3463	3467	3469	3491	3499	3511	3517	3527
3529	3533	3539	3541	3547	3557	3559	3571	3581	3583	3593	3607
3613	3617	3623	3631	3637	3643	3659	3671	3673	3677	3691	3697
3701	3709	3719	3727	3733	3739	3761	3767	3769	3779	3793	3797
3803	3821	3823	3833	3847	3851	3853	3863	3877	3881	3889	3907
3911	3917	3919	3923	3929	3931	3943	3947	3967	3989	4001	4003
4007	4013	4019	4021	4027	4049	4051	4057	4073	4079	4091	4093
4099	4111	4127	4129	4133	4139	4153	4157	4159	4177	4201	4211
4217	4219	4229	4231	4241	4243	4253	4259	4261	4271	4273	4283
4289	4297	4327	4337	4339	4349	4357	4363	4373	4391	4397	4409
4421	4423	4441	4447	4451	4457	4463	4481	4483	4493	4507	4513
4517	4519	4523	4547	4549	4561	4567	4583	4591	4597	4603	4621

4637	4639	4643	4649	4651	4657	4663	4673	4679	4691	4703	4721
4723	4729	4733	4751	4759	4783	4787	4789	4793	4799	4801	4813
4817	4831	4861	4871	4877	4889	4903	4909	4919	4931	4933	4937
4943	4951	4957	4967	4969	4973	4987	4993	4999	5003	5009	5011
5021	5023	5039	5051	5059	5077	5081	5087	5099	5101	5107	5113
5119	5147	5153	5167	5171	5179	5189	5197	5209	5227	5231	5233
5237	5261	5273	5279	5281	5297	5303	5309	5323	5333	5347	5351
5381	5387	5393	5399	5407	5413	5417	5419	5431	5437	5441	5443
5449	5471	5477	5479	5483	5501	5503	5507	5519	5521	5527	5531
5557	5563	5569	5573	5581	5591	5623	5639	5641	5647	5651	5653
5657	5659	5669	5683	5689	5693	5701	5711	5717	5737	5741	5743
5749	5779	5783	5791	5801	5807	5813	5821	5827	5839	5843	5849
5851	5857	5861	5867	5869	5879	5881	5897	5903	5923	5927	5939
5953	5981	5987	6007	6011	6029	6037	6043	6047	6053	6067	6073
6079	6089	6091	6101	6113	6121	6131	6133	6143	6151	6163	6173
6197	6199	6203	6211	6217	6221	6229	6247	6257	6263	6269	6271
6277	6287	6299	6301	6311	6317	6323	6329	6337	6343	6353	6359
6361	6367	6373	6379	6389	6397	6421	6427	6449	6451	6469	6473
6481	6491	6521	6529	6547	6551	6553	6563	6569	6571	6577	6581
6599	6607	6619	6637	6653	6659	6661	6673	6679	6689	6691	6701
6703	6709	6719	6733	6737	6761	6763	6779	6781	6791	6793	6803
6823	6827	6829	6833	6841	6857	6863	6869	6871	6883	6899	6907
6911	6917	6947	6949	6959	6961	6967	6971	6977	6983	6991	6997
7001	7013	7019	7027	7039	7043	7057	7069	7079	7103	7109	7121
7127	7129	7151	7159	7177	7187	7193	7207	7211	7213	7219	7229
7237	7243	7247	7253	7283	7297	7307	7309	7321	7331	7333	7349
7351	7369	7393	7411	7417	7433	7451	7457	7459	7477	7481	7487
7489	7499	7507	7517	7523	7529	7537	7541	7547	7549	7559	7561
7573	7577	7583	7589	7591	7603	7607	7621	7639	7643	7649	7669
7673	7681	7687	7691	7699	7703	7717	7723	7727	7741	7753	7757
7759	7789	7793	7817	7823	7829	7841	7853	7867	7873	7877	7879
7883	7901	7907	7919	7927	7933	7937	7949	7951	7963	7993	8009
8011	8017	8039	8053	8059	8069	8081	8087	8089	8093	8101	8111

8117	8123	8147	8161	8167	8171	8179	8191	8209	8219	8221	8231
8233	8237	8243	8263	8269	8273	8287	8291	8293	8297	8311	8317
8329	8353	8363	8369	8377	8387	8389	8419	8423	8429	8431	8443
8447	8461	8467	8501	8513	8521	8527	8537	8539	8543	8563	8573
8581	8597	8599	8609	8623	8627	8629	8641	8647	8663	8669	8677
8681	8689	8693	8699	8707	8713	8719	8731	8737	8741	8747	8753
8761	8779	8783	8803	8807	8819	8821	8831	8837	8839	8849	8861
8863	8867	8887	8893	8923	8929	8933	8941	8951	8963	8969	8971
8999	9001	9007	9011	9013	9029	9041	9043	9049	9059	9067	9091
9103	9109	9127	9133	9137	9151	9157	9161	9173	9181	9187	9199
9203	9209	9221	9227	9239	9241	9257	9277	9281	9283	9293	9311
9319	9323	9337	9341	9343	9349	9371	9377	9391	9397	9403	9413
9419	9421	9431	9433	9437	9439	9461	9463	9467	9473	9479	9491
9497	9511	9521	9533	9539	9547	9551	9587	9601	9613	9619	9623
9629	9631	9643	9649	9661	9677	9679	9689	9697	9719	9721	9733
9739	9743	9749	9767	9769	9781	9787	9791	9803	9811	9817	9829
9833	9839	9851	9857	9859	9871	9883	9887	9901	9907	9923	9929
9931	9941	9949	9967	9973							

A.3. TABLE OF ALL KNOWN MERSENNE PRIMES

k	Mersenne Prime 2^k-1	Number of Digits	Year Discovered
2	3	1	antiquity
3	7	1	antiquity
5	31	2	antiquity
7	127	3	antiquity
13	8191	4	1461
17	131071	6	1588
19	524287	6	1588
31	2147483647	10	1750
61	2305843009213693951	19	1883
89	618970019642690137449562111	27	1911
107	162259276829213363391578010288127	33	1913
127	170141183460469231731687303715884105727	39	1876
521	68647976601306097149 . . . 574028291115057151	157	1952
607	53113799281676709868 . . . 835393219031728127	183	1952
1279	10407932194664399081 . . . 710555703168729087	386	1952
2203	14759799152141802350 . . . 419497686697771007	664	1952
2281	44608755718375842957 . . . 133172418132836351	687	1952
3217	25911708601320262777 . . . 160677362909315071	969	1957
4253	19079700752443907380 . . . 034687815350484991	1281	1961
4423	28554254222827961390 . . . 231057902608580607	1332	1961
9689	47822027880546120295 . . . 992696826225754111	2917	1963
9941	34608828249085121524 . . . 426224883789463551	2993	1963
11213	28141120136973731333 . . . 391476087696392191	3376	1963
19937	43154247973881626480 . . . 741539030968041471	6002	1971
21701	44867916611904333479 . . . 410828353511882751	6533	1978

23209	40287411577898877818 . . . 343355523779264511	6987	1979
44497	85450982430363380319 . . . 867686961011228671	13395	1979
86243	53692799550275632152 . . . 857021709433438207	25962	1982
110503	52192831334175505976 . . . 951621083465515007	33265	1988
132049	51274027626932072381 . . . 138578455730061311	39751	1983
216091	74609310306466134368 . . . 336204103815528447	65050	1985
756839	17413590682008709732 . . . 603793328544677887	227832	1992
859433	12949812560420764966 . . . 414267243500142591	258716	1994
1257787	41224577362142867472 . . . 257188976089366527	378632	1996
1398269	81471756441257307514 . . . 532025868451315711	420921	1996
2976221	62334007624857864988 . . . 506256743729201151	895832	1997
3021377	12741168303009336743 . . . 422631973024694271	909526	1998
6972593	43707574412708137883 . . . 366526142924193791	2098960	1999
13466917	92494773800670132224 . . . 073855470256259071	4053946	2001
20996011	12597689545033010502 . . . 714065762855682047	6320430	2003
24036583	29941042940415717208 . . . 436921882733969407	7235733	2004
25964951	12216463006127794810 . . . 933257280577077247	7816230	2005
30402457	31541647561884608093 . . . 134297411652943871	9152052	2005
32582657	12457502601536945540 . . . 752880154053967871	9808358	2006
37156667	20225440689097733553 . . . 340265022308220927	11185272	2008
42643801	16987351645274162247 . . . 101954765562314751	12837064	2009
43112609	31647026933025592314 . . . 022181166697152511	12978189	2008
57885161	58188726623224644217 . . . 141988071724285951	17425170	2013

A.4. TABLE OF ALL KNOWN PERFECT NUMBERS

k	Perfect Number	Number of Digits	Year Discovered
2	6	1	antiquity
3	28	2	antiquity
5	496	3	antiquity
7	8128	4	antiquity
13	33550336	8	1456
17	8589869056	10	1588
19	137438691328	12	1588
31	2305843008139952128	19	1772
61	265845599 . . . 953842176	37	1883
89	191561942 . . . 548169216	54	1911
107	131640364 . . . 783728128	65	1914
127	144740111 . . . 199152128	77	1876
521	235627234 . . . 555646976	314	1952
607	141053783 . . . 537328128	366	1952
1279	541625262 . . . 984291328	770	1952
2203	108925835 . . . 453782528	1327	1952
2281	994970543 . . . 139915776	1373	1952
3217	335708321 . . . 628525056	1937	1957
4253	182017490 . . . 133377536	2561	1961
4423	407672717 . . . 912534528	2663	1961
9689	114347317 . . . 429577216	5834	1963
9941	598885496 . . . 073496576	5985	1963
11213	395961321 . . . 691086336	6751	1963
19937	931144559 . . . 271942656	12003	1971
21701	100656497 . . . 141605376	13066	1978

23209	811537765 ... 941666816	13973	1979
44497	365093519 ... 031827456	26790	1979
86243	144145836 ... 360406528	51924	1982
110503	136204582 ... 603862528	66530	1988
132049	131451295 ... 774550016	79502	1983
216091	278327459 ... 840880128	130100	1985
756839	151616570 ... 565731328	455663	1992
859433	838488226 ... 416167936	517430	1994
1257787	849732889 ... 118704128	757263	1996
1398269	331882354 ... 723375616	841842	1996
2976221	194276425 ... 174462976	1791864	1997
3021377	811686848 ... 022457856	1819050	1998
6972593	955176030 ... 123572736	4197919	1999
13466917	427764159 ... 863021056	8107892	2001
20996011	793508909 ... 206896128	12640858	2003
24036583	448233026 ... 572950528	14471465	2004
25964951	746209841 ... 791088128	15632458	2005
30402457	497437765 ... 164704256	18304103	2005
32582657	775946855 ... 577120256	19616714	2006
37156667	204534225 ... 074480128	22370543	2008
42643801	144285057 ... 377253376	25674127	2009
43112609	500767156 ... 145378816	25956377	2008
57885161	169296395 ... 270130176	34850340	2013

A.5. TABLE OF KAPREKAR NUMBERS

Kaprekar Number	Square of the Number		Decomposition
1	$1^2 =$	1	$1 = 1$
9	$9^2 =$	81	$8 + 1 = 9$
45	$45^2 =$	2,025	$20 + 25 = 45$
55	$55^2 =$	3,025	$30 + 25 = 55$
99	$99^2 =$	9,801	$98 + 01 = 99$
297	$297^2 =$	88,209	$88 + 209 = 297$
703	$703^2 =$	494,209	$494 + 209 = 703$
999	$999^2 =$	998,001	$998 + 001 = 999$
2,223	$2,223^2 =$	4,941,729	$494 + 1,729 = 2,223$
2,728	$2728^2 =$	7,441,984	$744 + 1,984 = 2,728$
4,879	$4,879^2 =$	23,804,641	$238 + 04,641 = 4,879$
4,950	$4,950^2 =$	24,502,500	$2,450 + 2,500 = 4,950$
5,050	$5,050^2 =$	25,502,500	$2,550 + 2,500 = 5,050$
5,292	$5,292^2 =$	28,005,264	$28 + 005,264 = 5,292$
7,272	$7,272^2 =$	52,881,984	$5,288 + 1,984 = 7,272$
7,777	$7,777^2 =$	60,481,729	$6,048 + 1,729 = 7,777$
9,999	$999^2 =$	99,980,001	$9,998 + 0,001 = 9,999$
17,344	$17,344^2 =$	300,814,336	$3,008 + 14,336 = 17,344$
22,222	$22,222^2 =$	493,817,284	$4,938 + 17,284 = 22,222$
38,962	$38,962^2 =$	1,518,037,444	$1,518 + 037,444 = 38,962$
77,778	$77,778^2 =$	6,049,417,284	$60,494 + 17,284 = 77,778$
82,656	$82,656^2 =$	6,832,014,336	$68,320 + 14,336 = 82,656$
95,121	$95,121^2 =$	9,048,004,641	$90,480 + 04,641 = 95,121$
99,999	$99,999^2 =$	9,999,800,001	$99,998 + 000001 = 99.999$
142,857	$142,857^2 =$	20408122449	$20,408 + 122,449 = 142,857$
148,149	$148,149^2 =$	21948126201	$21,948 + 126,201 = 148,149$
181,819	$181,819^2 =$	33058148761	$33,058 + 148,761 = 181,819$
187,110	$187,110^2 =$	35010152100	$35,010 + 152,100 = 187,110$

The next Kaprekar numbers are: 208495, 318682, 329967, 351352, 356643, 390313, 461539, 466830, 499500, 500500, 533170, 857143, . . .

A.6. TABLE OF ARMSTRONG NUMBERS

No.	Digits	Armstrong Number	No.	Digits	Armstrong Number
0	1	0	45	17	35641594208964132
1	1	1	46	17	35875699062250035
2	1	2	47	19	1517841543307505039
3	1	3	48	19	3289582984443187032
4	1	4	49	19	4498128791164624869
5	1	5	50	19	4929273885928088826
6	1	6	51	20	63105425988599693916
7	1	7	52	21	128468643043731391252
8	1	8	53	21	449177399146038697307
9	1	9	54	23	21887696841122916288858
10	3	153	55	23	27879694893054074471405
11	3	370	56	23	27907865009977052567814
12	3	371	57	23	28361281321319229463398
13	3	407	58	23	35452590104031691935943
14	4	1634	59	24	174088005938065293023722
15	4	8208	60	24	188451485447897896036875
16	4	9474	61	24	239313664430041569350093
17	5	54748	62	25	1550475334214501539088894
18	5	92727	63	25	1553242162893771850669378
19	5	93084	64	25	3706907995955475988644380
20	6	548834	65	25	3706907995955475988644381
21	7	1741725	66	25	4422095118095899619457938
22	7	4210818	67	27	121204998563613372405438066
23	7	9800817	68	27	121270696006801314328439376
24	7	9926315	69	27	128851796696487777842012787
25	8	24678050	70	27	174650464499531377631639254
26	8	24678051	71	27	177265453171792792366489765
27	8	88593477	72	29	14607640612971980372614873089
28	9	146511208	73	29	19008174136254279995012734740
29	9	472335975	74	29	19008174136254279995012734741
30	9	534494836	75	29	23866716435523975980390369295
31	9	912985153	76	31	1145037275765491025924292050346
32	10	4679307774	77	31	1927890457142960697580636236639
33	11	32164049650	78	31	2309092682616190307509695338915
34	11	32164049651	79	32	17333509997782249308725103962772
35	11	40028394225	80	33	186709961001538790100634132976990
36	11	42678290603	81	33	186709961001538790100634132976991
37	11	44708635679	82	34	1122763285329372541592822900204593
38	11	49388550606	83	35	12639369517103790328947807201478392
39	11	82693916578	84	35	12679937780272278566303885594196922
40	11	94204591914	85	37	1219167219625434121569735803609966019
41	14	28116440335967	86	38	12815792078366059955099770545296129367
42	16	4338281769391370	87	39	115132219018763992565095597973971522400
43	16	4338281769391371	88	39	115132219018763992565095597973971522401
44	17	21897142587612075			

A.7. TABLE OF AMICABLE NUMBERS

	First Number	Second Number	Year of Discovery
1	220	284	ca. 500 BCE-
2	1184	1210	1860
3	2620	2924	1747
4	5020	5564	1747
5	6232	6368	1747
6	10744	10856	1747
7	12285	14595	1939
8	17296	18416	ca. 1310/1636
9	63020	76084	1747
10	66928	66992	1747
11	67095	71145	1747
12	69615	87633	1747
13	79750	88730	1964
14	100485	124155	1747
15	122265	139815	1747
16	122368	123152	1941/42
17	141664	153176	1747
18	142310	168730	1747
19	171856	176336	1747
20	176272	180848	1747
21	185368	203432	1966
22	196724	202444	1747
23	280540	365084	1966
24	308620	389924	1747
25	319550	430402	1966
26	356408	399592	1921
27	437456	455344	1747
28	469028	486178	1966
29	503056	514736	1747
30	522405	525915	1747
31	600392	669688	1921
32	609928	686072	1747
33	624184	691256	1921
34	635624	712216	1921
35	643336	652664	1747
36	667964	783556	1966
37	726104	796696	1921
38	802725	863835	1966
39	879712	901424	1966
40	898216	980984	1747

41	947835	1125765	1946
42	998104	1043096	1966
43	1077890	1099390	1966
44	1154450	1189150	1957
45	1156870	1292570	1946
46	1175265	1438983	1747
47	1185376	1286744	1929
48	1280565	1340235	1747
49	1328470	1483850	1966
50	1358595	1486845	1747
51	1392368	1464592	1747
52	1466150	1747930	1966
53	1468324	1749212	1967
54	1511930	1598470	1946
55	1669910	2062570	1966
56	1798875	1870245	1967
57	2082464	2090656	1747
58	2236570	2429030	1966
59	2652728	2941672	1921
60	2723792	2874064	1929
61	2728726	3077354	1966
62	2739704	2928136	1747
63	2802416	2947216	1747
64	2803580	3716164	1967
65	3276856	3721544	1747
66	3606850	3892670	1967
67	3786904	4300136	1747
68	3805264	4006736	1929
69	4238984	4314616	1967
70	4246130	4488910	1747
71	4259750	4445050	1966
72	4482765	5120595	1957
73	4532710	6135962	1957
74	4604776	5162744	1966
75	5123090	5504110	1966
76	5147032	5843048	1747
77	5232010	5799542	1967
78	5357625	5684679	1966
79	5385310	5812130	1967
80	5459176	5495264	1967
81	5726072	6369928	1921
82	5730615	6088905	1966
83	5864660	7489324	1967
84	6329416	6371384	1966

85	6377175	6680025	1966
86	6955216	7418864	1946
87	6993610	7158710	1957
88	7275532	7471508	1967
89	7288930	8221598	1966
90	7489112	7674088	1966
91	7577350	8493050	1966
92	7677248	7684672	1884
93	7800544	7916696	1929
94	7850512	8052488	1966
95	8262136	8369864	1966
96	8619765	9627915	1957
97	8666860	10638356	1966
98	8754130	10893230	1946
99	8826070	10043690	1967
100	9071685	9498555	1946
101	9199496	9592504	1929
102	9206925	10791795	1967
103	9339704	9892936	1966
104	9363584	9437056	ca. 1600/1638
105	9478910	11049730	1967
106	9491625	10950615	1967
107	9660950	10025290	1966
108	9773505	11791935	1967

A.8. PYTHAGOREAN TRIPLES WITH A PAIR OF PALINDROMIC NUMBERS

3	4	5
6	8	10
363	484	605
464	777	905
3993	6776	7865
6776	23232	24200
313	48984	48985
8228	69696	70180
30603	40804	51005
34743	42824	55145
29192	60006	66730
25652	55755	61373
52625	80808	96433
36663	616616	617705
48984	886688	888040
575575	2152512	2228137
6336	2509052	2509060
2327232	4728274	5269970
3006003	4008004	5010005
3458543	4228224	5462545
80308	5578755	5579333
2532352	5853585	6377873
5679765	23711732	24382493
4454544	29055092	29394580
677776	237282732	237283700
300060003	400080004	500100005
304070403	402080204	504110405
276626672	458515854	535498930
341484143	420282024	541524145
345696543	422282224	545736545
359575953	401141104	538710545
277373772	694808496	748127700

635191536	2566776652	2644203220
6521771256	29986068992	30687095560
21757175712	48337273384	53008175720
27280108272	55873637855	62177710753
30000600003	40000800004	50001000005
30441814403	40220802204	50442214405
34104840143	42002820024	54105240145

NOTES

CHAPTER 1: NUMBERS AND COUNTING

1. Marvin Minsky, *The Society of Mind* (New York: Simon & Schuster, 1988), p. 192.

2. Paul Auster, *The Music of Chance* (New York: Viking, 1990), p. 73.

3. Bertrand Russell, *Introduction to Mathematical Philosophy* (New York: Macmillan, 1920), 2nd ed., chapter 2. Retrieved from http://www.gutenberg.org/ebooks/41654.

4. It is much more difficult to "count" infinite sets. A definition of the cardinality of infinite sets is beyond the scope of this book.

5. This notion comes from a lengthy discussion in chapter 1 of Georges Ifrah, *The Universal History of Numbers: From Prehistory to the Invention of the Computer* (New York: John Wiley & Sons, 2000).

CHAPTER 2: NUMBERS AND PSYCHOLOGY

1. K. C. Fuson, "Research on Learning and Teaching Addition and Subtraction of Whole Numbers," in *Analysis of Arithmetic for Mathematics Teaching*, ed. G. Leinhardt, R. Putnam, and R. A. Hattrup (Hillsdale, NJ: Lawrence Erlbaum Associates, 1992), p. 63.

CHAPTER 3: NUMBERS IN HISTORY

1. Georges Ifrah, *The Universal History of Numbers: From Prehistory to the Invention of the Computer* (New York: John Wiley & Sons, 2000).

2. Ibid., p. 538.

3. Ibid., p. 414.

CHAPTER 4: DISCOVERING PROPERTIES OF NUMBERS

1. Stanislas Dehaene, *The Number Sense: How the Mind Creates Mathematics*, rev. and updated ed. (New York: Oxford University, 2011), p. 104.

2. Aristotle, *Metaphysics*, trans. W. D. Ross (The Internet Classics Archive), book 1, part 5, http://classics.mit.edu/Aristotle/metaphysics.1.i.html.

3. Euclid, *Euclid's* Elements, trans. and ed. Thomas L. Heath (New York: Dover, 1956), book 7, http://www.perseus.tufts.edu/hopper/text?doc=Perseus:text :1999.01.0086.

4. Georges Ifrah, *The Universal History of Numbers: From Prehistory to the Invention of the Computer* (New York: John Wiley & Sons, 2000), pp. 5–6.

5. Aristotle, *Metaphysics*.

6. Giovanni Reale, *A History of Ancient Philosophy*, trans. and ed. John R. Catan (Albany: State University of New York, 1987), pp. 63–64.

CHAPTER 5: COUNTING FOR POETS

1. James D. McCawley, *The Phonological Component of a Grammar of Japanese* (The Hague: Mouton, 1968).

CHAPTER 9: NUMBER RELATIONSHIPS

1. Leonardo Fibonacci, *The Book of Squares (Liber Quadratorum)*, trans. L. E. Sigler (Orlando, FL: Academic Press, 1987).

CHAPTER 10: NUMBERS AND PROPORTIONS

1. Aristotle, *Metaphysics*, trans. W. D. Ross (The Internet Classics Archive), book 1, part 5, http://classics.mit.edu/Aristotle/metaphysics.1.i.html.

2. Herodotus, *The Histories*, trans. and ed. A. D. Godley (Cambridge, MA: Harvard University Press, 1920), book 2, chapter 124, http://www.perseus.tufts.edu/hopper/text?doc=Perseus:text:1999.01.0126.

CHAPTER 11: NUMBERS AND PHILOSOPHY

1. Charles Hermite, *Correspondance d'Hermite et de Stieltjes*, vol. 2, ed. B. Baillaud and H. Bourget (Paris: Gauthier-Villars, 1905), p. 398.

2. Charles Hermite, quoted in "Notice Historique sur Charles Hermite," in *Éloges académiques et discours*, by G. Darboux (Paris: Hermann, 1912), p. 142.

3. G. H. Hardy, *A Mathematician's Apology*, 19th ed. (Cambridge, UK: Cambridge University Press, 2012), pp. 123–24.

4. E. B. Davies, "Let Platonism Die," *Newsletter of the European Mathematical Society*, June 2007, p. 24.

5. Ibid., p. 25.

6. Barry Mazur, "Mathematical Platonism and Its Opposites," *Newsletter of the European Mathematical Society*, June 2008, p. 19. (Italics in original.)

7. Reuben Hersh, "On Platonism," *Newsletter of the European Mathematical Society*, June 2008, p. 17. (Italics in original.)

8. Richard Dedekind, "The Nature and Meaning of Numbers: Preface to the First Edition, 1887," in *Essays on the Theory of Numbers*, trans. Wooster Woodruff Beman (Chicago: Open Court, 1901), p. 14, http://www.gutenberg.org/ebooks/21016. (Italics in original.)

9. Bertrand Russell, *The Principles of Mathematics*, vol. 1 (Cambridge, UK: Cambridge University Press, 1903), p. xliii.

10. Ibid., p. 111.

11. Eric Temple Bell, *Mathematics: Queen and Servant of Science* (New York: McGraw-Hill, 1951), p. 21.

12. Bertrand Russell, *Introduction to Mathematical Philosophy* (London: George Allen & Unwin, 1919), p. 22, http://www.gutenberg.org/ebooks/41654.

13. Ibid., p. 11.

14. Hermann Weyl, "Mathematics and the Laws of Nature," in *The Armchai Science Reader*, ed. Isabel Gordon and Sophie Sorkin (New York: Simon and Schuste 1959), p. 300.

15. Paul Benacerraf, "What Numbers Could Not Be," in *Philosophy c Mathematics: Selected Readings*, ed. Paul Benacerraf and Hilary Putnam (Cambridg UK: Cambridge University Press, 1964), p. 290.

16. Ibid., p. 291.

17. Heike Wiese, *Numbers, Language, and the Human Mind* (Cambridge, UK Cambridge University Press, 2003), p. 79.

18. Albert Einstein, "Geometry and Experience," in *The Collected Papers c Albert Einstein*, vol. 7, *The Berlin Years*, ed. M. Janssen et al. (Princeton, NJ: Princeto University Press, 2002), p. 385. See also *The Digital Einstein Papers*, http://einsteir papers.press.princeton.edu.

19. Eugene P. Wigner, "The Unreasonable Effectiveness of Mathematics i the Natural Sciences," *Communications in Pure and Applied Mathematics* 13, no. (February 1960), New York: John Wiley & Sons, 1960.

20. Stewart Shapiro, "Philosophy of Mathematics and Its Logic: Introduction, in *The Oxford Handbook of Philosophy of Mathematics and Logic*, ed. Stewart Shapi (Oxford, UK: Oxford University Press, 2007), p. 5.

21. Einstein, "Geometry and Experience," p. 385.

22. Steven Weinberg, *Dreams of a Final Theory* (New York: Random Hous 1993), p. 167.

23. Thomas Forster, "Does Mathematics Need a Philosophy?" (notes for pape presented to the Trinity Mathematical Society on October 21, 2013, https://www.src .ucam.org/tms/talks-archive/).

24. Mazur, "Mathematical Platonism and Its Opposites," p. 20.

INDEX

Ohm, Martin, 299
one-to-one principle. *See* bijection
    principle
On the Calculation with Hindu Numerals
    (al-Khwārizmī), 101
ordering of numbers, 25, 32, 37, 39, 68,
    331. *See also* counting tags; sequence
order-irrelevance principle. *See* invariance
ordinal principle and ordinal numbers, 15,
    18, 20–21, 30, 37–39, 63, 64, 66, 71, 340
    and cardinal numbers, 31–33, 38, 64,
    66, 68, 340
*Oxford Handbook of Philosophy of Math-
    ematics and Logic, The* (Shapiro), 352

Pacioli, Luca, 299
Paganini, B. Nicolò, 248
pairing, 18–19, 110–11
palindromic numbers, 206–12, 228, 237,
    275
Pascal, Blaise, 167, 177
Pascal triangle, 153–81
    and Fibonacci numbers, 150, 156,
    157, 166, 167, 180
    and Pingala's problems, 154–57
    See also Meru Prastara
pattern recognition, 55
patterns, 210–11, 228
    in Pascal triangle, 170–77
    of paving slabs in a garden path, 141
    in Pythagorean triples, 265–67
    and verse meters in poetry, 134–39,
    142
    ways of climbing a staircase, 141–42
Peano axioms, 340–45, 348, 350, 364
    counter examples to, 342, 344
pebbles and counting, 28–31, 37, 39–40,
    97–98, 110

and sticks for larger quantities, 40,
    42–44, 72, 97
pentagonal numbers, 123–24, 126–27
pentagram, 305–306
perfect cube, 272–73
perfect numbers, 231–34, 247
Philolaus of Croton, 108, 109
philosophy and numbers, 329–66
    Pythagorean philosophy of numbers,
    107–10
    See also formalism; intuitionism;
    logicism; Platonism; structuralism
pi (π), 49, 306–17
    decimal expansion of, 307, 309–10,
    311, 312–13, 316
    and dimensions of Great Pyramid,
    322–26, 328
*Pi: A Biography of the World's Most
    Mysterious Number* (Posamentier and
    Lehmann), 317
Piaget, Jean, 52, 65–66
Pica, Pierre, 58
pigeonhole principle, 29
pinecones, number of spirals on, 167–68
Pingala, 133, 135
    solving Pingala's problems, 136–39,
    142–50, 154–57
Pirahã Indians, 58–59
placement of numbers, 183–220
place-value systems, 44–46, 48, 74, 77–84,
    91–95, 102–103
    and the abacus, 97–98
Planck length (shortest observable dis-
    tance), 313
Plato and Platonism, 331–32, 333, 334,
    365–66
plethron as a unit of measure, 322
poetry and counting, 129–60

4.5

Praise for *People Love Dead Jews*

NATIONAL JEWISH BOOK AWARD WINNER

NEW YORK TIMES NOTABLE BOOK OF 2021

CHICAGO PUBLIC LIBRARY BEST BOOK OF THE YEAR

PUBLISHERS WEEKLY BEST BOOK OF THE YEAR

ALA NOTABLE BOOK

KIRKUS PRIZE FINALIST

NATAN NOTABLE BOOK WINNER

"Dara Horn proposes a disturbingly fresh reckoning with an ancient hatred, refusing all categories of victimhood and sentimentality. She offers a passionate display of the self-renewing vitality of Jewish belief and practice. Because antisemitism is a Christian problem more than a Jewish one, Christian readers need this book. It is urgently important."

—James Carroll, author of *The Truth at the Heart of the Lie*

"So necessary. . . . *People Love Dead Jews* is an outstanding book with a bold mission. It criticizes people, artworks and public institutions that few others dare to challenge."

—Yaniv Iczkovits, *New York Times Book Review*

"This is a beautiful book, and in its particular genre—nonfiction meditations on the murder of Jews, particularly in the Holocaust, and the place of the dead in the American imagination—it can have few rivals. In fact, I can't think of any. . . . [O]ne-of-a-kind gorgeousness." —Martin Peretz, *Wall Street Journal*

"This is one of those unexpected, memorable books. . . . [R]iveting, gorgeously written."
—Pamela S. Nadell, *Washington Post*

"How can a book filled with anger, a book about anti-Semitism and entitled *People Love Dead Jews*, be delectable at the same time? The novelist Dara Horn has done it, combining previously published pieces in a work that is far greater than the sum of its parts."
—Elliott Abrams, *Commentary*

"Weaving together history, social science, and personal story, she asks readers to think critically about why we venerate stories and spaces that make the destruction of world Jewry a compelling narrative while also minimizing the current crisis of antisemitism. . . . *People Love Dead Jews* offers no definitive solution to the paradox it unfolds. Horn leaves the reader with several interwoven explanations, each of which lead us to confront the dark reality that Jewish deaths make for a compelling educational narrative, while facing the antisemitism of the present demands a commitment to equality that the world remains unable to embrace."
—Jonathan Fass, *Jewish Book Council*

"*People Love Dead Jews* is, of all things, a deeply entertaining book, from its whopper of a title on. Horn's sarcasm is bracing, reminding us that the politics of Jewish memory often becomes an outrageous marketing of half-truths and outright lies. . . . Horn is a masterful essayist. . . . She has the instincts of a stand-up comic with something deadly serious on her mind."
—David Mikics, *Tablet*

"A superb new essay collection. . . . Horn comes at her subject with a deep grasp of history and a personal commitment to the living Jewish tradition, with an acerbic sense of humour that pops out now and then—and also, refreshingly and necessarily, with anger." —Matti Friedman, *UnHerd*

"Horn is clearly exhausted about thinking about dead Jews, and about antisemitism, and you can feel her emotion through the page. But she channels the emotion to weave together a large [number] of stories—from Russian Jews living in China to Daf Yomi—and what results is a compelling series of essays." —Emily Burack, *Alma*

"The questions and ideas raised by Horn in *People Love Dead Jews* are—like the Yiddish stories she writes about—endless and defiant of neat solutions. But there is comfort to be found, in the most Jewish ways, in her humour and clear-eyed critical thinking." —Keren David, *Jewish Chronicle*

"[A] searing essay collection. . . . Enlivened by Horn's sharp sense of humor and fluid prose, this penetrating account will provoke soul-searching by Jews and non-Jews alike."
—*Publishers Weekly*, starred review

"Brilliantly readable. . . . Readers will be enthralled throughout by the fierce logic of Horn's arguments, novelty of research, black humor, and sharp phrasing. . . . A riveting, radical, essential revision of the stories we all know—and some we don't."
—*Kirkus Reviews*, starred review

"There is an immediacy to her writing that makes it seem as though everything she addresses is happening at once, even though the incidents described may be separated by centuries. . . . [W]restling with Horn's ideas makes for a rich experience. . . . Profound."   —*Booklist*, starred review

"A moving, meditative, well-written book. . . . Horn's writing is personable and engaging from start to finish."
                                                        —*Library Journal*

"Novelist Horn's piercing intellect and caustic wit enliven these meditations. . . . Surveying Holocaust memorials, media coverage of anti-Semitic crimes, Jewish heritage sites in the Chinese city of Harbin, and other topics, Horn punctures shibboleths and provokes genuine soul-searching."
                                    —*Publishers Weekly*, "Best Books of 2021"

"Dara Horn's thoughtful, incisive essays constitute a searing investigation of modern-day antisemitism, in all its disguises and complications. No matter where Horn casts her acute critical eye—from the ruins of the Jewish community in Harbin, China, to the tragedy at Pittsburgh's Tree of Life synagogue—the reports she brings back are at once surprising and enlightening and necessary."
                            —Ruth Franklin, author of *Shirley Jackson* and
                                                        A Thousand Darknesses

"Dara Horn has an uncommon mastery of the literary essay, and she applies it here with a relentless, even furious purpose.

Horn makes well-worn debates—on Anne Frank and Hannah Arendt, for instance—newly provocative and urgent. Her best essays are by turns tragic and comic, and her magnificent mini biography of Varian Fry alone justifies paying the full hardcover price."

—Tom Reiss, Pulitzer Prize–winning author of
The Black Count

"To see what is in front of one's nose needs a constant struggle, George Orwell told us. Dara Horn has engaged that struggle, and in *People Love Dead Jews* she explains why so many prefer the mythologized, dead Jewish victim to the living Jew next door. It's gripping, and stimulating, and it's the best collection of essays I have read in a long, long time."

—Mark Oppenheimer, author of *Squirrel Hill: The Tree of Life Synagogue Shooting and the Soul of a Neighborhood*

ALSO BY DARA HORN

In the Image

The World to Come

All Other Nights

A Guide for the Perplexed

Eternal Life

PEOPLE

LOVE

DEAD

JEWS

Reports from a Haunted Present

DARA HORN

W. W. NORTON & COMPANY
Independent Publishers Since 1923

Copyright © 2021 by Dara Horn

All rights reserved
Printed in the United States of America
First published as a Norton paperback 2022

Essays in this book originally appeared, in different form, in the following
publications: "Everyone's (Second) Favorite Dead Jew" and "Dead Jews of the
Desert" in *Smithsonian*; "Frozen Jews," "Fictional Dead Jews," and passages from
"Executed Jews" in *Tablet*; "Dead American Jews, Part One" and "Dead American
Jews, Part Two" in the *New York Times*; "Executed Jews" in *Jewish Review of Books*;
"Legends of Dead Jews" in *Azure*, as well as in the anthology *Esther in America* (Stuart
Halpern, ed., Maggid Books, 2021, reprinted with permission from Maggid Books);
and "Blockbuster Dead Jews" in the *Atlantic*. "On Rescuing Jews and Others" was
originally published by *Tablet* as an Amazon Kindle Single entitled *The Rescuer*.

For information about permission to reproduce selections from this book, write to
Permissions, W. W. Norton & Company, Inc., 500 Fifth Avenue,
New York, NY 10110

For information about special discounts for bulk purchases, please contact
W. W. Norton Special Sales at specialsales@wwnorton.com or 800-233-4830

Manufacturing by LSC Communications, Harrisonburg
Production manager: Beth Steidle

Library of Congress Cataloging-in-Publication Data

Names: Horn, Dara, 1977– author.
Title: People love dead Jews : reports from a haunted present / Dara Horn.
Description: First edition. | New York : W. W. Norton & Company, 2021. |
Includes bibliographical references and index.
Identifiers: LCCN 2021012209 | ISBN 9780393531565 (hardcover) |
ISBN 9780393531572 (epub)
Subjects: LCSH: Jews—History. | Jews—Public opinion. |
Jews—Persecutions—Public opinion. | Antisemitism—History. |
Death—Political aspects. | Horn, Dara, 1977–
Classification: LCC DS117 .H66 2021 | DDC 909/.04924—dc23
LC record available at https://lccn.loc.gov/2021012209

ISBN 978-1-324-03594-7 pbk.

W. W. Norton & Company, Inc., 500 Fifth Avenue, New York, N.Y. 10110
www.wwnorton.com

W. W. Norton & Company Ltd., 15 Carlisle Street, London W1D 3BS

1 2 3 4 5 6 7 8 9 0

R0463811692

For Maya, Ari, Eli, and Ronen,
who know how to live

CONTENTS

Introduction

IN THE
HAUNTED
PRESENT

SOMETIMES YOUR BODY IS SOMEONE ELSE'S HAUNTED house. Other people look at you and can only see the dead.

I first discovered this at the age of seventeen in the most trivial of moments, at an academic quiz bowl tournament in Nashville, Tennessee—where, as the only girl from my New Jersey high school, I shared a hotel room with two girls from Mississippi. We were strangers and competitors pretending to be friends. One night we stayed up late chatting about our favorite childhood TV shows, about how we had each believed that Mr. Rogers was personally addressing us through the screen. We laughed together until one girl said, "It's like Jesus. Even if he didn't know my name when he was dying on the cross, I still know he loved me, and if he knew my name, he would have loved me too." The other girl squealed, "I know, right? It's just like Jesus!" Then the two of them, full of messianic joy, looked at me.

I said nothing—a very loud nothing. The girls waited, uncomfortable, until one braved the silence. "It seems like people up north are much less religious," she tried. "How often do you go to church?"

It so happened that I was very religious. My family attended synagogue services weekly, or even more often than that; my parents were volunteer lay leaders in our congregation, and I had a job chanting publicly from the Torah scroll for the children's congregation every Saturday morning, which effectively meant that I knew large swaths of the Five Books of Moses in the original Hebrew by heart. On Sundays, I spent four hours learning ancient Jewish legal texts at a program for teenagers at a rabbinical school in New York, and from eight to ten p.m. every Tuesday and Thursday, I studied Hebrew language in a local adult-education class. My public school closed for Rosh Hashanah and Yom Kippur, but my siblings and I also skipped school for holidays like Sukkot, Simchat Torah, Passover, and Shavuot. I read works of Jewish philosophy for fun, tracking medieval and modern arguments about the nature of God. I often privately began and ended my days with traditional Hebrew prayers.

All of this and more required an enormous amount of countercultural effort, education, and commitment on the part of my family that vastly exceeded merely "going to church." But I sensed that this—"this" being the central pillar of my experience as a human being—was irrelevant to the question these girls were asking me. I mumbled something about a synagogue and tried to think of a way to steer us back to Mr. Rogers. But now the girls were staring at me, gaping in disbelief.

"You," one of the girls stammered, "you—you have blond hair!"

The second girl inspected me, squinting at my face in a way that made me wonder if I had acne. "And what color are your eyes?"

"Blue," I said.

The first girl said, "I thought Hitler said you all were dark."

In retrospect I can imagine many ways I might have felt about this statement, but at the time I was only baffled. I pictured my hand on the quiz-bowl buzzer I'd been pounding all week, and provided the correct answer: "Hitler was full of shit."

After a pause that lasted an eternity, one girl meekly offered, "I guess you're kind of right." *Kind of.* The other girl doubled down, demanding an explanation for my eye color if I were "from the Middle East." But I was done being nice, if being nice meant defending my own face. I left the room, confused.

That night I blurted to my mother from a hotel pay phone, "I don't get it. These girls made it to the nationals. These are the smart people! And they're getting their information from Hitler?"

My mother sighed, a long, tired sigh. "I know," she said, without elaborating. "I know."

My mother was the age then that I am now. And now I know too.

Those girls were not stupid, and probably not even bigoted. But in their entirely typical and well-intentioned education, they had learned about Jews mainly because people had killed Jews. Like most people in the world, they had only encountered dead Jews: people whose sole attribute was that they had been murdered, and whose murders served a clear purpose, which was *to teach us something*. Jews were people who, for moral and educational purposes, were supposed to be dead.

It took me many years to understand that those girls were not entirely wrong to look at me and think only of a terrifying past. I often felt haunted too.

When I was a child, I had a question that burned within me, and Judaism seemed to answer it. My question was about the nature of time.

I was obsessed with the unspoken and unnerving problem of being trapped in an eternal and inescapable present. As I got into bed each night, I would lie in the dark and wonder: *This day that just ended is gone now. Where did it go?* If I were a character in one of my novels, I would give the character a motivation for that constant longing. But I am not a fictional character. I had few words then for this deep sense of loss, nor any clear reason for it, other than that my mother came from a long line of women who had all died young, and her mourning for her own mother was something I observed and absorbed, without knowing what she or I was missing. The secondary nature of this grief only underscored the inexpressible feeling I had of arriving too late. With nothing to mourn, I nonetheless felt as though something were constantly flowing out of my life, just beneath its relentlessly cheerful surface. When I began writing as a child, my driving force was not the urge to invent stories but the urge to stop time, to preserve those disappearing days. I kept journals that were more like reporter's notebooks, taking minutes on even the most boring events for no reason other than to lock them down on paper. It did not occur to me that most people were not concerned with this problem. It did not occur to me because in my family's religious practice, I found many thousands of years' worth of people who shared my obsession with this problem—and who had, to my child's mind, succeeded in solving it.

One of America's many foundational legends is that it doesn't matter who your parents are, or who their parents were, or where you came from—that what matters is what you do now with the opportunities this country presents to you, and this is what we call the American dream. The fact that this legend is largely untrue does not detract from its power; legends are not reports on reality but expressions of a culture's values and aspirations. Judaism, too, has many foundational legends, and all of them express exactly the opposite of this idea. Ancient rabbinic tradition insists that it was not merely our ancestors who were liberated from Egyptian slavery, but that *we ourselves* were also personally freed by God. When God gave the Israelites the laws of the Torah at Mount Sinai, this tradition teaches, it was not merely that generation of Israelites who were present, but all of their future descendants—both biological and spiritual—stood with them at Sinai. In America, time was supposed to be a straight line where only the future mattered; in Judaism, it was more like a spiral of a spiral, a tangled old telephone cord in which the future was the present, which was essentially the past.

This profound difference between these two sides of my identity was not abstract or subtle; it was obvious even to a child. In public school, my classmates and I pledged allegiance to the flag and aspired to form a more perfect union, fully invested in America's future. But when we learned about the past—Pilgrims and Native Americans, Patriots and Tories, Yankees and Rebels—there was no "we." In Hebrew school and in the traditional texts I read in synagogue and at home, it was just the opposite. The Hebrew Bible was never discussed in historical context, because we were the historical context. It was our present, and in my family's religious life, it was treated that way. The creation of the world recurred every week at our Sabbath table, where we chanted Hebrew biblical passages about God resting on the seventh day and sang long medieval Hebrew poems about divine creativity in multipart harmony. On Passover, we ate the same matzah we'd been eating for millennia, still unable to find the time to let our bread rise during our flight to freedom. Every New Year, Abraham once more drew his knife to his son Isaac's throat, holding our future hostage, fate and free will bound together in a double helix that caught us in its grip.

When I went to Israel for the first time at the age of nine, I was stunned to discover that there was an actual answer to my question about where those disappearing days had gone: they were underground. The first time I entered Jerusalem's Old City, I walked down a flight of stairs that began on the current street level; at the bottom, I was stunned to step onto the paving stones from the street level during the Roman period, as though I had traveled through time instead of rock. The city itself was a kind of *tel*, an archaeological mound with layers of past centuries piled one on the other, some of which were preserved and exposed. As I grew older, I discovered that people had these layers too—that all people had those vanished days within them, whether or not they knew it. There was an alternative to being trapped in the present: a deep consciousness of memory that transcended any one person or lifetime. I dove into this possibility, body and soul. My studies of Hebrew— the language itself had layers, from airheaded TV shows all the way down to biblical bedrock—led to studies of Yiddish, which led to a doctorate in both, which flowed into my fiction. I wrote my way down into that tunnel through time, burrowing deeper into a past that was in fact the present, a breathing reality just beneath the surface of the current moment, until I was no longer afraid to fall asleep.

This, to me, was what being Jewish meant, the gift it gave me in the wonderland of a country that long ago gave my family a future. But as I slowly came to understand, this was not what it meant to people who weren't Jewish, or even to many Jews with little education in the culture. What Jewish identity meant to those people, it turned out, was simply a state of nonbeing: not being Christian or Muslim or whatever else other people apparently were (in Britain, for instance, more people identify as Jedis than as Jews), being alienated, being marginalized, or best of all, being dead. As thousands of Holocaust books and movies and TV shows and lectures and courses and museums and mandatory school curricula made abundantly clear, dead Jews were the most popular of all.

For most of my adult life, I had no reason to recall that moment in the Nashville hotel. I had filed it deep in my brain, in the same mental sock drawer where I kept the high schoolers from the adjacent town who cheered for my school's soccer team to "go to the gas," or the student in the first college class I ever taught who refused to read an assigned 1933 Hebrew novel because Hebrew was "racist," or the roommate who sobbed uncontrollably while informing me that I was going to hell. (I reassured her that at least I would know a lot of people there.) These incidents were oddities, weird and even laughable. They weren't my normal, or the normal of anyone I knew.

More than twenty-five years later, they still aren't my normal, though they are now the normal of more than a few people I know. But in recent years I have had the misfortune of discovering the deep vein of normalcy that runs beneath these oddities, which is shared by seemingly good-faith cultural enterprises like Holocaust museums, canonical Western literature, and the elaborate restoration of Jewish historical sites as far away as China. I began to notice a certain gaslighting about the Jewish past and present that I had never seen before, even when it was right in front of me. I had mistaken the enormous public interest in past Jewish suffering for a sign of respect for living Jews. I was very wrong.

This fact should have been obvious to me from the beginning of my writing career, when my most acclaimed early published piece, the one nominated for a major award, wasn't the one about Jewish historical sites in Spain but rather the one about death camps. I made a point of resisting this reality, asking people at my public talks if they could name three death camps, and then asking the same people if they could name three Yiddish authors—the language spoken by over 80 percent of death-camp victims. What, I asked, was the point of caring so much about how people died, if one cared so little about how they lived? At the time, I did not appreciate how deep the obsession with dead Jews went, how necessary it was to so many people's unarticulated concept of civilization, to their unarticulated concept of themselves. But as our current century wore on and public conversations about Israel became increasingly toxic—far beyond any normal political concern—and as public conversations about observant Jews took on the same tone, I came to recognize the mania for dead Jews as something deeply perverse, and all the more so when it wore its goodwill on its sleeve. I dealt with this perversity in the most honorable way possible: by avoiding it.

For a writer and scholar of Jewish history and literature, this was challenging, because it meant avoiding the subjects my readers and students clearly loved most. Still, I tried. I wrote novels about Jewish spies during the Civil War, about a medieval Hebrew archive in Cairo, about Soviet Yiddish Surrealists, about a woman born in ancient Jerusalem who couldn't manage to die. In my university courses and lectures, I emphasized the unprecedented revival of Hebrew, the evolving patterns of Israeli fiction, the growth of modern Yiddish poetry and drama out of traditional art forms, the complex internal religious debates that shaped secular writers' works generations later. I fought hard to keep everything as autonomous as possible, making sure to tell the stories of how Jews had lived and what they had lived for, rather than how they had died. As I insisted to my Nashville roommates long ago, I was not that dark.

But the past kept seeping into the present. By the end of 2018, after a massacre of Jews in our more perfect union that hardly came from nowhere, the only thing my readers, students, colleagues, and editors wanted me to talk about was dead Jews. I became the go-to person for the emerging literary genre of synagogue-shooting op-eds—a job I did not apply for, but one that I accepted out of fear of what someone less aware of history might write instead. Even outside of those news-headline incidents, I found myself asked, again and again, for my opinions on dead Jews. Perhaps I was expected to approach the subject with a kind of piety, an attitude that would generate some desperately needed hope and grace. After all, I was a living Jew (a writer, a religious person, even a Hebrew and Yiddish scholar), so I was clearly equipped to say something decorous and inspiring, something sad and beautiful that would flatter everyone involved.

I couldn't do it. I was too angry. My children were growing up in an America very different from the one I'd grown up in, one where battling strangers' idiocies consumed large chunks of brain space and where the harassment and gaslighting of others—encounters like those I'd once buried in my mental sock drawer—were not the exception but the rule. My efforts to prove a negative—that we weren't all dark—had failed, overwhelmed by the reality of being part of a ridiculously small minority that nonetheless played a behemoth role in other people's imaginations.

So instead of avoiding and rejecting this haunted-house world, where my family's identity was defined and determined by the opinions and projections of others, I decided to lean directly into that distorted public looking glass and report what I found there: to unravel, document, describe, and articulate the endless unspoken ways in which the popular obsession with dead Jews, even in its most apparently benign and civic-minded forms, is a profound affront to human dignity. I wish I did not feel the need to do this. But I want my children, and your children, to know.

This book explores the many strange and sickening ways in which the world's affection for dead Jews shapes the present moment. I hope you will find it as disturbing as I do.

December 2020

PEOPLE

LOVE

DEAD

JEWS

Chapter 1

EVERYONE'S
(SECOND) FAVORITE
DEAD JEW

PEOPLE LOVE DEAD JEWS. LIVING JEWS, NOT SO MUCH.

This disturbing idea was suggested by an incident in 2018 at the Anne Frank House, the blockbuster Amsterdam museum built out of Frank's "Secret Annex," or in Dutch, "Het Achterhuis [The House Behind]"—a series of tiny hidden rooms where the teenage Jewish diarist lived with her parents, her sister, and four other persecuted Jews for over two years before being captured by Nazis and deported to Auschwitz in 1944. Here's how much people love dead Jews: Anne Frank's diary, first published in Dutch in 1947 via her surviving father, Otto Frank, has been translated into seventy languages and has sold more than 30 million copies worldwide, and the Anne Frank House now hosts well over a million visitors each year, with reserved tickets selling out months in advance. But when a young employee at the Anne Frank House tried to wear his yarmulke to work, his employers told him to hide it under a baseball cap. The museum's goal was "neutrality," one spokesperson explained to the British newspaper *Daily Mail*, and a live Jew in a yarmulke might "interfere" with the museum's "independent position." The museum finally relented after deliberating for four months, which seems like a rather long time for the Anne Frank House to ponder whether it was a good idea to force a Jew into hiding.

One could call this a simple mistake, except that it echoed a similar incident the previous year, when visitors noticed a discrepancy in the museum's audio-guide displays. Each audio-guide language was represented by a national flag—with the exception of Hebrew, which was represented only by the language's name in its alphabet. The display was eventually corrected to include the Israeli flag.

These public-relations mishaps, clumsy though they may have been, were not really mistakes, nor were they even the fault of the museum alone. On the contrary: these instances of concealed Jewish identity are the key to the runaway success of Anne Frank's diary and fame. This sort of hiding was an essential part of the diary's original publication, in which several direct references to Jewish practice were edited away. They were also part of the psychological legacy of Anne Frank's parents and grandparents, German Jews for whom the price of admission to Western society was assimilation, hiding their differences by accommodating and ingratiating themselves to the culture that ultimately sought to destroy them. That price lies at the heart of Anne Frank's endless appeal. After all, Anne Frank had to hide her identity so much that she was forced to spend two years in a closet rather than breathe in public. And that closet, hiding place for a dead Jewish girl, is what millions of visitors want to see.

———

Surely there is nothing left to say about Anne Frank, except that there is everything left to say about her: all the books she never lived to write. For she was unquestionably a talented writer, possessed of both the ability and the commitment that real literature requires. Quite the opposite of how the influential Dutch historian Jan Romein described her work in April of 1946, in his article in the newspaper *Het Parool* that spurred her diary's publication—a "diary by a child, this *de profundis* stammered out in a child's voice"—Frank's diary was not the work of a naif, but rather of a writer already planning future publication. Frank had begun the diary casually, but soon sensed its potential. Upon hearing a radio broadcast in March of 1944 calling on Dutch civilians to preserve diaries and other personal wartime documents, she immediately began to revise two years of previous entries, with a title (*Het Achterhuis*, or *The House Behind*) already in mind, along with pseudonyms for the hiding place's residents. Nor were her revisions simple corrections or substitutions. They were thoughtful edits designed to draw the reader in, intentional and sophisticated. Her first entry in the original diary, for instance, begins with a long description of her birthday gifts (the blank diary being one of them), an entirely unself-conscious record by a thirteen-year-old girl. The first entry in her revised version, on the other hand, begins with a deeply self-aware and ironic pose: "It's an odd idea for someone like me to keep a diary; not only because I have never done so before, but because it seems to me that neither I—nor for that matter anyone else—will be interested in the unbosomings of a thirteen-year-old schoolgirl."

The innocence here is all affect, carefully achieved. Imagine writing this as your second draft, with a clear vision of a published manuscript; this is hardly the mind of a "stammering" child. In addition to the diary, Frank also worked hard on her stories, or as she proudly put it, "my pen-children are piling up." Some of these were scenes from her life in hiding, but others were entirely invented: stories of a poor girl with six siblings, or a dead grandmother protecting her orphaned grandchild, or a novel-in-progress about star-crossed lovers featuring multiple marriages, depression, a suicide, and prophetic dreams. Already wary of a writer's pitfalls, she noted, "It isn't sentimental nonsense for it's modeled on the story of Daddy's life."

"I am the best and sharpest critic of my own work," she wrote a few months before her arrest. "I know myself what is and what is not well written."

What is and what is not well written: it is likely that Frank's opinions on this subject would have evolved if she had had the opportunity to age. Reading the diary as an adult, one sees the limitations of a teenager's perspective, and longs for more. In one entry, Frank describes how her father's business partners—now her family's protectors—hold a critical corporate meeting in the office below the family's hiding place. Her father, she, and her sister discover that they can hear what is said by lying down with their ears pressed to the floor. In Frank's telling, the episode is a comic one; she gets so bored that she falls asleep. But adult readers cannot help but ache for her father, a man who clawed his way out of bankruptcy to build a business now stolen from him, reduced to lying facedown on the floor just to overhear what his subordinates might do with his life's work. When Frank complains about her insufferable middle-aged roommate Fritz Pfeffer (Albert Dussel, per Frank's pseudonym) taking his time on the toilet, adult readers might empathize with him as the only single adult in the group, permanently separated from his non-Jewish life partner whom he could not marry due to antisemitic laws. Readers Frank's age connect with her budding romance with fellow hidden resident Peter van Pels (renamed Peter van Daan), but adults might wonder how either of the married couples in the hiding place managed their own relationships in confinement with their children. More broadly, readers Frank's age relate to her constant complaints about grown-ups and their pettiness, but adults are equipped to appreciate these grown-ups' psychological devastation, how they endured not only their physical deprivation, but the greater blow of being reduced to a childlike state of dependence on the whims of others.

Frank herself sensed the limits of the adults around her, writing critically of her own mother's and Peter's mother's apparently trivial preoccupations—and in fact these women's circumstances, not only their wartime deprivation but their prewar lives as housewives, were a chief driver for Frank's ambitions. "I can't imagine that I would have to lead the same sort of life as Mummy and Mrs. v.P. [van Pels] and all the women who do their work and are then forgotten," she wrote. "I must have something besides a husband and children, something that I can devote myself to!" In the published diary, this passage is immediately followed by the famous words, "I want to go on living even after my death!"

By plastering this sentence on Frank's book jackets, publishers have implied that through her posthumous fame, the writer's dreams were achieved. But when we consider the writer's actual ambitions, it is obvious that her dreams were in fact destroyed—and that the writer who would have emerged from Frank's experience would not be anything like the writer Frank herself originally planned to become. Imagine this obituary of a life unlived:

Anne Frank, noted Dutch novelist and essayist, died this past Wednesday at her home in Amsterdam. She was 92.

A survivor of Auschwitz and Bergen-Belsen, Frank's acclaim was hard-won. In her twenties, Frank struggled to find a publisher for her first book, The House Behind, a memoir of her experiences in hiding and in Nazi concentration camps. Disfigured by a brutal beating, Frank rarely granted interviews; her later work, The Return, describes how her father did not recognize her upon their reunion in 1945.

Frank supported herself as a journalist, and in 1961 she earned notoriety for her fierce reporting on the Israeli capture of Nazi henchman Adolf Eichmann, an extradition via kidnapping that the European elite condemned. After covering Eichmann's Jerusalem trial for the Dutch press, Frank found the traction to publish Margot, a novel that imagined her sister living the life she once dreamed of, as a midwife in the Galilee. A surreal work that breaks the boundaries between novel and memoir, and leaves ambiguous which of its characters are dead or alive, the Hebrew translation of Margot became a runaway bestseller, while an English-language edition eventually found a small but appreciative audience in the United States.

Frank's subsequent books and essays brought her renown as a clear-eyed prophet carefully attuned to hypocrisy. Her reputation for relentless conscience, built on her many investigative articles on subjects ranging from Soviet oppression to Arab-Israeli wars, was cemented by her internationally acclaimed 1984 book Every House Behind, written after her father's death. Beginning with an homage to her father's unconditional devotion, the book progresses into a searing and accusatory work that reimagines her childhood hiding place as a metaphor for Western civilization, whose façade of high culture concealed a demonic evil. "Every flat, every house, every office building in every city," she wrote, "they all have a House Behind."

Her readers will long remember the words from her
first book, quoted from a diary she kept at 15: "I don't
believe that the big men are guilty of the war, oh no,
the little man is just as guilty, otherwise the peoples of
the world would have risen in revolt long ago! There's
in people simply an urge to destroy, an urge to kill,
to murder and rage, and until all mankind with-
out exception undergoes a great change, wars will be
waged, everything that has been built up, cultivated
and grown will be cut down and disfigured, and man-
kind will have to begin all over again."

Her last book, a memoir, was titled To Begin Again.

The problem with this hypothetical, or any other hypothetical, about Frank's nonexistent adulthood isn't just the
impossibility of knowing how Frank's life and career might
have developed. The problem is that the entire appeal of Anne
Frank to the wider world—as opposed to those who knew
and loved her—lay in her lack of a future.

There is an exculpatory ease to embracing this "young girl,"
whose murder is almost as convenient for her many enthusias-
tic readers as it was for her persecutors, who found unarmed
Jewish children easier to kill off than the Allied infantry. After
all, an Anne Frank who lived might have been a bit upset at
her Dutch betrayers, still unidentified, who received a reward
for each Jew they turned in of approximately $1.40. An Anne
Frank who lived might not have wanted to represent "the children of the world"—particularly since so much of her diary
is preoccupied with a desperate plea to be taken seriously, to

not be perceived as a child. Most of all, an Anne Frank who lived might have told people about what she saw at Westerbork, Auschwitz, and Bergen-Belsen, and people might not have liked what she had to say.

And here is the most devastating fact of Frank's posthumous success, which leaves her real experience forever hidden: we know what she would have said, because other people have said it, and we don't want to hear it.

The line most often quoted from Frank's diary are her famous words, "I still believe, in spite of everything, that people are truly good at heart." These words are "inspiring," by which we mean that they flatter us. They make us feel forgiven for those lapses of our civilization that allow for piles of murdered girls—and if those words came from a murdered girl, well, then, we must be absolved, because they must be true. That gift of grace and absolution from a murdered Jew (exactly the gift that lies at the heart of Christianity) is what millions of people are so eager to find in Frank's hiding place, in her writings, in her "legacy." It is far more gratifying to believe that an innocent dead girl has offered us grace than to recognize the obvious: Frank wrote about people being "truly good at heart" before meeting people who weren't. Three weeks after writing those words, she met people who weren't.

Here's how much some people dislike living Jews: they murdered 6 million of them. This fact bears repeating, as it does not come up at all in Anne Frank's writings. Readers of her diary are aware that the author was murdered in a genocide, but this does not mean that her diary is a work about genocide. If it were, it is unlikely that it would have been any-where near as universally embraced.

We know this, because there is no shortage of writings from victims and survivors who chronicled this fact in vivid detail, and none of those documents have achieved anything like Frank's diary's fame. Those that have come close have only done so by observing those same rules of hiding, the ones that insist on polite victims who don't insult their persecu-tors. The work that came closest to achieving Frank's interna-tional fame might be Elie Wiesel's *Night*, a memoir that could be thought of as a continuation of Frank's diary, recounting the tortures of a fifteen-year-old imprisoned in Auschwitz. As the scholar Naomi Seidman has discussed, Wiesel first pub-lished his memoir in Yiddish, under the title *And the World Was Silent*. The Yiddish book told the same story told in *Night*, but it exploded with rage against his family's murderers and, as the title implies, the entire world whose indifference (or active hatred) made those murders possible. With the help of the French Catholic Nobel laureate François Mauriac, Wie-sel later published a French version under the new title *La Nuit*—a work that repositioned the young survivor's rage into theological angst. After all, what reader would want to hear about how his society had failed, how he was guilty? Better to blame God. This approach earned Wiesel a Nobel Peace Prize, as well as, years later, selection for Oprah's Book Club, the American epitome of grace. It did not, however, make teen-age girls read his book in Japan, the way they read Frank's. For that he would have had to hide much, much more.

What would it mean for a writer not to hide this horror? There is no mystery here, only a lack of interest. You have probably never heard of another young murdered Jewish chronicler of the same moment, Zalmen Gradowski. Like Frank's, Gradowski's work was written under duress, and discovered only after his death—except that Gradowski's work was written in Auschwitz.

Gradowski, a young married man whose entire family was murdered, was one of the Jewish prisoners in Auschwitz's Sonderkommando: those forced to escort new arrivals into the gas chambers, haul the newly dead bodies to the crematoria, extract any gold teeth, and then burn them. He reportedly maintained his religious faith, reciting the Kaddish (mourner's prayer) each evening for the souls of the thousands of people whose bodies he burned that day—including Peter van Pels's father, who was gassed upon the group's arrival in Auschwitz on September 6, 1944. Gradowski recorded his experiences in Yiddish and buried the documents, which were discovered after the war; he himself was killed on October 7, 1944, in a Sonderkommando revolt he had organized that lasted only one day.

"I don't want to have lived for nothing like most people," Frank wrote in her diary. "I want to be useful or give pleasure to the people around me who don't yet know me, I want to go on living even after my death!" Gradowski, too, wrote with a purpose. But Gradowski's goal wasn't personal or public fulfillment. His was truth: searing, blinding prophecy, Jeremiah lamenting a world aflame.

"It may be that these, the lines that I am now writing, will be the sole witness to what was my life," Gradowski writes.

"But I shall be happy if only my writings should reach you, citizen of the free world. Perhaps a spark of my inner fire will ignite within you, and even should you sense only part of what we lived for, you will be compelled to avenge us—avenge our deaths! Dear discoverer of these writings! I have a request for you: this is the real reason why I write, that my doomed life may attain some meaning, that my hellish days and hopeless tomorrows may find a purpose in the future." And then Gradowski tells us what he has seen.

Gradowski's chronicle walks us, step by devastating step, through the murders of five thousand people, a single large "transport" of Czech Jews who were slaughtered on the night of March 8, 1944—a group that was unusual only because they had already been detained in Auschwitz for months, and therefore knew what was coming. Gradowski tells us how he escorted the thousands of women and young children into the disrobing room, marveling at how "these same women who now pulsed with life would lie in dirt and filth, their pure bodies smeared with human excrement." He describes how the mothers kiss their children's limbs, how sisters clutch each other, how one woman asks him, "Say, brother, how long does it take to die? Is it easy or hard?" Once the women are naked, Gradowski and his fellow prisoners escort them through a gauntlet of SS officers who had gathered for this special occasion—a night gassing arranged intentionally on the eve of Purim, the biblical festival celebrating the Jews' narrow escape from a planned genocide. He recalls how one woman, "a lovely blond girl," stopped in her death march to address the officers: "'Wretched murderers! You look at me with your thirsty, bestial eyes. You glut yourselves on my nakedness. Yes, this is what you've been waiting for. In your civilian lives you could never even have dreamed about it. [. . .] But you won't enjoy this for long. Your game's almost over, you can't kill all the Jews. And you will pay for it all.' And suddenly she leaped at them and struck Oberscharfuhrer Voss, the director of the crematoria, three times. Clubs came down on her head and shoulders. She entered the bunker with her head covered with wounds [. . .] she laughed for joy and proceeded calmly to her death." Gradowski describes how people sang in the gas chambers, songs that included "Hatikvah" (The Hope), now the national anthem of Israel. And then he describes the mountain of open-eyed naked bodies that he and his fellow prisoners had to pull apart and burn: "Their gazes were fixed, their bodies motionless. In the deadened, stagnant stillness there was only a hushed, barely audible noise—a sound of fluid seeping from the different orifices of the dead. [. . .] Frequently one recognizes an acquaintance." In the specially constructed ovens, he tells us, the hair is first to catch fire, but "the head takes the longest to burn; two little blue flames flicker from the eyeholes—these are the eyes burning with the brain. [. . .] The entire process lasts twenty minutes—and a human being, a world, has been turned to ashes. [. . .] It won't be long before the five thousand people, the five thousand worlds, will have been devoured by the flames."

Gradowski was not poetic; he was prophetic. He did not gaze into this inferno and ask why. He knew. Aware of both the long recurring arc of destruction in Jewish history, and of the universal fact of cruelty's origins in feelings of worthlessness, he writes: "This fire was ignited long ago by the barbarians and murderers of the world, who had hoped to drive darkness from their brutal lives with its light."

One can only hope that we have the courage to hear this truth without hiding, to face the fire and to begin again.

Chapter 2

FROZEN
JEWS

ONE OF MY STRANGE AND VIVID MEMORIES FROM MY
first trip to Israel, when I was nine years old, is of a brief
cartoon I watched at the Diaspora Museum in Tel Aviv.
The cartoon described the travels of Benjamin of Tudela, a
twelfth-century Spanish Jewish merchant who documented
his six-year journey traversing the known world, across the
Mediterranean to Turkey, Israel, Egypt, Babylonia, and Persia,
and reporting on India and China, staying with Jewish communities in each place and sharing crowded boats and wagons
in between. The Diaspora Museum has since been revamped
and rebranded as the Museum of the Jewish People, but in 1986
it was a dark and openly depressing place, its dour displays
about Jewish communities around the world all leading to a
"Scrolls of Fire" atrium describing how the hapless Jews in
these communities were either expelled or burned alive.

But the cartoon was bright and curious. Benjamin was a ridiculous bowling-pin figure with googly eyes, bobbing across the screen and cheerfully reporting on thriving Jewish communities around the world—the Jews in France who inexplicably lived in a castle, the Jews in Babylonia who had their own googly-eyed king, the Jews in Yemen who joined local Arab armies and stampeded with them in a cloud of dust, the Jews in Syria who pacified wiggly-eyebrowed assassins by offering free silk scarves. For reasons I could not articulate at the age of nine, I was utterly enchanted.

I feel that same enchantment now when I am seduced by the travel industry's branding of the world as an amazing place full of welcoming people who, beneath it all, are actually the same. In reality, the more time I have spent in any of the fifty-plus countries I have visited as a tourist, the more I notice the differences between myself and the inhabitants, and the more alienated, uncomfortable, and anxious I become. Yet colorful photos of exotic places on TripAdvisor lure me every time.

So I was eager to make my way to a city called Harbin in a remote province of northeastern China, south of Siberia and north of North Korea, where the temperature hovers around minus 35 Celsius for much of the year, and where every winter, over ten thousand workers construct an entire massive city out of blocks of ice. The Harbin Ice Festival dwarfs similar displays in Canada and Japan by orders of magnitude, its enormous ice buildings laced through with LED lighting and sometimes replicating famous monuments at or near life-size. It attracts over 2 million visitors a year; it needs to be seen to be believed. As I considered a trip to Harbin, my mindless travel-industry scrolling took me to a list of other local tourist attractions, including synagogues.

Yes, synagogues. Plural. And then I discovered something deeply strange: the city of Harbin was built by Jews.

———

Jews have lived in China for more than a thousand years, which is as long as they have lived in Poland. But the story of the Jews of Harbin, and of Harbin itself, begins with the railroad—because before the railroad, Harbin did not exist.

Like most Chinese cities you've never heard of, Harbin today is larger than New York, with a population around 16 million. But as late as 1896, there was only a cluster of small fishing villages around a bend in a river. That year Russia received a concession from China to build part of the Trans-Siberian Railroad through Manchuria—the traditional name for the vast, frigid, and, at that time, barely populated region of northeastern China. Building this route would shave precious time off the trip from Moscow to Vladivostok. The route would also include a branch line deeper into China, requiring a large administrative center at the junction—essentially, a town. Mikhail Gruliov, a Jew who had converted to Russian Orthodoxy in order to become a general in the Russian Army, selected the site that became Harbin.

With an enormous investment to protect, railroad officials quickly realized that they could not depend on local warlords or Siberian peasants to create this not-yet-existent town. They needed experienced Russian-speaking entrepreneurs. But who would ever want to move to Manchuria? The Russian minister of finance, Sergei Yulyevich Witte, hit on a genius idea: the Jews.

Russia's crippling antisemitic laws and violent pogroms were already driving hundreds of thousands of Jews to America, including my own ancestors. Witte argued to the regime in St. Petersburg that to get capital and talent to Manchuria, one only had to tell the Jews that they could live free of antisemitic restrictions—without learning a new language or becoming bottom-feeders in New York's sweatshops—if they moved there.

The regime reluctantly agreed. So did hundreds, and then thousands, of Russian Jews.

The first Jews arrived in 1898 and incorporated an official community in 1903; in only five years, the plan was working splendidly. A 1904 *National Geographic* article written by a U.S. consul to Manchuria reported, wide-eyed, that "one of the greatest achievements in city construction that the world has ever witnessed is now going on in the heart of Manchuria," and that "the capital for most of the private enterprises is furnished by Siberian Jews." These Jewish entrepreneurs created Harbin's first hotels, banks, pharmacies, insurance companies, department stores, publishing houses, and more. By 1909, twelve of the forty members of Harbin's city council were Jewish. These initial entrepreneurs were later joined by new Jewish veterans of the 1904–1905 Russo-Japanese War, then by Jewish refugees fleeing the 1905 Russian pogroms, then by even more refugees fleeing World War I and the Russian Civil War.

At its peak, Harbin's Jewish community numbered around

twenty thousand. The "Old" Synagogue was built in 1909, and by 1921 there was enough demand for a "New" Synagogue a few blocks away, as well as a kosher slaughterer, ritual bath, and matzah bakery, not to mention a Jewish elementary and secondary school, a hospital, a charity kitchen, a free loan association, an old-age home, multiple magazines and newspapers, performances of Jewish music and theater, and Zionist clubs that were the center of many young people's lives—featuring not only competitive athletics (these clubs owned their own sports facilities and even yachts) but also rigorous study of Hebrew language and Zionist ideas. Harbin hosted major international Zionist conferences that drew Jews from all over Asia. Zionist parades were held in the streets.

You already know this story has to end badly. Like almost every place Jews have ever lived, Harbin was great for the Jews until it wasn't—but in Harbin, the usual centuries-long rise-and-fall was condensed into approximately thirty years. The flood of refugees from the 1917 Russian Revolution included many non-Jewish "White" Russians (anti-Communist royalists), whose virulent antisemitism was soon institutionalized in a Fascist party within Harbin's government, and who burned the Old Synagogue in 1931. That was also the year the Japanese occupied Manchuria, noticed rich Jews there, and decided they wanted their money. Conveniently, White Russian thugs were ready to help.

The Japanese gendarmerie embarked on a partnership with White Russian criminals, targeting Jewish business owners and their families for extortion, confiscation, kidnapping, and murder. Later they manipulated the Jewish community for political purposes, sending Abraham Kaufman, a respected physician and the community's elected leader, off to two separate audiences with the Japanese emperor, and forcing him to publish official statements from Harbin's Jewish community announcing their love for Nazi-allied Japan. When the Soviets took over in 1945, they rounded up the city's remaining Jewish leaders, including Dr. Kaufman, and sent them to gulags. Dr. Kaufman endured eleven years in a gulag and then five years in exile in Kazakhstan before he was allowed to join his family in Israel. He was the luckiest; no one else survived. Then again, dying in a gulag was less dramatic than the fate of some Jews under the Japanese. While retreating from the Manchurian town of Hailar, the Japanese military beheaded its Jewish residents.

By 1949, Chinese Maoists controlled Harbin. The thousand-plus Jews still in town were gradually stripped of their businesses and livelihoods, while Israel's government made secret contact with Harbin's remaining Jews and began arranging for them to leave—a process that mostly involved submitting to extortion. As Walter Citrin, an Israeli official responsible for facilitating Jewish emigration from Communist China, explained, "It is obvious that the Communist government is keen to clear the country of the foreign element. However . . . the authorities make things very difficult as long as the person who wants to leave is still in funds, and let the person go only after making quite sure that his personal funds are exhausted." The last Jewish family left town in 1962. After that, only one Jew remained in the city, a woman named Hannah Agre, who refused to leave. Leaning into the crazy-old-lady motif, she moved into a tiny room in the Old Synagogue (by then the building, its interior subdivided, was being used as government office space) and died there in 1985, the official Last Jew of Harbin.

She wasn't quite the last, though. Today there is one Jew in Harbin, an Israeli in his seventies named Dan Ben-Canaan. Ben-Canaan was covering the Far East for Israeli news media when he was invited to teach at a local university, and he settled permanently in Harbin in 2002. Ben-Canaan is a busy man, not only because of his university responsibilities and his work editing local English-language news programs, but because his enormous research into Harbin's Jewish past has made him indispensable to the local government as they restore Jewish sites—so he is also basically employed as the semiofficial One Jew of Harbin.

Ben-Canaan spends enough of his time being the One Jew of Harbin that when I first spoke with him over Skype, he had his one-liner ready: "I'm the president of the community here, which consists of me and me alone. It's great because I don't have anyone to argue with." Ben-Canaan's interest in Harbin's Jewish history, stemming from his days as a journalist, intensified when he learned that Harbin's government owned the Jewish community's official archives—and kept them under lock and key. "I tried to get them to reopen the archives, and they refused," he told me. "I've been given two reasons for it. One is that it contains politically sensitive material, and the other is that they're afraid of being sued for property restitution. There were some wealthy Jews here whose property was worth millions." The lack of access motivated Ben-Canaan to re-create the archives himself by collecting photographs, memorabilia, and testimony from more than eight hundred former Harbin Jews and their descendants around the world. As a result, as he put it, "I've become an address" for Harbin's Jewish history. When the provincial government decided—for reasons that only gradually became clear to me—to spend $30 million to restore, renovate, or reconstruct its synagogues and other Jewish buildings, they hired him.

The One Jew of Harbin spoke with me for nearly two hours, because that was how long it took him to describe the Jewish sites whose refurbishment he had supervised. There was apparently a lot to see. When I asked if I might meet him in Harbin in January, he laughed, explaining that he spends his winters in southern China. "Winter here is not like winter in other places," he warned me. "You can't just walk around outside. Come in the spring or summer instead." But I'd been lured by the city of ice. So he connected me with one of his former students who now worked as a tour guide, and I was on my way.

———

There is a tourist-industry concept, popular in places largely devoid of Jews, called "Jewish Heritage Sites." The term is a truly ingenious piece of marketing. "Jewish Heritage" is a phrase that sounds utterly benign, or to Jews, perhaps ever so slightly dutiful, suggesting a place that you surely ought to visit—after all, you came all this way, so how could you not? It is a much better name than "Property Seized from Dead or Expelled Jews." By calling these places "Jewish Heritage Sites," all those pesky moral concerns—about, say, why these "sites" exist to begin with—evaporate in a mist of goodwill. And not just goodwill, but goodwill aimed directly at you, the Jewish tourist. These non-Jewish citizens and their benevolent government have chosen to maintain this cemetery or renovate this synagogue or create this museum purely out of their profound respect for the Jews who once lived here (and who, for unstated reasons, no longer do)—and out of their sincere hope that you, the Jewish tourist, might someday arrive. But still, you cannot help but feel uncomfortable, and finally helpless, as you engage in the exact inverse of what Benjamin of Tudela once did: instead of traveling the world and visiting Jews, you are visiting their graves.

Harbin was enjoying a heat wave when I arrived, a balmy ten below with a wind chill of minus eighteen. I only needed to wear a pair of thermals, a shirt, a sweater, a fleece, a parka, a balaclava, a neck warmer, a hat, gloves, three pairs of socks, and three pairs of pants to go outdoors.

My first stop was the city's Jewish cemetery, billed by tour companies as the largest Jewish cemetery in the Far East—except that it's not a cemetery, since cemeteries contain dead bodies, and this one doesn't have any. In 1958, Harbin's local government was redesigning the city and decided that the Jewish cemetery, home to around 3,200 dead Jews, had to go. The city offered families the option of moving their dead relatives' graves to the site of a large Chinese cemetery called Huang-shan, an hour's drive outside the city, for the price of about $50 per grave. Many Jewish families were long gone by then, so only 812 graves were moved—and, as it turned out, only the gravestones, since city authorities saw no reason to move the bodies too. The human remains from the old cemetery are now in what the Chinese call "deep burial"—that is, the space containing them has been paved over and turned into an amusement park. "It is nice for them to be there," my tour guide—whom I'll call Derek to keep him out of trouble—said of the dead Jews under the rides. "They are always with happy people now."

The drive to Huangshan took about an hour through industrial wastelands and frozen fields, culminating in a grandiose toll plaza with enormous Russian-style onion domes and then several miles more of abandoned warehouses, with a few bundled people by the roadside selling stacks of fake money to burn as offerings—because Huangshan is really a vast Chinese cemetery, filled with endless rows of identical shiny white tombstones on mini plots containing cremated remains. After driving past tens of thousands of dead Chinese people, we found the entrance to the cemetery's Jewish section, paid our fee, and entered the gates.

The Jewish section was compact and stately, with roughly seven hundred gravestones elaborately carved in Hebrew and Russian, along with many modern metal plaques sponsored by former Harbin Jews whose relatives' original stones hadn't been moved. Many of the original grave markers had ceramic inserts with photographic portraits of the deceased, which would have been intriguing if every single one hadn't been shattered or removed. The damage was clearly deliberate, which might explain why a cemetery employee kept following us around. The idea that Jewish cemetery desecration was currently in vogue in Harbin was a tad depressing, but to my surprise, this snowy Jewish Heritage Site didn't feel at all lonely or bereft. In fact, it was rather glam.

Inside the gate was a plaza with a massive granite Star of David sculpture, next to a two-story-high domed synagogue building festooned with more Stars of David. The synagogue's doors were locked, but through its windows I could see that the building was a shell, with nothing inside but some scattered tools and junk. When I asked what the building was for, Derek laughed. "They built it for Olmert's visit," he explained. "Now it's just used by the cemetery workers to stay warm." Ehud Olmert, a former Israeli prime minister who served prison time for corruption, had roots in Harbin. His father was born there, and his grandfather, or at least his grandfather's gravestone, was in Huangshan—a gravestone that had now been outdone by a twelve-foot-high black marble obelisk. The obelisk, crowned with yet another Jewish star, was carved with greetings written in English in Olmert's handwriting and painted in gold: "Thank you for protecting the memory of our family, and restoring dignity into [sic] the memory of those who were part of this community and [illegible] a reminder of a great Jewish life which a long time ago was part of Harbin." The words were a dashed-off scribble, suggesting that Olmert didn't quite expect them to be set in stone. His grandfather's gravestone had been replaced with a black-and-gold marble one to match the obelisk, outshining the plebeians with their smashed ceramic photos. Near his grave stood a trash can designed to look like a soccer ball.

Olmert's visit to Harbin in 2004 as Israel's deputy prime minister had been a big deal, but the (fake) synagogue built in his honor at the (also fake) cemetery was just one part of an enormous and expensive project on the part of the local provincial government to restore Jewish Heritage Sites. The government's explicit goal is to attract Jewish money, in the form of both tourism and investment by foreign Jews.

In our conversation, the One Jew of Harbin had only praise for these efforts, in which he has been deeply involved. "The restoration cost $30 million—it's unheard-of here. Everything was of the highest quality," Ben-Canaan told me, adding that Harbin's Jewish Heritage Sites have the same official designation as Chinese landmarks like the Forbidden City. But one of the many sources on Harbin he shared with me was a long 2007 news article from a Chinese magazine by a journalist named Su Ling, whom he described as one of China's rare investigative reporters. The article, titled "Harbin Jews: The Truth," traced a very particular history: not Harbin's Jewish Heritage, but the Heilongjiang provincial government's attempts to capitalize on that heritage.

The story began innocently enough, with a social-scientist-cum-real-estate-agent named Zhang Tiejiang, who discovered the prior Jewish ownership of many historic homes that he was supposed to demolish for a city planning project in 1992. Taking an interest, he studied the Jewish graves in Huang-shan cemetery, translating their Russian text with the help of a computer program. His timing was auspicious: 1992 was the year China established diplomatic relations with Israel, and in 1999 China's premier made his first official visit to Jerusalem. Also auspicious: Heilongjiang Province, long reliant on declining industries like coal mining, had hit an economic slump. In

1999 Zhang Tiejiang published his idea in an article for a state news agency titled "Suggestions for the Study of Harbin Jews to Quicken Heilongjiang Economic Development."

This article made its way to the higher-ups in the Chinese government in Beijing, who dispatched an official to Heilongjiang's Academy of Social Sciences to "intensify the study of the history of Harbin Jews." A Center for Jewish Studies was established, with a massive budget enabling unqualified people producing minimal research to enjoy trips abroad. "Develop[ing] the travel industry and attracting business investments," the center's original website announced, was "the tenet of our existence and purpose." In years following, the government's $30 million produced far more tangible results, including not only the cemetery refurbishment but also the transformation of the New Synagogue into a Jewish museum, the reconstruction of the Old Synagogue and the Jewish secondary school, and the labeling of formerly Jewish-owned buildings as landmarks in the city's historic heart.

This attempt to "attract business investments" by researching Jewish history seems, to put it gently, statistically unsound. Among the tens of millions of tourists to China each year, forty thousand annual Israeli visitors and even fewer Jewish tourists from elsewhere amount to a rounding error. And the idea that Israeli or other Jewish-owned companies would be moved to invest in Heilongjiang Province out of nostalgia for its Jewish heritage seems unlikely at best. The only way to understand this thinking is to appreciate the role Jews play in the Chinese imagination.

Most Chinese people know next to nothing about Jews or

Judaism. But in a 2009 essay reviewing trends in Jewish studies in China, Lihong Song, a professor of Jewish studies at Nanjing University, pointed out a common pattern in what they do know. "My students' first association with Jews is that they are 'rich and smart,'" he noted. "The shelves of Chinese bookstores," Song explained, "are lined with bestsellers on Jewish subjects." What Jewish subjects might those be? Well, some of those bestselling titles are *Unveiling the Secrets of Jewish Success in the World Economy, What's Behind Jewish Excellence?, The Financial Empire of the Rothschilds, Talmudic Wisdom in Conducting Business,* and of course, *Talmud: The Greatest Jewish Bible for Making Money.* Song claimed that this was not antisemitic, but rather "some sort of Judeophilia."

At a 2007 "International Forum on Economic Cooperation between Harbin and the World's Jews," held in Harbin with dozens of invited Jewish guests who ranged from the Israeli ambassador to a group of Hungarian Jewish dentists, Harbin's mayor welcomed participants by citing esteemed Jews such as J. P. Morgan and John D. Rockefeller (neither of whom was Jewish). He then announced that "the world's money is in the pockets of the Americans, and the Americans' money is in the pockets of the Jews. This is the highest acclaim and praise to Jewish wisdom."

————

Former Harbin Jews often remembered Harbin as a kind of paradise. "They owned the town," Irene Clurman, a daughter of former Harbin Jews, told me, describing the nostalgia that many "Harbintsy"—ex-Harbiners—expressed for their beloved city. "It was a semicolonial situation; they had Chinese servants and great schools and fur coats." Or in the words of her grandmother Roza (later Ethel) Clurman in a 1986 interview, "Harbin was a dream."

Roza Clurman's husband—Irene Clurman's grandfather—was kidnapped, tortured, and murdered in Harbin during the Japanese antisemitic reign of terror, after which his lucrative business (he introduced indoor plumbing to Manchuria) and his high-end rental building were confiscated, leaving his family with nothing. And the Clurman family's horror stories had begun much earlier: Roza Clurman was five during the 1905 Odessa pogrom, hiding in an attic for days on end while the neighborhood was ransacked and her neighbors murdered. The move to Harbin didn't quite prevent her family from being targeted, given that her husband also wound up murdered. But "my grandmother absolutely had a nostalgia for Harbin," Irene Clurman insisted. In her interview, Roza Clurman admitted that "everything changed" in Harbin, but she spent far more time describing its glory: the steaks the family ate, their household staff, the children's private lessons.

The ascent from pogroms to private lessons was dizzyingly fast, obscuring the community's equally precipitous decline. One Harbintsy descendant, Jean Ispa, told me how her father, an orphan, made his way to Harbin alone solely to study music, since Russian conservatories didn't take Jewish students. Running away from an orphanage, he collected scrap metal to buy a ticket to Harbin, where he was promptly jailed for entering the country illegally—and where musicians in the Jewish community bailed him out. "He was sixteen when he made this journey," Ispa told me in wonder. "He

gave concerts in Harbin. I even have the programs he played."
Another Harbin exile, Alexander Galatzky, was eight during
the pogroms of the 1919–1920 Russian Civil War, when he and
his mother repeatedly barricaded themselves in their apartment in Ukraine and listened to the screams of their neighbors being murdered and raped. When the ship fare his father
sent from New York was stolen, their only hope was to go east
to Manchuria, where his father planned to meet them. In rem-
iniscences he wrote down for his family, Galatzky described
boarding a cattle car to leave Ukraine: "Mother has a bundle
of old clothes with her. The soldier on guard of the cattle car
is trying to take it from her. She clutches at it, crying, kissing
the soldier's hand. We have no money or valuables and the old
clothes can be bartered for food en route. Without them we
would starve." After a life like that, Manchuria was paradise.

Of course, one could tell the same story about Russian
Jews who emigrated to New York. But in Harbin, where Russian Jews created their own Russian-Jewish bubble, their sense
of ownership and pride was greater—and that pride made the
story of their community's destruction into a footnote. Of the
Harbintsy descendants I interviewed, most mentioned friends
or relatives who were kidnapped, tortured, or murdered during the Japanese occupation. All had their family's hard-earned
assets seized by Manchuria's various regimes. But in the next
sentence they would tell me, again, how Harbin was "a golden
age." An entire organization in Israel, Igud Yotzei Sin (Association of Chinese Exiles), exists solely to connect homesick
"Chinese Jews" around the world with one another through
networking, social events, scholarships, and trilingual news-

letters which run to hundreds of pages. Until recent years, members gathered weekly in Tel Aviv to play mah-jongg, drink tea, and reminisce about the wonders of Harbin. Teddy Kaufman, who ran the organization until his death in 2012, published a memoir entitled *The Jews of Harbin Live On in My Heart*, extolling the Jewish paradise. His father was the community president who'd wound up in a gulag.

Harbin's Jewish "golden age" lasted less than one generation. Even before the Japanese occupation, things were unpleasant enough that leaving was, for many, a foregone conclusion. Alexander Galatzky, the boy whose mother bartered old clothes to feed him on the Trans-Siberian Railroad, kept diaries as a teenager from 1925 to 1929 that his daughter Bonnie Galat recently had translated. The diaries revealed an assumption that most teenagers don't live with: everyone planned to leave, and the only question was where to go. He counted off his friends' departures—to Palestine, to Russia, to Australia, to America—and waxed nostalgic about leaving, as he capitalized in his diary, "FOR GOOD." "My old classmate Misha leaves for Paris today," he wrote, describing one of many permanent goodbyes. "For good, I think . . . It's a scary word, 'FOR GOOD.' Biro left, and Pinsky, and I think I'm leaving next year too . . . and not with Mom and Dad, but alone." Galatzky's fears came true; the following year he left for Paris via train and ship through Shanghai, Ceylon, and Suez. Later he wound up supporting his parents, after they fled Harbin with little more than old clothes.

Many came to recall the community's destruction as if it were almost expected, like snow or rain. Alex Nahumson, who was born in Harbin and emigrated in 1950 at the age of three with his family, reports only "very happy memories" discussed by his parents. "The Chinese never did anything bad to us, just the Russians and the Japanese," he told me by phone in Hebrew from his home in Israel—despite the fact that his family's assets were plundered by the Maoist regime. "When my parents talked about Harbin, they only talked about their dacha [country home], the theater, the opera," he averred. When I brought up the kidnappings during the Japanese occupation, he verbally shrugged. "That's just crime," he insisted. "Crime happens everywhere." His parents survived all of these regime changes, he said cheerfully, "between the raindrops"—a Hebrew expression for evading repeated disaster. Losing everything they had was inevitable, like the weather. As the Russian Jewish writer Sholem Aleichem once put it, Jewish wealth is like snow in March, melting and washing away. Later in our conversation, Nahumson mentioned, almost casually, that his own grandfather was kidnapped and tortured by the Japanese.

———

It is hard to describe what, exactly, was wrong with Harbin's New Synagogue Jewish Museum—or as it said on my ticket, the "Construction Art Museum," a name that comes from the building's current ownership by the Harbin Municipal Construction Department. One feels the overwhelming need to applaud this (mostly) Jewish museum's mere existence, to carefully delineate its many strengths, to thank the locals for their bountiful goodwill. For it did have enormous strengths, and the goodwill was abundant. Still, from the moment I arrived at the large domed building and entered its wide-open space with an enormous Star of David decorating the floor—it only occurred to me later how ridiculous this detail was, since the floor would have been covered with seats when the synagogue was in use—I felt that creeping "Jewish Heritage" unease, the unarticulated sense that despite all the supposed goodwill, something was clearly off. But then my actual Jewish heritage kicked in, consisting of centuries of epigenetic instincts reminding me that I am only a guest. I swallowed my discomfort and started snapping pictures.

The Jewish history exhibition filled the second floor—the women's gallery of the synagogue. There, in vast arrays of photographs, I observed smiling, well-dressed people building synagogues, celebrating weddings, attending Zionist meetings, patronizing a library, posing in scout uniforms, working in a hospital, rescuing neighbors from a flood, and skating on the river. The displays were informative enough, even if their translated captions sometimes disintegrated into word salad. Beneath one portrait of a man wearing a prayer shawl and a tall clerical hat, for instance, the English caption read, "Judean assembly mark in harbin choir leading singer gram benefit maxwell minister radical." I asked Derek what the original Chinese caption meant. He smiled apologetically and said, "I'm not sure."

It was all admirably thorough, if a little garbled. But toward the far end of the gallery, on the part of the floor that had been constructed over the alcove where the ark for Torah scrolls once stood (the actual alcove for the ark is now a foyer leading to a restroom), I entered a set of little rooms whose contents puzzled me.

The first room was dominated by a large wooden desk, with a life-size white plaster sculpture of a bald and bearded Western man seated before an ancient paperless typewriter. The brass plaque in front of him read, "Real workplace of Jewish industrialist in Harbin." Confused by the word "real," I asked Derek if this was supposed to be a specific person. He glanced at the plaque and explained, "It is showing a Jew in Harbin. He is doing business."

In subsequent rooms, more tableaux of frozen Jews unfolded. There were life-size plaster Jews frozen at a grand piano, a life-size plaster Jew frozen in a chair with knitting needles, and two child-sized plaster Jews frozen on a bed, playing eternally with plaster blocks. This, the brass plaque informed me, was "The Display of the Jews' Family in Harbin." The plaque continued: "At the first half of the 20th century, not only was the display of the Jews' family simple, but also practical and the children lived a colorful life there." The children's blocks, like the children, were devoid of color. Later I discovered the unnamed inspiration for this display: Harbin's annual Snow Sculpture Park, full of figures carved from blocks of manufactured snow.

After the rooms full of frozen Jews, the photographs of mostly dead Jews resumed, dominated by "real Jewish industrialists" who "brought about numerous economic miracles" in Harbin, including the founders of Harbin's first sugar refinery, first soybean-export business, first candy factory, and China's first brewery. The wall text explained how Harbin "offered the Jews an opportunity for creating new enterprises and providing a solid foundation for their later economic activities in

Europe and America." This was true, I suppose, if one thinks of Harbin as a kind of business-school exercise, rather than a place where actual Jews created actual capital that was subsequently seized, transforming them overnight into penniless refugees, if they were lucky.

One enterprise prominently featured in the museum, for instance, was the Skidelsky Coal Mine Corporation. The Skidelskys were among the "Siberian Jews" who provided the initial capital for Harbin—although "initial capital" is an understatement. In an account of his family's holdings in *Prospect* magazine, Robert Skidelsky, a member of the British House of Lords and a Harbin native, described how his great-grandfather Leon Skidelsky held the contract in 1895—prior to Harbin's founding—to build the Trans-Siberian Railroad from Manchuria to Vladivostok. The Skidelskys were one of only ten Jewish families allowed to live in Vladivostok, since the railroad desperately needed them. They held long-term concessions on three thousand square kilometers of timber in Siberia and Manchuria, and enough long-term mining concessions to make them one of the region's largest employers. They continued supplying the railroad as it changed hands from the Russians to the Chinese to the Japanese. In 1924, Leon's son Solomon even charmed a local warlord into selling him a thirty-year lease on a mine, by repeatedly and deliberately losing to him in poker.

In 1945, Solomon Skidelsky was still nine years shy of running out the lease when the Soviets sent him and his brother to die in a gulag, and Communists—first Soviet and then Chinese—seized the mines. Decades later, Lord

Skidelsky filed his claim. "In 1984," Lord Skidelsky recounted, "I received a cheque for 24,000 English pounds in full settlement of a claim for compensation that amounted to £11 million." When he visited Harbin in 2006, local TV crews trailed him and presented him with flowers. The flowers were worth somewhat less than £11 million.

When I expressed my sense that the museum was telling only part of a story, Derek raised an issue that Ben-Canaan brought up with me repeatedly, that the museum focused exclusively on wealthy people—thus underscoring the idea that Jews are rich. "Obviously there were poor Jews here too," Derek pointed out. "The building across the street was the Jewish Free Kitchen."

It was only as I was leaving, through the enormous mezuzah-less door, that I looked back at what was once the sanctuary and understood what, exactly, was wrong. Above the vast Star of David floor, the museum was dominated by an enormous blown-up photograph of a 1930s farewell banquet, its rows of Harbin Jews in their tuxedos gathered to say goodbye to yet another Jewish family fleeing, as Alexander Galatzky had put it, "FOR GOOD." Suddenly the Jewish Heritage miasma melted away, and the realization hit me: Nothing in this museum explained why this glorious community no longer exists.

———

Harbin is a rather hideous city, its Soviet-style apartment blocks stretching as far as the eye can see. But the city's historic heart has been restored so thoroughly that if not for the

Chinese crowds and street signs, one could imagine being in prewar Europe. The restoration included turning the historic tree-lined Central Avenue into a pedestrian mall that doubles as an outdoor architectural museum, where each original building—80 percent of which were once Jewish-owned—is labeled with a plaque describing its past. Unfortunately the restoration also included installing loudspeakers that constantly blast high-volume Western music. When I arrived, they were playing "Edelweiss": *Bless my homeland forever*. The music made it hard to think.

Derek pointed out the various restored buildings on Central Avenue and elsewhere in the neighborhood: the Jewish-owned pharmacy, the Jewish Free Kitchen, the Jewish People's Bank, and many private homes, all now occupied by other enterprises. The "Heritage Architecture" plaques affixed to each historic building couldn't have been more direct: "This mansion," a typical one read, "was built by a Jew."

The most impressive Central Avenue building "built by a Jew" was the Modern Hotel, a building whose story captures the Harbin Jewish community's roller coaster of triumph and horror. The Modern Hotel was built by the Jewish entrepreneur Joseph Kaspe, and from the moment it opened in 1909, it was the height of Manchurian chic. The Modern wasn't merely a high-class establishment frequented by celebrities and diplomats. Its premises also included China's first movie theater, and the hotel frequently hosted theatrical performances, lectures, and concerts with seating for hundreds. Kaspe also created other Modern-labeled luxury products like jewelry and high-end food. In other words, the Modern was a brand.

When the Japanese occupied Harbin, they immediately set their sights on the Modern. But Joseph Kaspe was one step ahead of them. His wife and two sons had moved to Paris, where they had acquired French citizenship—so Kaspe put the Modern in his son's name and raised the French flag over the hotel. He assumed the Japanese wouldn't risk an international incident just to steal his business. He was wrong.

In 1932, Kaspe invited his older son Semion, a celebrated pianist, back to Manchuria for a concert tour. On the last night of his tour, Semion was kidnapped. Instead of paying the bankruptcy-inducing ransom, Joseph Kaspe went to the French consulate. It didn't help; the kidnappers upped the ante, mailing Kaspe his son's ear. After three months, Semion's body was found outside the city. When Kaspe saw his son's maimed and gangrenous corpse, he went insane. Friends shipped him off to Paris, where he died in 1938. His wife was deported and died at Auschwitz five years later. His younger son escaped to Mexico, where he died in 1996, refusing to ever discuss Harbin.

The Modern Hotel is still in operation today, though at a few stars lower than the Holiday Inn where I stayed down the street. The large pink stone building with its glamorous arched windows and turrets still dominates Central Avenue, its girth expanding for an entire city block, Cyrillic letters spelling out "MODERN" running down one corner of its facade. Outside, I saw a long line of people winding its way down the street toward one end of the hotel, the hordes queuing in minus-ten degrees. The line, Derek explained, was for the Modern's famous ice cream. "In Harbin, we love eating cold foods at cold temperatures," he said with a grin. It's true; the streets of Harbin are lined with snack stands selling skewers of frozen fruit. The Kaspes figured this out and created China's first commercially produced ice cream, sold today under the chic English-language brand name "Modern 1906." (The three-year error on the brand's founding date clearly troubled no one but me.) Passing up the frozen treats, I went inside.

The Modern Hotel's lobby was shabby and nondescript, except for an exhibit celebrating the hotel's illustrious history. It began with a bronze bust of Joseph Kaspe, with wall text in Chinese and English describing the accomplishments of the Modern Corporation and its founder, "The Jew of Russian Nationality Mr. Alexander Petrovich Kaspe." (The "Alexander" was inexplicable; Joseph Kaspe's actual first name appeared in Russian on the bust.) As the wall text explained, this impressive Jew founded a "flagship business in Harbin integrated with hotel, cinema, jewelry store, etc." "In recent years," the text continued, "the cultural brand of Modern is continuously consolidated and developed." It then listed the numerous businesses held by this storied company— including the Harbin Ice Festival, which belonged to the Modern Corporation until the provincial government took it over a few years ago. "Currently," the wall text gloated, "Modern Group . . . is riding on momentum, and is shaping a brand-new international culture industry innovation platform." Mr. Kaspe's descendants would indeed be proud of this heritage, if any of them had inherited it.

But after all, the Modern Hotel clearly honors its Jewish Heritage! There, on its walls, were enlarged photos of

Joseph Kaspe's family, including his murdered son, sexy in his white tie and tails, frozen over his piano. There, under glass, were Real Historic Items from the Kaspe family, including silver candlesticks, an old-timey telephone, and a samovar! And there, in one particularly dusty glass case near the floor, were "the Kaspe collection of household utensils of Judaism sacrificial offerings," including an actual Passover Seder plate!

I squatted down for a closer look at this display and saw that there were two plates inside it. The Seder plate had a bronzy Judaica motif suspiciously familiar from my own American Jewish childhood. I squelched my skepticism until I saw that it was carved all around with English words. The second plate, a ceramic one, sported an Aztec-esque design, with the word "Mexico" painted across the bottom—a 1980s airport souvenir. At that point it became clear that any Real Historic Items from the Kaspes had long disappeared into some regime's pockets, and that this display had been sourced from eBay.

I put my balaclava back on and went out into the cold again, past the hundreds of Chinese people clamoring for Kaspe's ice cream, and headed to the Old Synagogue, which is now a concert hall. The result of a multimillion-dollar renovation project for which the One Jew of Harbin served as an adviser, the building is part of an entire "Jewish block" that includes the music school next door, which was once the Jewish secondary school. Ben-Canaan was meticulous about the project, gathering and examining old photographs and descriptions to exactly replicate the ark with its granite Ten Commandments motif, the pillars, the gallery that was once the women's section, and the seats with their prayer-book stands. His only concession, he told me, was to make the bimah (the platform before the ark) wide enough to accommodate a chamber orchestra. When the person manning the ticket booth refused to let me peek inside, I bought a ticket for that night's string quartet.

The Old Synagogue's interior shocked me. I don't know what I was expecting, but what I didn't expect was to be standing in a synagogue no different from every single urban early-twentieth-century synagogue I've ever entered around the world, from my own former synagogue in New York City to others as far as London and Moscow and Cape Town and Buenos Aires and Melbourne—all those buildings around the world where you walk into the sanctuary (usually after passing an armed guard) and could literally be in any synagogue anywhere, which is exactly the point. The One Jew of Harbin did a marvelous job—so marvelous that as I walked into the large hall and saw the massive ark looming before me, with its familiar Hebrew inscription imploring me to *Know Before Whom You Stand*, I instinctively listened for what part of the service I was walking in on, how late I was this time, whether they were up to the Torah reading yet. My thoughts about how far back I should sit finally gave way to logic, and I looked at the seat number on my ticket.

But when I took my seat in the third row, I still could not shut down my muscle memory. My hands went straight to the slot in the seat in front of me, reaching for a prayer book that wasn't there. I almost couldn't stop myself from reciting all the words I've recited in rooms like this, the words I've repeated my entire life, the same words recited by all the people who have gathered in rooms like this over the past twenty centuries, in Yavneh and Pumbedita and Aleppo and Rome and Marrakesh and Philadelphia and Mumbai and São Paulo and Harbin, facing Jerusalem. I was awed, googly-eyed. In that moment I suddenly knew, in a vast sense that expanded far beyond space and time, before whom I stand.

Then a Chinese string quartet walked up to the bimah in front of the ark, and instead of bowing before the ark, they bowed before me. The lights dropped, and they played, spectacularly well, Brahms's "Hungarian Dance Number 5," and Tchaikovsky's "Romeo and Juliet," and inexplicably, "Cotton-Eyed Joe."

And suddenly I was very, very tired.

———

Somewhere in between the synagogues, the Belle Epoque–style bookstore named for Nikolai Gogol, the pool carved out of the frozen river with people swimming in minus-thirty degrees, and the hundreds of dead Jews, I found myself in a "Siberian Tiger Park," where seven hundred of the world's remaining tigers loll behind high chain-link fences or pace in isolation cells, in what resembles a tiger reeducation camp. There, after riding a bus painted with tiger stripes through bare icy yards full of catatonic-looking tigers, I was encouraged to buy slabs of raw meat—since, as Derek explained, the facility only provided the animals with meager rations, with the assumption that tourists would make up the difference. The pull of curiosity tainted by guilt—a feeling remarkably similar to my "Jewish Heritage" unease—brought me to a woman selling buckets of raw pork slabs, which visitors feed to the tigers with tongs through the chain link. The woman selling the slabs also offered a crate of live chickens that I could alternatively have purchased as tiger food; this would have involved buying a live chicken and thrusting it into the tiger enclosure via a dedicated chicken chute. For the first time in my life, I bought pork.

As I struggled to pick up slippery pieces of meat with the tongs, I remembered a moment in the Talmud (*The Greatest Jewish Bible for Making Money*) when the rabbis claim that the last thing created during the week of creation was the world's first pair of tongs, since tongs can only be forged with other tongs—a story whose haunting image of human limits transcends its lack of logic. When I succeeded in wielding the meat, the otherwise catatonic tigers pounced against the fence at me in a cartoon-like fury, rattling the Soviet-style barriers as they battled one another for the strips of flesh. I watched these almost mythic captives through a thick haze of pity and fear, helplessly flinging inadequate scraps at something powerful, beautiful, and trapped. Much later, I came across a *National Geographic* article claiming that this "park" was in fact a tiger farm, where these endangered animals—only seven of which still exist in the northeastern Chinese wild, thus outnumbering Jews in the region by 7,000 percent— were bred and slaughtered for trophies and traditional medicines. It all felt like an elaborate con. Or if not quite a con, a *display*.

The Harbin Ice Festival was the greatest display of all, surpassing my most fevered expectations. It was much, much larger and more elaborate than I had imagined from the photos and videos that had lured me to Harbin. I'd been amply warned by online strangers about how difficult the festival would be to endure, since it requires long periods outside, at night, in punishing temperatures. But once I was there, I was shocked by how easy it was. All I needed to add to my Harbin ensemble was a second sweater, a fourth pair of pants, a second pair of gloves, a fourth pair of socks, handwarmers stuck into my gloves and boots, and ice cleats, and I was good to go. I had been told that I wouldn't be able to bear the cold for more than forty minutes. In the company of approximately ten thousand others who were also visiting that evening, a number relatively insignificant in the vastness of the festival, I stayed for three hours.

Among the ice castles and ice fortresses clustered around a snow Buddha the size of a high school, I recognized shimmering tacky neon versions of places I'd visited in real life, cataloging them in my brain like Benjamin of Tudela: the Wild Goose Pagoda of Xian, the Summer Palace outside Beijing, the gate to the Forbidden City, Chartres Cathedral, the Campanile tower near Venice's original Jewish ghetto, the Colosseum built by Jewish slaves brought from Jerusalem to Rome. I wandered around and through these flashing structures, their colors changing every few seconds as the LED wiring blinked within each ice block, passing over bridges and through moon gates and up staircases and down slides that wound their way through castles of ice. China is a place full of enormous, gaudy, extravagantly impersonal monuments made possible through cheap labor, from a two-thousand-year-old tomb filled with ten thousand terra cotta warriors in Xian to the medieval Great Wall outside Beijing to the 1994 Oriental Pearl Tower in Shanghai. The Harbin Ice Festival was the gargantuan fluorescent opposite of intimate or subtle. It was mind-blowing, and mindless. It was the most astounding man-made thing I had ever seen.

What was most shocking about the Ice Festival was the bizarre fact that all of it was temporary. In another month or two, the vast city would begin to melt. But unlike what I ignorantly assumed, the ice city does not simply vanish on its own. Instead, when the melting begins, ten thousand workers return to hack apart the millions of ice blocks, remove their electrical wiring, and then haul them out and dump them in the river. Like all cities, there is nothing natural about its creation, and also nothing natural about its destruction.

Nothing simply disappears. As I left Harbin, I thought of Hannah Agre, the Last Jew of Harbin—the crazy old lady who refused to leave the city, who died alone in 1985, twenty-three years after the last Jewish family left, in an office space that she had rejiggered into an apartment on the second floor of the Old Synagogue. It occurred to me, as I passed through the industrial wastelands and endless high-rises on my way to the Harbin airport, that maybe she wasn't so crazy. Maybe she didn't like being told to leave. Maybe she was physically enacting what all the other Harbintsy spent the rest of their lives trying to do, as they gathered in San Francisco and Tel Aviv to play mah-jongg and share photos of their samovars and fur coats. Maybe she wanted to keep the castle her family had built, preserved in ice.

By the time I reached the airport, the Harbin Holiday Inn's breakfast buffet of dragon fruit and lychee nuts was a distant memory, and I was hungry. Fortunately, right next to my gate there was a hip-looking eatery, with historic black-and-white photos framed on trendy brick walls. Its sign read: "Modern 1906."

I almost couldn't believe it, but yes, here it was once more: Joseph Kaspe's business. As if responding to my private disbelief, a giant flat screen on a hip brick wall flashed a photo of Kaspe's family, then one of Kaspe's face. I stared at the photos before they blinked away, looking at this murdered family and then at Kaspe, the man who built a city only to lose his son, his property, and his mind. I suddenly felt shaken by the "success" of this business that has apparently persisted uninterrupted for over a century, by the sheer chutzpah of this open bragging about a corporate "heritage," by the enduring quality of stolen goods. It was twenty below outside, but I bought an ice cream in a flavor labeled "Original." The sweet frozen cream melted in my mouth, gone before I even put away my Chinese change.

I was in the last row of the Air China plane leaving Harbin, the only Westerner on board. There was an intense smell of barbecued pork as someone in the row in front of me celebrated the Year of the Pig. I thought of Alexander Galatzky leaving Harbin "FOR GOOD," boarding the train to Shanghai and then the boat to Ceylon and on through the Suez Canal, a journey embarked upon nine years after he first traversed the world at the age of nine on the Trans-Siberian Railroad, with his mother and her bag of old clothes. A cheerful animated panda on the screen in front of me explained the many safety features of the aircraft, including what to do if we should require, as the awkward English translation put it, "Emergency Ditching." I thought of the Clurmans, the Kaspes, the Nahumsons moving between the raindrops, ditching as needed, ditching as expected, ditching a foregone conclusion. I watched the animation and remembered Benjamin of Tudela, the chipper cartoon of the perilous journey around the world, where every Jewish community was documented and counted and marveled at, full of cheery animated people who never felt the need to ditch, where cities never melted away.

Within two minutes of takeoff, Harbin was no longer visible. Outside my window, I saw only snow-dusted farmland and the gleam of sunlight on the frozen river. The land was vast and empty. The enormous city was gone.

Chapter 3

DEAD
AMERICAN JEWS,
PART ONE

"THERE ARE NO WORDS."

This was what I heard most frequently from good people stunned by the news one Saturday morning in October of 2018: eleven people murdered at a Pittsburgh synagogue, the largest massacre of Jews in American history. But there are words for this, entire books full of words: the books the murdered people were reading at the hour of their deaths. News reports described these victims as praying, but Jewish prayer is not primarily personal or spontaneous. It is communal reading: public recitations of ancient words, scripts compiled centuries ago and nearly identical in every synagogue in the world. A lot of those words are about exactly this.

When I told my children what had happened, they didn't

ask why; they knew. "Because some people hate Jews," they said. How did these American children know that? They shrugged. "It's like the Passover story," they told me. "And the Hanukkah story. And the Purim story. And the Babylonians. And the Romans." My children are descendants of Holocaust survivors, but they didn't need to go that far forward in history. The words were already there.

The people murdered in Pittsburgh were mostly old, because the old are the pillars of Jewish life, full of days and memories. They are the ones who come to synagogue first, the ones who know the words by heart. The oldest victim at the Tree of Life synagogue was Rose Mallinger, age ninety-seven.

The year Mallinger was born was the tail end of the mass migration of over a million Eastern European Jews to America. Many brought with them memories of pogroms, of men invading synagogues with weapons, of blood on holy books. This wasn't shocking, because it was already described in those books. On Yom Kippur in synagogue, these Jews read the stories of rabbis murdered by the Romans, including Rabbi Haninah ben Teradion, who was wrapped in a Torah scroll set aflame. Before dying, he told his students, "The parchment is burning, but the letters are flying free!" My synagogue's old High Holiday prayer book, a classic edition edited by Rabbi Morris Silverman that dominated twentieth-century American synagogues, hints at what these stories meant to American Jews of Ms. Mallinger's age. Its 1939 English preface asks: "Who can forget, even after decades, the sight of his father huddled in the great prayer shawl and trying in vain to conceal the tears which flowed down his cheeks during the recital of this poem?" By the time I was a kid reciting those poetic stories, no one was crying. Instead, my siblings and I smirked at the excessive gory details, the violence unfamiliar enough to be absurd. But Rabbi Haninah must have been right, because we were still reading from that same scroll, the same words Jews first taught the world: *Do not oppress the stranger. Love your neighbor as yourself.*

People Ms. Mallinger's age were in their twenties when word spread about mass murders of Jews in Europe. In synagogue on Rosh Hashanah, they read the old words begging God for compassion, "for the sake of those killed for your holy name," and "for the sake of those slaughtered for your uniqueness." My husband's grandparents came here after those massacres, their previous spouses and children slaughtered like the people in the prayer. They kept reciting the prayer, and for their new American family it reverted to metaphor.

In the decades that followed, Jews from other places joined American synagogues, many bringing memories that American Jews had forgotten. Those memories were waiting for them in the synagogue's books. On the holiday of Purim, they recited the Book of Esther, about an ancient Persian leader's failed attempt at a Jewish genocide. It's a time for costumes and levity, for shaking noisemakers to blot out the evildoer's name. One year my brother dressed as the ayatollah, and the Persians in our congregation laughed. Another year someone dressed as Gorbachev; the Russians loved it. The evildoers seemed defeated.

In 2000, when Ms. Mallinger was seventy-nine, a Jewish senator was his party's nominee for vice president. A year later the White House hosted its first official Hanukkah party. About a decade later I attended one myself. In the White House we recited ancient words thanking God for rescuing us from hatred. To older Jews, this felt miraculous: My parents and grandfather gawked at my photos, awestruck. But at the party I met younger Jewish leaders who often attended these events. To them, this was normal. The ancient hatred was a memory, words on a page.

Or maybe it wasn't. In 2001, after terrorists attacked American cities, concrete barriers sprouted in front of my family's synagogue, police cruisers parked in the lot. This felt practical in a nation on edge; we assumed it affected everyone. As my children were born and grew, the barriers and guards became their normal. When I took my children to an interfaith Thanksgiving service at a church down the street from our synagogue, one of them asked me why no one was guarding the door.

In the years that followed, the internet suddenly allowed anyone to say whatever they wanted, rewarding the most outrageous from every political stripe. Soon, comments sections became an open sewer, flowing with centuries-old garbage— and as social media exploded, those comments evolved into open vitriol, as abstract hate scaled up to direct verbal attacks on Jewish institutions and individuals. To young Jews this felt confusing. To old Jews it must have felt familiar, a memory passed down and echoed in the holy books.

When Ms. Mallinger was ninety-seven, she and ten other Jews were murdered in their synagogue. There are words for this too, a Hebrew phrase for 2,500 years' worth of people murdered for being Jews: *kiddush hashem*, death in sanctification of God's name.

My children were right: This story is old, with far too many words. Yet they were wrong about one thing. In the old stories, those outside the community rarely helped or cared; our ancestors' consolation came only from one another and from God. But in this horrific new reality, perhaps our old words might mean something new.

When they return to synagogue, mourners are greeted with more ancient words: "May God comfort you among the mourners of Zion and Jerusalem." In that verse, the word used for God is *hamakom*—literally, "the place." *May the place comfort you.*

May the people in this place comfort you: the first responders who rush to your rescue, the neighbors who overwhelm evil with kindness, the Americans of every background who inspire more optimism than Jewish history allows. May this country comfort you, with its infinite promise. As George Washington vowed in his 1790 letter to a Rhode Island synagogue, America shall be a place where "every one shall sit in safety under his own vine and fig tree, and there shall be none to make him afraid." Those words aren't his. They're from the Hebrew prophet Micah, on the shelves of every synagogue in the world.

In synagogue as always, we read from the scroll we call the Tree of Life, and the place will comfort us. As we put the book away, we repeat the words from Lamentations: "Renew our days as of old."

Chapter 4

EXECUTED
JEWS

ALA ZUSKIN PERELMAN AND I HAD BEEN IN TOUCH online before I finally met her in person, and I still cannot quite believe that she exists. Years ago, I wrote a novel about Marc Chagall and the Yiddish-language artists whom he once knew in Russia, all of whom were eventually murdered by the Soviet regime. While researching the novel, I found myself sucked into the bizarre story of these people's exploitation and destruction: how the Soviet Union first welcomed these artists as exemplars of universal human ideals, then used them for its own purposes, and finally executed them. I named my main character after the executed Yiddish actor Benjamin Zuskin, a comic performer known for playing fools. After the book came out, I heard from Ala in an email written in halting English: "I am Benjamin Zuskin's daughter." That winter I was speaking at a literary conference in Israel, where Ala lived, and she and I arranged to meet. It was like meeting a character from a book.

My hosts had generously put me up with other writers in a beautiful stone house in Jerusalem. We were there during Hanukkah, the celebration of Jewish independence. On the first night of the holiday, I walked to Jerusalem's Old City and watched as people lit enormous Hanukkah torches at the Western Wall. I thought of my home in New Jersey, where in school growing up I sang fake English Hanukkah songs created by American music education companies at school Christmas concerts, with lyrics describing Hanukkah as being about "joy and peace and love." Joy and peace and love describe Hanukkah, a commemoration of an underdog military victory over a powerful empire, about as well as they describe the Fourth of July. I remembered challenging a chorus teacher about one such song, and being told that I was a poor sport for disliking joy and peace and love. (Imagine a "Christmas song" with lyrics celebrating Christmas, the holiday of freedom. Doesn't everyone like freedom? What pedant would reject such a song?) I sang those words in front of hundreds of people to satisfy my neighbors that my tradition was universal—meaning, just like theirs. The night before meeting Ala, I walked back to the house through the dense stone streets of the Old City's Jewish Quarter, where every home had a glass case by its door, displaying the holiday's oil lamps. It was strange to see those hundreds of glowing lights. They were like a shining announcement that this night of celebration was shared by all these strangers around me, that it was universal. The experience was so unfamiliar that I didn't know what to make of it.

The next morning, Ala knocked on the door of the stone house and sat down in its living room, with its view of the Old City. She was a small dark-haired woman whose perfect posture showed a firmness that belied her age. She looked at me and said in Hebrew, "I feel as if you knew my father, like you understood what he went through. How did you know?"

The answer to that question goes back several thousand years.

———

The teenage boys who participated in competitive athletics in the gymnasium in Jerusalem 2,200 years ago had their circumcisions reversed, because otherwise they wouldn't have been allowed to play. In the Hellenistic empire that had conquered Judea, sports were sacred, the entry point to being a person who mattered, the ultimate height of cool—and sports, of course, were always played in the nude. As one can imagine, ancient genital surgery of this nature was excruciating and potentially fatal. But the boys did not want to miss out.

I learned this fun fact in seventh grade, from a Hebrew school teacher who was instructing me and my pubescent classmates about the Hanukkah story—about how Hellenistic tyranny gained a foothold in ancient Judea with the help of Jews who wanted to fit in. This teacher seemed overly jazzed to talk about penises with a bunch of adolescents, and I suspected he'd made the whole thing up. At home, I decided to fact-check. I pulled a dusty old book off my parents' shelf, Volume One of Heinrich Graetz's opus *History of the Jews*.

In nineteenth-century academic prose, Graetz explained how the leaders of Judea demonstrated their loyalty to the occupying Hellenistic empire by building a gymnasium and recruiting teenage athletes—only to discover that "in uncovering their bodies they could immediately be recognized as Judeans. But were they to take part in the Olympian games, and expose themselves to the mockery of Greek scoffers? Even this difficulty they evaded by undergoing a painful operation, so as to disguise the fact that they were Judeans." Their Zeus-worshiping overlords were not fooled. Within a few years, the regime outlawed not only circumcision but all of Jewish religious practice, and put to death anyone who didn't comply.

Sometime after that, the Maccabees showed up. That's the part of the story we usually hear.

Those ancient Jewish teenagers were on my mind that Hanukkah when Ala came to tell me about her father's terrifying life, because I sensed that something profound united them—something that doesn't match what we're usually taught about what bigotry looks or feels like. It doesn't involve "intolerance" or "persecution," at least not at first. Instead, it looks likes the Jews themselves are choosing to reject their own traditions. It is a form of weaponized shame.

Two distinct patterns of antisemitism can be identified by the Jewish holidays that celebrate triumphs over them: Purim and Hanukkah. In the Purim version of antisemitism, exemplified by the Persian genocidal decrees in the biblical Book of Esther, the goal is openly stated and unambiguous: Kill all the Jews. In the Hanukkah version of antisemitism, whose appearances range from the Spanish Inquisition to the Soviet regime, the goal is still to eliminate Jewish civilization. But in the Hanukkah version, this goal could theoretically be accomplished simply by destroying Jewish civilization, while leaving the warm, de-Jewed bodies of its former practitioners intact.

For this reason, the Hanukkah version of antisemitism often employs Jews as its agents. It requires not dead Jews but cool Jews: those willing to give up whatever specific aspect of Jewish civilization is currently uncool. Of course, Judaism has always been uncool, going back to its origins as the planet's only monotheism, featuring a bossy and unsexy invisible God. Uncoolness is pretty much Judaism's brand, which is why cool people find it so threatening—and why Jews who are willing to become cool are absolutely necessary to Hanukkah antisemitism's success. These "converted" Jews are used to demonstrate the good intentions of the regime—which of course isn't antisemitic but merely requires that its Jews publicly flush thousands of years of Jewish civilization down the toilet in exchange for the worthy prize of not being treated like dirt, or not being murdered. For a few years. Maybe.

I wish I could tell the story of Ala's father concisely, compellingly, the way everyone prefers to hear about dead Jews. I regret to say that Benjamin Zuskin wasn't minding his own business and then randomly stuffed into a gas chamber, that his thirteen-year-old daughter did not sit in a closet writing an uplifting diary about the inherent goodness of humanity, that he did not leave behind sad-but-beautiful aphorisms pondering the absence of God while conveniently letting his fellow humans off the hook. He didn't even get crucified for his beliefs. Instead, he and his fellow Soviet Jewish artists—extraordinarily intelligent, creative, talented, and empathetic adults—were played for fools, falling into a slow-motion psychological horror story brimming with suspense and twisted self-blame. They were lured into a long game of appeasing and accommodating, giving up one inch after another of who they were in order to win that grand prize of being allowed to live.

Spoiler alert: they lost.

———

I was in graduate school studying Yiddish literature, itself a rich vein of discussion about such impossible choices, when I became interested in Soviet Jewish artists like Ala's father. As I dug through library collections of early-twentieth-century Yiddish works, I came across a startling number of poetry books illustrated by Marc Chagall. I wondered if Chagall had known these Yiddish writers whose works he illustrated, and it turned out that he had. One of Chagall's first jobs as a young man was as an art teacher at a Jewish orphanage near Moscow, built for children orphaned by Russia's 1919–1920 civil war pogroms. This orphanage had a rather renowned faculty, populated by famous Yiddish writers who trained these traumatized children in the healing art of creativity.

It all sounded very lovely, until I noticed something else. That Chagall's art did not rely on a Jewish language—that it had, to use that insidious phrase, "universal appeal"— allowed him a chance to succeed as an artist in the West. The rest of the faculty, like Chagall, had also spent years in western Europe before the Russian revolution, but they chose to return to Russia because of the Soviet Union's policy of endorsing Yiddish as a "national Soviet language." In the 1920s and '30s, the USSR offered unprecedented material support to Yiddish culture, paying for Yiddish-language schools, theaters, publishing houses, and more, to the extent that there were Yiddish literary critics who were salaried by the Soviet government. This support led the major Yiddish novelist Dovid Bergelson to publish his landmark 1926 essay "Three Centers," about New York, Warsaw, and Moscow as centers of Yiddish-speaking culture, asking which city offered Yiddish writers the brightest prospects. His unequivocal answer was Moscow, a choice that brought him back to Russia the following year, where many other Jewish artists joined him.

But Soviet support for Jewish culture was part of a larger plan to brainwash and coerce national minorities into submitting to the Soviet regime—and for Jews, it came at a very specific price. From the beginning, the regime eliminated anything in the celebrated Jewish "nationality" that didn't suit its needs. Jews were awesome, provided they weren't practicing the Jewish religion, studying traditional Jewish texts, using Hebrew, or supporting Zionism. The Soviet Union thus pioneered a versatile gaslighting slogan, which it later spread through its client states in the developing world and which remains popular today: it was not antisemitic, merely anti-Zionist. (In the process of not being antisemitic and merely being anti-Zionist, the regime managed to persecute, imprison, torture, and murder thousands of Jews.) What's left of Jewish culture once you surgically remove religious practice, traditional texts, Hebrew, and Zionism? In the Soviet Empire, one answer was Yiddish, but Yiddish was also suspect for its supposedly backward elements. Nearly 15 percent of its words came directly from biblical and rabbinic Hebrew, so Soviet Yiddish schools and publishers, under the guise of "simplifying" spelling, implemented a new and quite literally antisemitic spelling system that eliminated those words' ancient Near Eastern roots. Another answer was "folklore"— music, visual art, theater, and other creative work reflecting Jewish life—but of course most of that cultural material was also deeply rooted in biblical and rabbinic sources, or reflected common religious practices like Jewish holidays and customs, so that was treacherous too.

No, what the regime required were Yiddish stories that showed how horrible traditional Jewish practice was, stories in which happy, enlightened Yiddish-speaking heroes rejected both religion and Zionism (which, aside from its modern political form, is also a fundamental feature of ancient Jewish texts and prayers traditionally recited at least three times daily). This de-Jewing process is clear from the repertoire of the government-sponsored Moscow State Yiddish Theater, which could only present or adapt Yiddish plays that denounced traditional Judaism as backward, bourgeois, corrupt, or even more explicitly—as in the many productions involving ghosts and graveyard scenes—as dead. As its actors would be, soon enough.

The Soviet Union's destruction of Jewish culture commenced, in a calculated move, with Jews positioned as the destroyers. It began with the Yevsektsiya, committees of Jewish Bolsheviks whose paid government jobs from 1918 through 1930 were to persecute, imprison, and occasionally murder

Jews who participated in religious or Zionist institutions—categories that included everything from synagogues to sports clubs, all of which were shut down and their leaders either exiled or "purged." This went on, of course, until the regime purged the Yevsektsiya members themselves.

The pattern repeated in the 1940s. As sordid as the Yevsektsiya chapter was, I found myself more intrigued by the undoing of the Jewish Antifascist Committee, a board of prominent Soviet Jewish artists and intellectuals established by Joseph Stalin in 1942 to drum up financial support from Jews overseas for the Soviet war effort. Two of the more prominent names on the JAC's roster of talent were Solomon Mikhoels, the director of the Moscow State Yiddish Theater, and Ala's father Benjamin Zuskin, the theater's leading actor. After promoting these people during the war, Stalin decided these loyal Soviet Jews were no longer useful, and charged them all with treason. He had decided that this committee he himself had created was in fact a secret Zionist cabal, designed to bring down the Soviet state. Mikhoels was murdered first, in a 1948 hit staged to look like a traffic accident. Nearly all the others—Zuskin and twelve more Jewish luminaries, including the novelist Dovid Bergelson, who had proclaimed Moscow as the center of the Yiddish future—were executed by firing squad on August 12, 1952.

Just as the regime accused these Jewish artists and intellectuals of being too "nationalist" (read: Jewish), today's long hindsight makes it strangely tempting to read this history and accuse them of not being "nationalist" enough—that is, of being so foolishly committed to the Soviet regime that they were unable to see the writing on the wall. Many works on this subject have said as much. In *Stalin's Secret Pogrom*, the indispensable English translation of transcripts from the JAC "trial," Russia scholar Joshua Rubenstein concludes his lengthy introduction with the following:

> As for the defendants at the trial, it is not clear what they believed about the system they each served. Their lives darkly embodied the tragedy of Soviet Jewry. A combination of revolutionary commitment and naïve idealism had tied them to a system they could not renounce. Whatever doubts or misgivings they had, they kept to themselves, and served the Kremlin with the required enthusiasm. They were not dissidents. They were Jewish martyrs. They were also Soviet patriots. Stalin repaid their loyalty by destroying them.

This is completely true, and also completely unfair. The tragedy—even the term seems unjust, with its implied blaming of the victim—was not that these Soviet Jews sold their souls to the devil, though many clearly did. The tragedy was that integrity was never an option in the first place.

———

Ala was almost thirteen years old when her father was arrested, and until that moment she was immersed in the Soviet Yiddish artistic scene. Her mother was also an actor in the Moscow State Yiddish Theater; her family lived in the same building as the murdered theater director Solomon Mikhoels, and moved in the same circles as other Jewish actors and writers. After seeing her parents perform countless times, Ala had a front-row seat to the destruction of their world. She attended Mikhoels's state funeral, heard about the arrest of the brilliant Yiddish author Der Nister from an actor friend who witnessed it from her apartment across the hall, and was present when secret police ransacked her home in conjunction with her father's arrest. In her biography of her father, *The Travels of Benjamin Zuskin*, she provides for her readers what she gave me that morning in Jerusalem: an emotional recounting, with the benefit of hindsight, of what it was really like to live through the Soviet Jewish nightmare.

It's as close as we can get, anyway. Her father Benjamin Zuskin's own thoughts on the topic are available only from state interrogations extracted under unknown tortures. (One typical interrogation document from his three and a half years in the notorious Lubyanka Prison announces that that day's interrogation lasted four hours, but the transcript is only half a page long—leaving to the imagination how interrogator and interrogatee may have spent their time together. Suffice it to say that another JAC detainee didn't make it to the trial alive.) His years in prison began when he was arrested in December of 1948 in a Moscow hospital room, where he was being treated for chronic insomnia brought on by the murder of his boss and career-long acting partner, Mikhoels; the secret police strapped him to a gurney and carted him to prison in his hospital gown while he was still sedated.

But in order to truly appreciate the loss here, one needs to know what was lost—to return to the world of the great

Yiddish writer Sholem Aleichem, the author of Benjamin Zuskin's first role on the Yiddish stage, in a play fittingly titled *It's a Lie!*

Benjamin Zuskin's path to the Yiddish theater and later to the Soviet firing squad began in a shtetl comparable to those immortalized in Sholem Aleichem's work. Zuskin, a child from a traditional family who was exposed to theater only through traveling Yiddish troupes and clowning relatives, experienced that world's destruction: his native Lithuanian shtetl, Ponievezh, was among the many Jewish towns forcibly evacuated during the First World War, catapulting him and hundreds of thousands of other Jewish refugees into modernity. He landed in Penza, a city with professional Russian theater and Yiddish amateur troupes. In 1920, the Moscow State Yiddish Theater opened, and by 1921, Zuskin was starring alongside Mikhoels, the theater's leading light.

In the one acting class I have ever attended, I learned only one thing: acting isn't about pretending to be someone you aren't, but rather about emotional communication. Zuskin, who not only starred in most productions but also taught in the theater's acting school, embodied that concept. His very first audition was a one-man sketch he created, consisting of nothing more than a bumbling old tailor threading a needle— without words, costumes, or props. It became so popular that he performed it to entranced crowds for years. This physical artistry animated his every role. As one critic wrote, "Even the slightest breeze and he is already air-bound."

Zuskin specialized in playing figures like the Fool in *King Lear*—as his daughter puts it in her book, characters who "are supposed to make you laugh, but they have an additional dimension, and they arouse poignant reflections about the cruelty of the world." Discussing his favorite roles, Zuskin once explained that "my heart is captivated particularly by the image of the person who is derided and humiliated, but who loves life, even though he encounters obstacles placed before him through no fault of his own."

The first half of Ala's book seems to recount only triumphs. The theater's repertoire in its early years was largely adapted from classic Yiddish writers like Sholem Aleichem, I. L. Peretz, and Mendele Moykher Seforim. The book's title is drawn from Zuskin's most famous role: Senderl, the Sancho Panza figure in Mendele's *Don Quixote*–inspired work, *Travels of Benjamin the Third*, about a pair of shtetl idiots who set out for the Land of Israel and wind up walking around the block. These productions were artistically inventive, brilliantly acted, and played to packed houses both at home and on tour. *Travels of Benjamin the Third*, in a 1928 review typical of the play's reception, was lauded by the *New York Times* as "one of the most originally conceived and beautifully executed evenings in the modern theater."

One of the theater's landmark productions, I. L. Peretz's surrealist masterpiece *At Night in the Old Marketplace*, was first performed in 1925. The play, set in a graveyard, is a kind of carnival for the graveyard's gathered ghosts. Those who come back from the dead are misfits like drunks and prostitutes, and also specific figures from shtetl life—yeshiva idlers, synagogue beadles, and the like. Leading them all is a *badkhn*, or wedding jester—divided in this production into two mirror-characters played by Mikhoels and Zuskin—whose repeated chorus among the living corpses is "The dead will rise!" "Within this play there was something hidden, something with an ungraspable depth," Ala writes, and then relates how after a performance in Vienna, one theatergoer came backstage to tell the director that "the play had shaken him as something that went beyond all imagination." The theatergoer was Sigmund Freud.

As Ala traces the theater's trajectory toward doom, it becomes obvious why this performance so affected Freud. The production was a zombie story about the horrifying possibility of something supposedly dead (here, Jewish civilization) coming back to life. The play was written a generation earlier as a Romantic work, but in the Moscow production, it became a means of denigrating traditional Jewish life without mourning it. That fantasy of a culture's death as something compelling and even desirable is not merely reminiscent of Freud's death drive, but also reveals the self-destructive bargain implicit in the entire Soviet-sponsored Jewish enterprise. In her book, Ala beautifully captures this tension as she explains the *badkhn*'s role: "He sends a double message: he denies the very existence of the vanishing shadow world, and simultaneously he mocks it, as if it really does exist."

This double message was at the heart of Benjamin Zuskin's work as a comic Soviet Yiddish actor, a position that required him to mock the traditional Jewish life he came from while also pretending that his art could exist without it. "The chance to make fun of the shtetl which has become a thing of the past charmed me," he claimed early on, but later, according to his daughter, he began to privately express misgivings. The theater's decision to stage *King Lear* as a way of elevating itself disturbed him, suggesting as it did that the Yiddish repertoire was inferior. His own integrity came from his deep devotion to *yidishkayt*, a sense of essential and enduring Jewishness, no matter how stripped-down that identity had become. "With the sharp sense of belonging to everything Jewish, he was tormented by the theater forsaking its expression of this belonging," his daughter writes. Even so, "no, he could not allow himself to oppose the Soviet regime even in his thoughts, the regime that gave him his own theater, but 'the heart and the wit do not meet.'"

In Ala's memory, her father differed from his director, partner, and occasional rival, Mikhoels, in his complete disinterest in politics. Mikhoels was a public figure as well as a performer, and his leadership of the Jewish Antifascist Committee, while no more voluntary than any public act in a totalitarian state, was a role he played with gusto, traveling to America in 1943 and speaking to thousands of American Jews to raise money for the Red Army in their battle against the Nazis. Zuskin, on the other hand, was on the JAC roster, but seems to have continued playing the fool. According to both his daughter and his trial testimony, his role in the JAC was almost identical to his role on a Moscow municipal council, limited to playing chess in the back of the room during meetings.

In Jerusalem, Ala told me that her father was "a pure soul." "He had no interest in politics, only in his art," she said, describing his acting style as both classic and contemporary, praised by critics for its timeless qualities that are still evident today in his film work. But his talent was the most nuanced and sophisticated thing about him. Offstage, he was, as she put it in Hebrew, a *"tam"*—a biblical term sometimes translated as fool or simpleton, but which really means an innocent. (It is the first adjective used to describe the title character in the Book of Job.) It is true that in trial transcripts, Zuskin comes out looking better than many of his co-defendants by playing dumb instead of pointing fingers. But was this ignorance, or a wise acceptance of the futility of trying to save his skin? As *King Lear*'s Fool put it, "They'll have me whipp'd for speaking true; thou'lt have me whipp'd for lying, and sometimes I am whipp'd for holding my peace." Reflecting on her father's role as a fool named Pinia in a popular film, Ala writes in her book, "When I imagine the moment when my father heard his death sentence, I see Pinia in close-up . . . his shoulders slumped, despair in his appearance. I hear the tone that cannot be imitated in his last line in the film—and perhaps also the last line in his life?—'I don't understand anything.'"

Yet it is clear that Zuskin deeply understood how impossible his situation was. In one of the book's more disturbing moments, Ala describes him rehearsing for one of his landmark roles, that of the comic actor Hotsmakh in Sholem Aleichem's *Wandering Stars*, a work whose subject is the Yiddish theater. He had played the role before, but this production was going up in the wake of Mikhoels's murder. Zuskin was already among the hunted, and he knew it. As Ala writes:

One morning—already after the murder of Mikhoels—I saw my father pacing the room and memorizing the words of Hotsmakh's role. Suddenly, in a gesture revealing a hopeless anguish, Father actually threw himself at me, hugged me, pressed me to his heart, and together with me, continued to pace the room and to memorize the words of the role. That evening I saw the performance . . . "The doctors say that I need rest, air, and the sea . . . For what . . . without the theater?" [Hotsmakh asks], he winds the scarf around his neck—as though it were a noose. For my father, I think these words of Hotsmakh were like the motif of the role and—I think—of his own life.

Describing the charges levied against Zuskin and his peers is a degrading exercise, for doing so makes it seem as though these charges are worth considering. They are not. It is at this point that Hanukkah antisemitism transformed, as it inevitably does, into Purim antisemitism. Here Ala offers what hundreds of pages of state archives can't, describing the impending horror of the noose around one's neck.

Her father stopped sleeping, began receiving anonymous threats, and saw that he was being watched. No conversation was safe. When a visitor from Poland waited near his apartment building to give him news of his older daughter Tamara (who was then living in Warsaw), Zuskin instructed the man to walk behind him while speaking to him and then to switch directions, so as to avoid notice. When the man asked Zuskin what he wanted to tell his daughter, Zuskin "approached the guest so closely that there was no space between them, and whispered in Yiddish, 'Tell her that the ground is burning beneath my feet.'" It is true that no one can know what

Zuskin or any of the other defendants really believed about the Soviet system they served. It is also true—and far more devastating—that their beliefs were utterly irrelevant.

Ala and her mother were exiled to Kazakhstan after her father's arrest, and learned of his execution only when they were allowed to return to Moscow in 1955. By then, he had already been dead for three years.

In Jerusalem that morning, Ala told me, in a sudden private moment of anger and candor, that the Soviet Union's treatment of the Jews was worse than Nazi Germany's. I tried to argue, but she shut me up. Obviously the Nazi atrocities against Jews were incomparable, a fact Ala later acknowledged in a calmer mood. But over four generations, the Soviet regime forced Jews to participate in and internalize their own humiliation—and in that way, Ala suggested, they destroyed far more souls. And they never, ever paid for it.

"They never had a Nuremberg," Ala told me that day, with a quiet fury. "They never acknowledged the evil of what they did. The Nazis were open about what they were doing, but the Soviets pretended. They lured the Jews in, they baited them with support and recognition, they used them, they tricked them, and then they killed them. It was a trap. And no one knows about it, even now. People know about the Holocaust, but not this. Even here in Israel, people don't know. How did you know?"

———

That evening I went out to the Old City again, to watch the torches being lit at the Western Wall for the second night of Hanukkah. I walked once more through the Jewish Quarter, where the oil lamps, now each bearing one additional flame, were displayed outside every home, following the tradition to publicize the Hanukkah miracle—not merely the legendary long-lasting oil but the miracle of military and spiritual victory over a coercive empire, the freedom to be uncool, the freedom not to pretend. Somewhere nearby, deep underground, lay the ruins of the gymnasium where de-circumcised Jewish boys once performed naked before approving crowds, stripped of their integrity and left with their private pain. I thought of Benjamin Zuskin performing as the dead wedding jester, proclaiming, "The dead will rise!" and then performing again in a "superior" play, as *King Lear*'s Fool. I thought of the ground burning beneath his feet. I thought of his daughter, Ala, now an old woman, walking through Jerusalem.

I am not a sentimental person. As I returned to the stone house that night, along the streets lit by oil lamps, I was surprised to find myself crying.

Chapter 5

FICTIONAL

DEAD

JEWS

AS A NOVELIST I AM FORTUNATE TO GET A LOT OF MAIL from readers, some of it more pleasant than others. Years ago, one of my esteemed readers sent me the following email:

Dear Ms. Horn, I recently began reading your book The World to Come *[about a pogrom survivor]. After the scene of the horse being beaten, I threw the book across the room. With all the cruelty in the world, I find it more of a service to mankind to write a book for people to laugh, enjoy and be uplifted. Best wishes, Denise.*

I wrote, but did not send, a reply to Denise:

Dear Denise, Sorry about the horse. It was a reference to Crime and Punishment, *which is another book you might want to avoid. You should also steer clear of the Bible, which is likewise not a great book for people who want to laugh and enjoy. However, I do have some Garfield comics I can highly recommend for their service to mankind. Best wishes, Dara.*

It's easy to laugh at Denise, but her message reveals many readers' unspoken expectations about the purpose of literature. Sophisticated readers don't insist on Garfield and happy endings, but I've found that even educated readers who appreciate tragedy still secretly expect a "redemptive" ending—or as Denise put it, something "uplifting." I've thought about Denise's stupid email to me for many years, because it raises a very fundamental question: What are stories for? That question becomes even harder to answer when we consider that an astonishing proportion of what counts today as "Jewish" literature in English is basically Holocaust fiction. If the purpose of literature is to "uplift" us, is it even possible to write fiction that is honest about the most horrifying aspects of the Jewish past?

When I was in my twenties, I went and got a doctorate in comparative literature just to answer this question. The answer was depressing, for all the wrong reasons.

———

I entered graduate school in order to study Hebrew and Yiddish literature, but the scholars and critics I had to read in the comparative literature department's required courses were all basing their ideas on what I quickly learned were "normal" literatures: English, German, Russian, Spanish, French. That's when I noticed the problem that distinguishes literature in Jewish languages from most other literary canons, and essentially makes Jewish literature into a kind of anti-literature— one that should make everyone question what they want out of a work of literary art.

In 1965, the British literary critic Frank Kermode published his monumental work *The Sense of an Ending*, which essentially tries to explain the purpose of storytelling. Kermode points out how much readers desire coherent and satisfying endings, and then connects that desire to the history of Western religion. He compares literature with religion in what he calls its "desire for consonance," or the desire to live in a world that makes sense. As he puts it, "Everything is relevant if its relevance can be invented." Readers, he says, are like fundamentalists who see everything as a sign from God, because in literature, every detail actually does matter. Just as religion urges us to see our world as an inherently meaningful place where events always serve some larger purpose, writers create that purpose within stories by making every detail add up to something meaningful, thereby inventing a coherent world. As a graduate student who was simultaneously writing novels, I read this and patted myself on the back. See, I wasn't just procrastinating on my dissertation; I was inventing a coherent world!

But I very quickly saw the problem. This idea of religion imposing coherence on the world sounded absolutely nothing like the religion I knew best. Kermode's argument is based on the idea that Western religion is all about "endings." As he puts it, "The Bible is a familiar model of history. It begins at the beginning with the words 'In the beginning,' and it ends with a vision of the end, with the words, 'Even so, come, Lord Jesus.'"

Needless to say, this is not how the Hebrew Bible ends. The Tanakh, as Jews call the Hebrew Bible, has plenty of apocalyptic visions, but its final pages in the rather plodding Book of Chronicles don't exactly end with a bang. Even the Torah, the part of the Hebrew Bible that Jews publicly chant aloud from start to finish every year and then begin reading again, doesn't have much finality to it, other than the expected death of Moses. Instead, the Torah ends with a cliffhanger, stopping just before the Israelites' long-awaited arrival in the Promised Land. The characters never even make it home. It slowly dawned on me that Kermode's idea of religion giving us an "ending" isn't universal at all. It's Christian.

In graduate school, of course, we quickly left Kermode and his contemporaries behind and continued on to structuralism, deconstruction, and other more adventurous schools of thought. But as a writer rather than a scholar, I found that Kermode's idea stayed with me, because it felt true to the expectations of ordinary readers—including my own. The more I considered it, the more I realized just how pervasive, and how accurate, Kermode's idea of literature actually is for English-language readers. Think about what we expect from the endings of stories—not just Denise, but all of us. We expect the good guys to be "saved." If that doesn't happen, we at least expect the main character to have an "epiphany." And if that doesn't happen, then at least the author ought to give us a "moment of grace." All three are Christian terms. So many of our expectations of literature are based on Christianity—and not just Christianity, but the precise points at which Christianity and Judaism diverge. And then I noticed something else: the canonical works by authors in Jewish languages almost never give their readers any of those things.

I was studying modern Yiddish and Hebrew literature, and I began to see that the major works in these Jewish languages almost never involved characters getting saved, or having epiphanies, or experiencing moments of grace. In fact, as I read my way through the foundational works in these literatures, I saw that many of the canonical stories and novels in modern Yiddish and Hebrew literature actually didn't have endings at all.

One major point of entry for modern Jewish literature is the early-nineteenth-century fiction of Rabbi Nachman of Bratslav, a religious leader whose homiletic tales became a baseline inspiration for secular and religious Jewish writers alike. Nachman's stories were important enough to later secular Jewish writers that there are even plays and poems written *about* them, well into the twentieth century; even Franz Kafka was inspired by them and adapted them in his own work. Nachman's stories are fairy tales of a sort, written very much like the stories collected by the Brothers Grimm, which Nachman reportedly read in German with enormous interest. They have many of the same princes and knights and quests and magical agents. What Nachman's stories don't have, though, are happy endings—or any endings at all.

Nachman's story "The Loss of the Princess," for instance, has all the elements of a fairy tale: a banished princess hidden away in an inaccessible castle, a noble knight who sets out on a quest to rescue her, and plenty of riddles to solve and giants to battle on the way. But here's how Nachman's story, narrated aloud to his scribe, ends: "And how he freed her, Nachman did not tell. But he did free her." Another story of Nachman's, "Tale of the Seven Beggars," likewise has all the right fairy-tale elements: spunky orphans who get lost in a forest, and then seven roving beggars with stylized disabilities, a sort of Seven Dwarves, who offer the orphans generosity and wisdom. When the spunky orphans grow up and (of course) get married, each of the poetically disabled beggars provides his own *Canterbury Tales*–style story-within-a-story at their wedding feast. It's all rather conventionally satisfying—except that after the sixth beggar, the story simply stops. The seventh beggar never shows up. These missing endings seem like storytelling failures, but they're entirely deliberate. Nachman was making a religious point about living in a broken and unredeemed world.

As I read my way through modern Yiddish and Hebrew literature, I kept running into this pattern. Sholem Aleichem's *Tevye the Dairyman* stories are familiar to English-language audiences from their Broadway adaptation as *Fiddler on the Roof*. But *Fiddler on the Roof* left out a few details, including Tevye's wife, Golde, dropping dead; Tevye's son-in-law Motl dropping dead; and Tevye's daughter Shprintze drowning herself, none of which would have played well on Broadway. What's even less "uplifting" about the *Tevye* stories is their structure. They're like a TV series where each of Tevye's daughters' marriages is a different episode, and each one is more devastating than the last. But as the series progresses and twenty years pass, Tevye himself never changes. He never learns anything; he never realizes anything; he never has an epiphany or a moment of grace. And he's certainly never rescued or saved. Instead he just keeps enduring, which feels achingly realistic. His great power is that he remains exactly who he always was.

No matter whose work I read among these major Jewish-language writers, I kept running into this problem. The Hebrew Nobel laureate S. Y. Agnon wrote amazing novels that built entire worlds out of centuries of ancient Jewish texts that he brought to life in new and ironic ways, but the best ending you can get from him is that maybe after six hundred pages, somebody dies. The foundational nineteenth-century author Mendele Moykher Seforim wrote novels that essentially introduced stylistic literary sophistication to both Hebrew and Yiddish—but his most famous book, full of surreal and world-altering adventures, ends with the protagonist essentially saying, "And then I woke up." The Yiddish Nobel laureate Isaac Bashevis Singer ended almost every one of his novels not with a resolution but with the protagonist disappearing, whether by running away or by literally locking himself into a closet for the rest of his life. My favorite Hebrew novel of all time, A. B. Yehoshua's 1989 masterpiece *Mr. Mani*, is a fantastically inventive story that moves backwards in time through six generations of a Jerusalem family while tracing the family's recurring suicidal gene—until you get to the end, which is really the beginning, when the enduring mystery of the family's self-destruction "resolves" not with an answer, but with a question, one that casts new light over everything that came before, but which remains unanswered.

These stories, I came to understand, were presenting a challenge to the Western idea of the purpose of creativity. Stories with definitive endings don't necessarily reflect a belief that the world makes sense, but they do reflect a belief in the power of art to make sense of it. What one finds in Jewish storytelling, though, is something really different: a kind of realism that comes from humility, from the knowledge that one cannot be true to the human experience while pretending to make sense of the world. These are stories without conclusions, but full of endurance and resilience. They are about human limitations, which means that the stories are not endings but beginnings, the beginning of the search for meaning rather than the end—and the power of resilience and endurance to carry one through to that meaning. Tevye, after grieving for his wife, daughter, and son-in-law and being expelled from his home, finally leaves the reader with a line that would never work on Broadway: "Tell all our Jews everywhere that they shouldn't worry: our old God still lives!"

I eventually came to understand the profound insult inherent in the messages I was receiving, both directly and indirectly, from readers expecting uplifting Jewish literature full of moments of grace—not to me as a novelist, but to my ancestors who endured experiences like those I gave to my characters, and in a sense, to all those who have endured the most atrocious moments of Jewish history. Readers who demanded that "coherence" from literature about the modern Jewish experience were essentially insisting that Jewish suffering was only worth examining if it provided, in the words of my reader's memorable message, "a service to mankind." In retrospect I am stunned by how long it took me to understand just how hateful this was. Consider, as I only very slowly did, what this demand really entails. Dead Jews are supposed to teach us about the beauty of the world and the wonders of redemption—otherwise, what was the point of killing them in the first place? That's what dead Jews are for! If people were going to read about dead Jews, where was the service to mankind I owed them?

This is far from a fringe attitude among contemporary readers, as just about every bestselling Holocaust novel of our current century makes fantastically clear. Holocaust novels that have sold millions of copies both in the United States and overseas in recent years are all "uplifting," even when they include the odd dead kid. *The Tattooist of Auschwitz*, a recent international mega-bestseller touted for its "true story," manages to present an Auschwitz that involves a heartwarming romance. *Sarah's Key*, *The Book Thief*, *The Boy in Striped Pajamas*, and many other bestsellers, some of which have even become required reading in schools, all involve non-Jewish rescuers who risk or sacrifice their own lives to save hapless Jews, thus inspiring us all. (For the record, the number of actual "righteous Gentiles" officially recognized by Yad Vashem, Israel's national Holocaust museum and research center, for their efforts in rescuing Jews from the Holocaust is under 30,000 people, out of a European population at the time of nearly 300 million—or 0.01 percent. Even if we were to assume that the official recognition is an undercount by a factor of *a thousand*, such people remain essentially a rounding error.) In addition to their wonderful non-Jewish characters, these books are almost invariably populated by the sort of relatable dead Jews whom readers can really get behind: the mostly non-religious, mostly non-Yiddish-speaking ones whom noble people tried to save, and whose deaths therefore teach us something beautiful about our shared and universal humanity, replete with epiphanies and moments of grace. Statistically speaking, this was not the experience of almost any Jews who endured the Holocaust. But for literature in non-Jewish languages, that grim reality is both inconvenient and irrelevant.

What does a novel about the horrors of Jewish history in a Jewish language look like?

For English-language readers drowning in uplifting Holocaust fiction, here is one novel, among many, that demonstrates a more honest way to write fiction about atrocity: Chava Rosenfarb's *The Tree of Life*, a panoramic Yiddish-language trilogy about the Łódź Ghetto. To call it a masterpiece would be an understatement. It is the sort of work—long, immersive, engrossing, exquisite—that feels less like reading a book than living a life.

Make that ten lives. That's about how many major characters

we come to know intimately in Rosenfarb's sweeping epic, and we meet them all in vivid detail before the war begins, so we know who they are before sadists take over their lives. Some are sadists themselves, like Mordechai-Chaim Rumkowski, the infamous Nazi-selected "King of the Jews" who ruled the Łódź Ghetto with an iron fist; we first meet him before the war, as an orphanage director who sucks up to rich donors while sexually molesting his young female wards. Most are "ordinary" people—except there's no such thing as ordinary, as the vast variety of the Jews of Łódź makes clear. Prewar Łódź was one-third Jewish, and Rosenfarb brilliantly unfolds a panorama of the city in all its diversity by intertwining her complex characters' lives. The wealthy industrialist Samuel Zuckerman is obsessed with the history of the Jews of Łódź, an interest he shares with Itche Mayer, a poor Jewish carpenter in his employ—and into whose slum neighborhood Zuckerman himself moves when that slum becomes the ghetto. Zuckerman's family-man civility is disdained by Adam Rosenberg, another wealthy industrialist in his circle who thrives on cruelty and sexual conquest. We meet rationalists like the doctor Michal Levine, proud Polish patriots like the spinster teacher Dora Diamant, passionate Communists like the orphaned Esther, Socialists and Zionists among Itche Mayer's sons, and the slightly surreal "Toffee Man," a religious father of nine who periodically appears unbidden, offering other characters unexpected moments of hope. Nor are these characters reducible to representatives of a type or class. They are each embedded, as real people are, in networks of families, lovers, friends, and enemies; each is inspired by their own commitments and also plagued by private doubts. The integrity of these characters depends, as it does for all of us, on their inherent adulthood, their agency in their own choices. In the ghetto, none of that disappears; each character remains exactly who he or she was before, just in inhuman circumstances. The Holocaust was not a morality play, except perhaps for its perpetrators. And that's exactly what makes the ghetto's horrors real.

I would have thought these horrors would be impossible to convey, except that Rosenfarb brings you there. Despite our own culture's saturation in violent imagery, *The Tree of Life* is extremely difficult to read. There is no ruminating about God here, no contrived conversations with Nazis that show their humanity, nor even any brave rebellion—at least, not until the very end. Instead there is confusion, starvation, denial, and sheer sadistic horror. As you read, you are shocked to realize that no one in the book knows what you know. Instead they believe, when imprisonment and forced labor commence at the start of Volume Two, that this slavery and starvation is the central atrocity they are enduring. When deportations begin, some even opt in, reasoning that things cannot possibly be worse. It is only when familiar and sometimes bloodstained clothing begins returning to the slave-labor processing centers (in some cases with family photographs still in the pockets) that some characters realize what is happening—yet even they are quickly (and gladly) silenced by the forces of denial. Meanwhile, German soldiers shoot children in the streets for fun. Power politics among "influential" Jews quickly becomes a blood sport, with people stopping at nothing, including sexual servitude, to protect themselves and those they love—all, of course, to no avail. Soon characters we care about begin falling like dominoes, whether deported, starved, diseased, shot, or tortured; one major character winds up castrated. By Volume Three, the Germans demand that the Jews hand over all children under ten.

Rosenfarb herself (1923–2011), a renowned Yiddish poet who lived most of her life in Canada, survived the Łódź Ghetto and subsequently Auschwitz and Bergen-Belsen. *The Tree of Life*, published in Yiddish in 1971 and in English translation in 1985, could be mistaken for a survivor's testimony. The extreme detail with which Rosenfarb brings to life Jewish Łódź, its people and its passions, is itself an enormous achievement, a monument and a memorial to a destroyed community, written in the great (and alas, very long) tradition of Jewish literary lament. Yet *The Tree of Life* is not a work of testimony but a work of art. One character, the aspiring teenage poet Rachel Eibushitz, most closely resembles Rosenfarb herself, but this character is simply one of many and hardly the most important. Instead of memoir, Rosenfarb offers true imagination, bringing us into the minds of many different people and rendering even the most despicable figures with the utmost imaginative empathy.

Yet the greatest miracle for the reader is the chance to meet the city's artists, who come to life in Rosenfarb's words. One of the novel's most vivid characters is the poet Simkha Bunim Berkovitch, Rosenfarb's stand-in for her own poetic mentor Simkha Bunim Shayevitch, who was murdered at Dachau after the murders of his wife and two young children.

Berkovitch comes from a large Hasidic family, but loses his faith as he discovers his poetic talent. A poor factory worker, Berkovitch is a true artist, living only to create; his life means nothing to him without the ability to produce his poetry, and his only fortune in the ghetto is a menial job he obtains (with help) that allows him time and space to write. It's a drive all writers can understand. One of the book's most affecting early scenes involves Berkovitch's marriage to a woman who ultimately can't appreciate his art. A lesser novelist would play up this conflict, but Rosenfarb knows that artists are humans who live with contradictions. The uncompromised beauty of Berkovitch's family life is among the startling wonders of this novel—and its sudden destruction is among its most devastating.

Amid the unrelenting horror, Rosenfarb's characters render miracles. In one of the book's most astonishing scenes, a group of young people and another poet gather in the street, drawn by the poet's humming of a classical symphony; the poet leads them to the tiny room of Vladimir Winter, a middle-aged hunchback introduced as "the Rembrandt of the ghetto." The young people crowd the room, its walls covered with brown paper, as Winter orders the poet to recite his work. As he recites, Winter takes a box of crayons and begins illustrating the words, covering the walls with surreal drawings that incorporate the men and women in the room, imposing their faces on animals, casting their bodies into open meadows, dipping their hands into pools of water, winding their hair into clouds. When the poet finishes reciting, Winter continues drawing as the light fades outside, and one young woman begins to sing, continuing the creative trance. When all four walls are covered, the poet turns on the electric light and the visitors rise from the floor, looking around as if "falling into a dream. There was a land surrounding them, a land of painful beauty, of light and shadows, which enveloped them with the perfume of an unknown life." Winter then passes out from tubercular fever. The book's last volume ends—or rather, stops—at the gates of Auschwitz. It does not provide an inspirational quote.

That "unknown life," of course, was the creative worlds lost by the murders of these artists; Rosenfarb's work itself, for all its power, can only hint at their destroyed potential. For them there is no redemption except in this novel's pages—a redemption only possible through us, the readers. But we as readers cannot ask the book to uplift us, the way we expect, obscenely, for every other book about atrocity. Reading this monumental work requires an active commitment. It provides, one might say, a service to mankind: it broadens your life beyond your own imagining, allowing your life to include many other lives within it. It brings you down to the deepest level of existence, and offers what Rosenfarb herself describes in a poem called "Praise":

When the light fades
And the end approaches
And abruptly you see yourself standing
In a deep dark gate
Look back one more time
At that bubble of reality,

And praise it, that day
That drips out from being—
Unnoticed,
Vanished,
In the night of forgetting.

Chapter 6

LEGENDS

OF

DEAD JEWS

AMERICAN JEWS ARE A HIGHLY EDUCATED GROUP OF people—and not just educated, but great at asking annoying questions. In the most recent Pew Survey, 49 percent of American Jews claimed that a key part of their Jewish identity involved "being intellectually curious." In other words, American Jews see themselves as people who don't merely value their university degrees, but also their skepticism, their critical-thinking skills, and their refusal to take anything at face value.

So I didn't think it was a big deal a few years ago when I gave a public lecture at a Jewish institution and casually mentioned that the family story so many American Jews have heard, that their surnames were changed at Ellis Island, is a myth. At Ellis Island, which has been up and running as a National Park Service museum for over thirty years, this is routinely announced on public tours. More recently, we have entered an era of trendy genealogy, bolstered by cheap DNA testing that has led tens of thousands of Americans down the rabbit hole of ancestry research, with ample guidance from online forums, TV documentaries, family tree construction software, and accessible archival databases. With this public glut of information, I hardly thought my mention was news.

Wow, was I wrong.

After that talk, I was mobbed by people—angry people, in a scrum. These were well-read, highly educated American Jews, each of whom furiously explained to me that while maybe *most* people's names weren't changed at Ellis Island, *their* great-grandfather was the exception. None of these people offered any evidence, other than to assure me, "My great-grandfather wouldn't lie!"

I didn't lose any sleep over my Ellis Island mob. But then it happened again. I wrote an article for a Jewish publication in which I compared the "My name was changed at Ellis Island" story to similar historical material, such as Washington chopping down the cherry tree, the CIA killing Kennedy, and the lunar landing being faked to impress the Soviets. In the comments section, hundreds of people explained to me how I was totally wrong, because . . . well, instead of evidence, they then inserted a five-hundred-word anecdote about their great-grandmother, *so there.*

My angry hecklers have taught me a great deal about the power of founding legends, about mythmaking and its purpose.

But now I know I have to get the facts out of the way first. So, for the record: No, your family's name was not changed at Ellis Island, and your ancestors were not the exception. Here is how we know.

First of all, there was no language problem at Ellis Island. Immigration inspectors there were not rent-a-cops. These were highly trained people who were required to be fluent in at least three languages, and additional translators circulated to ensure competency—and in this context, the languages spoken by Jewish immigrants were far from obscure. Second, immigration processing at Ellis Island wasn't like checking ID at today's airports. These were long interviews, twenty minutes or more, because the purpose of this process was to weed out anyone who was likely to become, in the jargon of the time, "a public charge." So this was not a situation where some idiot behind a desk was just moving a line along.

Even if it were: *nobody at Ellis Island ever wrote down immigrants' names.* Immigrants' names were provided by ship's manifests, compiled at the port of origin. Ships' manifests in Europe were based on passports and other state-issued documents. Those compiling ships' manifests were very careful to get them right, because errors cost them money and potentially their jobs. Any immigrant who was improperly documented on board these vessels had to be sent back to Europe at the shipping company's expense.

Yet there is ample evidence of name changing: thousands of court records from the 1920s, '30s, '40s, and '50s of Jewish immigrants and their children filing petitions in New York City Civil Court in order to change *their own* family names.

In her book *A Rosenberg by Any Other Name*, the historian Kirsten Fermaglich tracks these court filings. For legal name changes, petitioners had to provide the court with their reasons for changing their names. And that's where we see the heartbreaking reality behind the funny stories about Ellis Island. In these legal petitions, as Fermaglich unemotionally reports, we meet thousands of American Jews, most of them born in the United States, explaining under oath that they are changing their names because they cannot find a job, or because their children are being humiliated or discriminated against at school, or because with their real names, no one will hire them for any white-collar position—because, essentially, American antisemitism has prevented their families' success.

In her analysis of thousands of name-change petitions, Fermaglich notes many clear patterns. One is that those with Jewish-sounding names overwhelmingly predominated such court filings. In 1932, for instance (nearly a decade after Ellis Island ceased routine immigrant processing), over 65 percent of name-change petitions in New York were filed by people with Jewish-sounding names. The next-largest group, those with Italian-sounding names, made up a mere 11 percent of filings. Granted, the Jewish population of New York that year was twice the size of the city's Italian American population—but not six times the size. Another pattern Fermaglich uncovered is that petitioners with Jewish-sounding names often filed name-change petitions *as families*; frequently the motivation cited for the name change involved the educational and professional prospects of the petitioners' children. In these petitions that Fermaglich rather dispassionately describes, we witness ordinary

American Jews in the debasing act of succumbing to discrimination instead of fighting it.

American antisemitism during the decades that followed the mass migration was, as Fermaglich puts it, "private" and therefore "insidious." In the earlier part of the twentieth century, such discrimination was not subtle, appearing in job advertisements with the warning "Christians Only" or at hotels and restaurants posting signs declaring "No Dogs or Jews Allowed." (My childhood piano teacher, a Juilliard alumnus and retired cocktail pianist, once told me how he was hired as a young man in the late 1940s to play the lobby of Florida's prestigious Kenilworth Hotel, where performing musicians were named on lobby signs. As he approached the hotel, he saw the dreaded sign reading "No Dogs or Jews Allowed," and wondered how he, Alan Wolfson, would manage to pass. He soon found himself playing a grand piano beside a marquee the hotel had provided, announcing: "Tonight's Performer: Alain de Wolfe.") By midcentury, these explicit markers had morphed into an elaborate glass ceiling that was an open secret, expressed at first through carefully worded advertising for employment or public accommodations ("sabbath observers need not apply"; "churches nearby") and later through byzantine job and school application forms that, as Fermaglich explains, demanded information not only about the applicant's birthplace and citizenship but also equally mandatory and entirely irrelevant information about the applicant's parents' and grandparents' birthplaces, parents' professions, mother's maiden name, and grandparents' surnames. Fermaglich points out the profound, "corrosive" effect of this type of intense and unacknowledged discrimination on the target population: "The unofficial nature of American antisemitism encouraged many Jews to resist discrimination by using bureaucratic name change petitions to reshape their personal identity rather than civil rights activism to change an unfair society."

Fermaglich is careful to note that the vast majority of Jewish name-changers did not actually take on new non-Jewish identities; most continued contributing to Jewish organizations and participating in Jewish communal life. Fermaglich presents this point optimistically, as a grand refutation of the popular assumption that such people rejected their Jewish roots. But to me, this fact demonstrates just how profoundly oppressive the situation must have been, if even those who chose to participate in organized Jewish communities felt that a name change was necessary. These people were not "self-hating Jews." They were simply staring down a reality that they could not deny. And as the wording of their petitions reveals, they also could not allow themselves to admit exactly what that reality was.

As I pored through Fermaglich's selections from this ream of archival material, what I found most heartbreaking was witnessing how these Jewish name-changers participated in the very humiliation that they were seeking to escape. They did so not merely by changing their names, but by censoring their own self-expression *during the very act of changing those names*—because in their court filings, as Fermaglich reports, virtually no petitioners identified antisemitism as their motivation.

Instead, the Jewish petitioners almost uniformly referred to how their names were "foreign-sounding" or "difficult to spell and pronounce"—even, Fermaglich notes, "when the name was spelled phonetically." "The name Greenberg is a foreign-sounding name and is not conducive to securing good employment," one very typical petition reads. Rose Lefkowitz declared her last name "difficult to pronounce." (Is there more than one way to pronounce "Lefkowitz"?) Louis Goldstein declared his name "un-American, uneuphonius, and an economic handicap"—a petition that was rejected by the judge, whose name was also Louis Goldstein. (Those who beat the odds in an unfair system, of course, are the ones most invested in claiming the system is fair; if they didn't need a workaround, there must not be anything to work around.) Max Hymowitz described how his son Emmanuel found their shared surname "cumbersome" and "an annoyance"; his father felt that changing their name would "substantiate and promote his son's comfort and interests, socially, educationally, economically, and patriotically." One couple, pleading on behalf of their family, testified that "The name of Tomshinsky is difficult to remember and properly spell, and because of this, petitioners and their children have been subject to embarrassment and your petitioners believe that it would be to the best interests of their children as they mature, to have the family name changed to the proposed name of Thomas." In fact, the only petitioners Fermaglich cites whose filings actually mention antisemitism are non-Jews seeking to change their Jewish-sounding names, so as not to be mistaken for Jews.

Of course, many names circulating in the United States during this period were "foreign-sounding" and "difficult to pronounce and spell"——for example, LaGuardia, Roosevelt, Juilliard, Lindbergh, DiMaggio, Vanderbilt, Earhart, Rockefeller, and Eisenhower. Yet as the remarkably low numbers of non-Jewish name-change petitioners in New York City demonstrate, such families and their forebears do not appear to have been "subject to embarrassment" or affected "socially, educationally, economically, and patriotically" by having names that were "difficult to pronounce and spell." Fermaglich interprets these Jewish petitioners' concealing of the actual problem to mean that, as she gently puts it, "Jews were uncomfortable talking about antisemitism, and may have even been ashamed of their experiences with antisemitism." The difficulty these American Jewish families were facing had nothing to do with spelling or pronunciation, but none of them could admit it. And thus the process of hiding one's name became embedded within the more elaborate process of hiding the reasons why.

This brings us to the reality behind the funny family stories of names that were "changed at Ellis Island." The Ellis Island legend is simply the final step in this multigenerational process of denying, hiding, and burying the reality that American Jews feared most—namely, the possibility that they were not welcome here.

So now we know the myth, and we know the reality. And now we can ask the more interesting questions: Why did so many American Jews' ancestors tell this story about their names being "changed at Ellis Island"? What purpose did it serve then, and why do educated skeptical people still want to believe it now?

Those people who accosted me at my talk and online weren't merely uninformed. They were responding to something enormously powerful and important. This mythological story about the Jews' arrival in America is shared by plenty of Americans from other ethnic groups, immortalized on film in the classic scene of Vito Corleone's arrival at Ellis Island in *The Godfather II*. But it is also part of a deep pattern in Jewish history, one that is much bigger than a single generation of immigrants and their children, one that goes back centuries.

—————

Nearly every diaspora Jewish community in world history has at least one founding legend, a story about its origins that members of that community accept as fact, no matter how ridiculous that story might be. The Jews of medieval Spain, for example, in some ways resembled American Jews today, a group that included many people who excelled professionally and politically in the society in which they lived. One of those accomplished medieval Spanish Jews was a twelfth-century man named Abraham ibn Daud. The leading philosopher of his generation, he also had a sideline writing groundbreaking books on astronomy—and in his spare time, he published *Sefer Ha-Qabbalah* (The Book of Tradition), the most widely accepted history of the Spanish Jewish community. In that book is a story about exactly how the Spanish Jewish community became the center of the Jewish world, and the story is so ridiculous that it's amazing that anyone in their right mind ever thought it was true.

Two centuries after the "fact," ibn Daud recounted this origin story: Four important rabbis from Babylonia, the center of Jewish scholarship for centuries, were traveling by ship on the Mediterranean in the year 990—and then, on the high seas, their ship was captured by a royal Spanish fleet. All four rabbis were taken as captives and sold as slaves in different places around the Mediterranean, and in each place, local Jewish communities bought the captives' freedom. One of the captives, Rabbi Moshe, wound up in the Spanish city of Cordoba. One day, the newly liberated Rabbi Moshe, now a penniless refugee, sat in the back of a Torah study class. When he began offering brilliant answers to the class's Torah questions, the community recognized his gifts and made him their new leader. And thus, says ibn Daud, the crown of Torah was transferred from Babylonia to Spain, making Spain the next link in the chain of Jewish tradition.

This story has some truthiness to it, but as the twentieth-century historian Gerson Cohen has thoroughly explained, it's impossible. The story claims that the leader of the royal fleet that hijacked the rabbis' ship was Abed al Rachman the Third. This would have sounded plausible to ibn Daud's readers, since Abed al Rachman the Third really was the king of Spain. Unfortunately, Abed al Rachman died thirty years before this story allegedly took place, among many other factual errors. Nearly every event that "happens" in the story is lifted directly from another source—the medieval version of an internet meme. The story about the unknown pauper

Rabbi Moshe wowing senior scholars in the study session, for instance, is suspiciously similar to a story in the Talmud about the rise of the Roman-era rabbi Hillel the Elder. So why would an obviously intelligent person record this story as official history? And why did centuries of smart Jews in Spain believe it?

One could ask the same question about the founding of the Jewish community in Poland. When the Jews arrived in Poland a thousand years ago, the story goes, the head of the Jewish community announced in Hebrew, "*Poh-lin*"—"Here we will dwell," and that name spread to the local people and stuck. In case that didn't make Poland Jewish enough, the Jews coming to Poland, afraid of persecution in this new land, hid during the day in caves in the Polish forests, studying Talmud. They then sneaked out at night to carve the names of the tractates they were studying onto the trunks of trees. The local Polish people noticed this and began revering these places as holy. Later, Jews who arrived in Poland discovered just how welcome they and their traditions were in this bountiful new land, because the names of all the tractates of the Talmud were already carved onto the trees of the Polish forest, waiting for them.

These ridiculous stories were not only in fashion in the generations following the Jewish migration to Poland centuries ago, but they were also repeated as fact by respected modern Jewish authors—including the wildly popular nineteenth-century Polish Yiddish writer I. L. Peretz, who essentially owned Yiddish publishing for fifty years, and also by the twentieth-century Polish-born Hebrew writer S. Y. Agnon, who won the Nobel Prize for Literature in 1966.

There are endless examples of such origin stories. Jewish communities in France claimed that Jews had lived there since the time of the First Temple in Jerusalem, nearly 3,000 years ago—it's not true, but it was a great alibi for explaining to their Christian neighbors why they weren't involved in killing Jesus. Jewish communities in parts of Algeria similarly claimed they had been living there since the time of the Second Temple, two thousand years ago—also not true (though it was true in other parts of North Africa), but it was a great alibi for telling Muslims that they were already there before the Islamic conquest. One of my favorite founding legends is a story about the very first Jewish "diaspora" community, the Israelites' biblical sojourn in Egypt. Rabbinic tradition claims that one reason the Israelites survived their time in Egypt was that they never changed their Jewish names. But anyone who's ever read the Torah knows that the very first Israelite in Egypt was Joseph, and the book of Genesis explicitly describes how he changed his name to an Egyptian one. Of course, Joseph's name change is nothing compared to the later biblical Book of Esther, in which the title character, who bears the Hebrew name Hadassah, becomes the queen of the Persian Empire and keeps her Jewish identity a secret—helped along by her new name, borrowed from the Persian goddess Ishtar. Jews have been changing their names to non-Jewish ones and lying about it for a long, long time.

There's a clear pattern to these legends, which are all about living in places where you are utterly vulnerable and cannot admit it. These stories express the Jewish community's two highest hopes and deepest fears. The first hope is that the Jews in this new place will remain part of the chain of Jewish tradition, and the second hope is that the local population will accept them. The fears, of course, are the inverse—of being cut off from that chain going back to Mount Sinai, and of being subject to the whims of the non-Jewish majority. These fears couldn't be more real, because being a diaspora community means being vulnerable. It is a highwire act of the highest order. There are political strategies for dealing with that vulnerability, but these founding legends are an emotional strategy, and their power is unmatched.

In ways that made sense to each community, these stories each created a fantasy of total acceptance in a non-Jewish setting, and of total continuity with Jewish tradition. In the Spanish story, the enslaved and victimized rabbi survives, untraumatized and unscathed, to become the communal leader—and the chain of Torah scholarship is unbroken between Babylonia and Spain. In the Polish story, the message is that the people of Poland love the Jews so much that they named their country after Hebrew words and even honored their holy texts. This was a fantasy that Polish Jews desperately needed to get them through the far less welcoming reality they lived in.

As minorities limited by sheer numbers in their ability to control the majority's determination of their fate, all diaspora Jewish communities are fundamentally vulnerable. This unforgiving fact also applied to the American Jewish community at the time when 2 million Jews arrived through Ellis Island, and perhaps even more so, to the American Jewish community at the time when those 2 million people's children became adults. Surviving and thriving in this reality required far more than bravery and resilience. It demanded creativity, imagination, and above all, an utterly irrational faith in the fantasy of acceptance. Believing in a fantasy takes conscious effort. It requires convincing oneself of the absolute necessity of believing, and then never relenting, ever.

These new Americans and their children, living in what they hoped was the first place in centuries where their families could enjoy full and free lives, soon discovered that when they applied for a job as Rosenberg no one would hire them, but when they applied as Rose, everyone would. Imagine the private humiliation of changing your name, of accepting the unspoken yet undeniable fact that this intergenerational marker of who you are is publicly considered revolting. Imagine the betrayal: At enormous risk and expense, you or your parents had fled other places to spare yourself and your children this very same humiliation, but now you suffer the slow, seeping, soul-shaking discovery that this new place is, in important and life-limiting ways, no different. Not merely accepting this new and devastating reality, you lie in court about your motivations for succumbing to it—or, even worse, you yourself believe your own lies, because the reality is too painful to acknowledge.

Now imagine telling your children, years later, about what you did.

Telling them the truth wouldn't only implicate you. It would also implicate America. You'd be telling your children that you thought you would be accepted here, but you were fooled, because this place is just like everywhere else—only more insidious, because the discrimination isn't written into the law, so you can't even publicly protest it. All you can do is submit to it, publicly agree with it, announce in court of your own free will that your name is "un-American," that the very essence of who you are is unacceptable. If you tell that story to your children, you'd be confirming two enormous fears: first, that this country doesn't really accept you, and second, that the best way to survive and thrive is to dump any outward sign of your Jewish identity and symbolically cut that cord that goes back to Mount Sinai—which, in the case of names like Levy and Cohen, Hebrew names denoting the descendants of the ancient Jewish priesthood, it actually does.

Now imagine that instead of telling your children these psychologically damaging things, you have another option. You can do what Jews have done for thousands of years, and create an origin story that turns those fears into hopes. You can tell your children that something funny happened at Ellis Island, something completely innocent that didn't hurt you but only helped you. If you tell that story, you've accomplished two things. First you've made America into a place where people maintain their Jewish identity without any interruption of the line from Sinai, and second, you've made America into a place so welcoming that happy non-Jews greet you at the door, and then make innocent mistakes that coincidentally help you to fit right in, at no cost to you or to the three-thousand-year-old tradition you want to maintain. This is the legend that the ancestors of today's American Jews created for their descendants. We were told that our ancestors were definitely not humiliated people facing the reality of American antisemitism, even though that is demonstrated by thousands of court records. No, our ancestors were brave and hopeful people whom America joyously welcomed through that golden door. The lethal attacks on American Jews in recent years have been so shocking and disorienting not merely because of their sheer violent horror, but because they contradict the story American Jews have told themselves for generations, which is that America has never been a place where antisemitism affected anyone's life. We don't simply prefer this founding legend. We need it. The story is more important than the history, because the story is the device that makes meaning.

—————

For some people, like my angry readers, it's upsetting to learn that the Ellis Island story is "only" a legend. But I find it empowering, because it reveals the enormous emotional resources available to our ancestors and to us. Our ancestors could have dwelled on the sordid facts, and passed down that psychological damage. Instead, they created a story that ennobled us, and made us confident in our role in this great country—which means we have that creative power too.

We can't change the past, but we can change its meaning. Doing so is an act of creativity, but it is foremost an act of bravery and love. To those who gave us that enduring legend of names changed at Ellis Island, I have only one thing to say: Thank you.

Chapter 7

DEAD
AMERICAN JEWS,
PART TWO

AT THE END OF THE HOLIDAY OF PASSOVER IN 2019, I WAS full of rage.

This is not how I usually feel after Passover. In my family, Passover is the holiday we look forward to all year. Inspired by the tradition of making people in each generation feel as if they personally were freed from Egyptian slavery, we act out the story with elaborate costumes, props, and special effects. That year, we even built a neon-painted black-lit Egyptian palace in our basement, along with four hundred yards of suspended blue yarn representing the parting of the Red Sea. It's silly, but it works: my children all feel, viscerally, as if they have left Egypt, that their lives are unfolding in the Promised Land.

That Passover, however, ended with an unhinged gunman opening fire in a synagogue outside San Diego, killing a sixty-year-old congregant and wounding several others, including an eight-year-old girl—six months after another unhinged gunman did the same thing in a Pittsburgh synagogue, killing eleven.

For most Americans, this was just another dismal news headline. For American Jews, though, it was something much, much worse: a confirmation that the Pittsburgh attack was not a one-off, that our cherished belief in America as an exception in Jewish history might be a delusion.

Most organizations I belong to emailed their members official statements of sorrow after the shooting. But unlike the statements from other groups, the Jewish organizations invariably included long detailed lists of their security protocols, concrete ways to prevent more bullets, reasons 1 through 6 why I shouldn't be afraid to take my children to a synagogue to study or to pray. After Pittsburgh, I knew what to tell my children to comfort them: that this wasn't like those ancient horror stories, that our neighbors love us, that America is different. After that Passover, I no longer knew what to tell them.

Passover has always been frightening. The very first Passover took place during the "night of vigil" before the Israelites fled Egypt when, we are taught, the Angel of Death struck down firstborn Egyptians and passed over the Israelites' homes. Since then, Passover has always been a vigil: For centuries, it has also been a time of antisemitic attacks, from medieval blood libels to modern pogroms to the massacre of thirty people at a Passover Seder in Israel in 2002.

Yet there is something even scarier about Passover than the sheer vulnerability of people gathered at prayer. The Bible's famous call for freedom is "Let my people go." But in the Bible, these words are nearly always followed by another phrase: "so they may serve me." The only purpose of this freedom is to enable the people to voluntarily accept divine laws—laws about welcoming strangers, loving one's neighbors, and accepting responsibility for creating a civic society of mutual obligation. For a nation of former slaves, Jewish tradition teaches, this sudden new agency was dramatic, even terrifying. These people discovered that freedom requires hard work: building a community, supporting the vulnerable, respecting others, educating children.

I'm enraged that I feel the need to apologize for writing about this, but I do: I'm so sorry to take up your time by writing—again—about a measly antisemitic attack where "only" one person was killed and "only" one child now has a leg full of shrapnel. There are so few Jews in the world; even in the United States, we are barely 2 percent of the population, a minority among minorities. Who cares if my children have to grow up praying in a lockdown? Statistically speaking, nothing that happens to Jews should be of any consequence to anyone else.

Except that it is.

Since ancient times, in every place they have ever lived, Jews have represented the frightening prospect of freedom. As long as Jews existed in any society, there was evidence that it in fact wasn't necessary to believe what everyone else believed, that those who disagreed with their neighbors could survive and even flourish against all odds. The Jews' continued distinctiveness, despite overwhelming pressure to become like everyone else, demonstrated their enormous effort to cultivate that freedom: devotion to law and story, deep literacy, and an absolute obsessiveness about consciously transmitting those values between generations. The existence of Jews in any society is a reminder that freedom is possible, but only with responsibility—and that freedom without responsibility is no freedom at all.

People who hate Jews know this. You don't need to read the latest screed by a hater to know that unhinged killers feel entitled to freedom without any obligations to others. Antisemitism is at heart a conspiracy theory, and one appeal of conspiracy theories is that they absolve their believers of accountability, replacing the difficult obligation to build relationships with the easy urge to destroy. The insane conspiracy theories that motivate people who commit antisemitic violence reflect a fear of real freedom: a fondness for tyrants, an aversion to ideas unlike their own, and most of all, a casting-off of responsibility for complicated problems. None of this is a coincidence. Societies that accept Jews have flourished. Societies that reject Jews have withered, fading into history's night.

I don't know what to tell my children about this horror, but I do know what to tell you. The freedoms that we cherish are meaningless without our commitments to one another: to civil discourse, to actively educating the next generation, to welcoming strangers, to loving our neighbors. The beginning of freedom is the beginning of responsibility. Our night of vigil has already begun.

Chapter 8

ON RESCUING
JEWS AND
OTHERS

ONE BALMY WINTER MORNING, I TOOK MYSELF ON A tour of homes in Los Angeles's Pacific Palisades, cruising along palm-lined streets called Napoli Drive, Amalfi Drive, Monaco Drive, and other names evoking the opposite side of the planet. I was the only tourist. The cartoonish palm trees among the European names reinforced my irrational fear of Los Angeles, a city that lacks so many of the things I was raised to consider normal—like seasons, or aging, or people who reserve the word "historic" for events that occurred prior to 1982. It is a place without markers of mortality, which made my tour particularly complicated. Instead of driving by the homes of the Kardashians, I was looking to solve the mystery of a group of people saved from the Holocaust by an American named Varian Fry.

Between 1940 and 1941, working out of a hotel room and later a small office in the French port city of Marseille, Varian Fry rescued hundreds of artists, writers, musicians, composers, scientists, philosophers, intellectuals, and their families from the Nazis, taking enormous personal risks to bring them to the United States. Fry was one of only a few American "righteous Gentiles," a man who voluntarily risked everything to save others, with no personal connection to those he saved. At the age of thirty-two, Fry had volunteered to go to France on behalf of the Emergency Rescue Committee, an ad-hoc group of American intellectuals formed in 1940 for the purpose of distributing emergency American visas to endangered European artists and thinkers. The U.S. Department of State, which initially supported the committee's mission, slowly turned against it in favor of its supposed allies in the "unoccupied" pro-Nazi French government—to the point of arranging for Fry's arrest and expulsion from France in 1941. During Fry's thirteen months in Marseille, he managed to rescue roughly two thousand people, including a handpicked list of the brightest stars of European culture—Hannah Arendt, Marcel Duchamp, Marc Chagall, Max Ernst, Claude Levi-Strauss, and André Breton, to name a few. Until recently, I had never heard of Fry, even though it is arguably because of him—and because of his equally brave colleagues, including several other non-Jewish Americans—that these artists and intellectuals not only survived but reshaped the culture of

America. But now I was driving through Los Angeles to see the former homes of some of these rescued luminaries—and to meet a filmmaker who is one of the few living Americans who has heard of Varian Fry.

"We pay tribute to the righteous in order to ignore them. There have been no high-caliber books written about the righteous, no rigorous, critical studies of what made these people do what they did." This is what I was told by Pierre Sauvage, the filmmaker who has spent much of the past two decades working on a documentary about Varian Fry. Bearded and bespectacled in a red polo shirt, Sauvage is convinced that the stories of Holocaust rescuers like Fry should be not merely inspirational but instructional—that by studying these exceptional people, we can learn to be more like them. It's a surprisingly lonely point of view. In 1984, Sauvage helped organize an international conference on the righteous, chaired by Elie Wiesel. "We brought all these righteous Gentiles to Washington," Sauvage recalled. "In the breaks between sessions, the righteous Gentiles were standing around being ignored by the scholars. No one spoke to them, no one engaged them. How can scholars not be fascinated by these people?"

Sauvage is the director (and proprietor) of the Varian Fry Institute, a nonprofit archive of "Fryana," as he calls it. On a warm winter morning in Los Angeles, he welcomed me to the "institute," which turned out to be a small office with floor-to-ceiling shelves of binders that revealed an obsession bordering on mania. Sauvage's collection of Fryana included everything from copies of Fry's letters to textbooks Fry wrote for a public affairs think tank to a poem Fry composed in

French not long before his death. But most of the Fryana was stored on computers, containing video files of what was easily several months of Sauvage's filmed interviews with nearly every person who ever worked with, talked to, knew of, or breathed near Varian Fry.

Sauvage's fascination with rescuers comes in part because he owes his life to them. He was born in 1944 in Le Chambon, France, a Huguenot village in the south central part of the country in which the entire town, following the leadership of its Protestant clergy, formed a silent "conspiracy of goodness," as Sauvage has called it, to shelter Jews from the Nazis. Sauvage's parents were among the thousands of Jews hidden by the righteous of Le Chambon. His 1989 film, *Weapons of the Spirit*, is a documentary about Le Chambon; it has become an educational staple that I watched in my high school French class. Sauvage's parents went to Le Chambon, he later discovered, after being rejected for rescue by Varian Fry.

Fry was honored by Israel's Yad Vashem Holocaust Museum in 1997, thirty years after his death, as one of the "Righteous Among the Nations"; there is also a street named after him in his hometown of Ridgewood, New Jersey, not far from where I live, as well as another street named after him in Berlin. But to Sauvage, this kind of recognition is meaningless when we make no attempt to learn what motivated him. "Many years ago in New York, I read about a guy who had fallen onto the subway tracks, and another man had jumped down to rescue him," Sauvage told me. "When he was asked why he did it, he said, 'What else could I do? There was a train coming.' For most people, that would be the reason *not* to do it. But this man's response was automatic. Fiction and drama have given us a distorted sense of how rescuers think. Writers need a narrative arc, so they show these people wrestling with themselves, agonizing over what to do. But rescuers actually don't hesitate or agonize. They immediately recognize what the situation calls for. When they say that what they did was no big deal, we think they are being modest. They aren't. They genuinely experienced it as no big deal."

From his research in Le Chambon, Sauvage developed his own theory about the righteous: that they are happy, secure people with a profound awareness of who they are. "I've never met an unhappy rescuer," he claimed. "These are people who are rooted in a clear sense of identity—who they are, what they love, what they hate, what they value—that gives them a footing to assess a situation." He described the inspiration the people of Le Chambon drew from their Protestant history and faith. Then he began showing me video footage of his interviews with Fry's colleagues, several exceedingly intelligent, colorful, and sincere Americans. All of them did indeed seem like happy people, with a deep sense of who they were.

The only person missing from his footage is Varian Fry.

I've long been uncomfortable with stories of Holocaust rescue, not least because of the painful fact that they are statistically insignificant—as are, for that matter, stories of Holocaust survival. But for me, the unease of these stories runs deeper. When I was twenty-three and just beginning my doctoral work in Yiddish, I barely understood the world I was entering; it was so distant from American culture, where even stories of the Holocaust are expected somehow to have happy endings.

In Holocaust literature written in Yiddish, the language of the culture that was successfully destroyed, one doesn't find many musings on the kindness of strangers, because there actually wasn't much of that. Instead one finds the overwhelming reality of the unavenged murder of innocents, along with cries of anguish, rage, and, yes, vengeance. Far more palatable to American audiences are stories of Christian rescuers, because while they have the imprimatur of true stories, they also conveniently follow the familiar arc of fiction, inserting heroes into a reality almost entirely populated by villains and victims.

But unlike the humble peasants of Le Chambon, Varian Fry felt oddly familiar to me. Not just because he was young and American, but because he was very much the kind of young American I know best. Like me, he grew up in a commuter suburb in northern New Jersey; he graduated from Harvard in 1931, sixty-eight years before I did. In photographs, he looks a lot like the guys with whom I went to college: thin, awkward, but handsome in a dorky way, his then-stylish glasses and carefully knotted ties a failed but endearing attempt at coolness. His personal letters, which I read in Columbia University's Rare Book Room, are well written and irreverent in a tone I recognize from my college friends—full of witty references to nerdy things ranging from the *Aeneid* ("I was surprised to find so many more / had joined us, ready for exile . . .") to Gilbert and Sullivan ("I am never disappointed in them [the rescued artists]—what never? Well, *hardly* ever!"). If he hadn't been dead for more than fifty years, I might have dated him.

What felt creepily familiar about him, too, were his

motivations. Unlike Le Chambon's pious French peasants, who spoke of living lives worthy of Christ, Varian Fry went to France with a far more secular goal: to save Western civilization. The Emergency Rescue Committee that Fry worked for in France had evolved from an activist organization called the American Friends of German Freedom, a group of prominent American intellectuals from many fields. The Emergency Rescue Committee was formed after France fell to the Nazis, out of the fear that European culture itself was about to be lost forever. American writers, curators, and scholars took great pains to compile an A-list of great brains in need of rescue.

It did not appear to occur to anyone at that time, as premier American minds argued over which premier European minds to include on the list, that there was a sort of eugenics to this anti-Nazi exercise as well—though later it would very much occur to Varian Fry. But when Fry volunteered, it was precisely the mission's elitist nature that excited him. In the introduction to his memoir, titled *Surrender on Demand*, Fry admitted it as one of his main reasons for going to France. "Among the refugees who were caught in France were many writers and artists whose work I had enjoyed: novelists like Franz Werfel and Lion Feuchtwanger; painters like Marc Chagall and Max Ernst; sculptors like Jacques Lipchitz," he wrote. "For some of these men, although I knew them only through their work, I had a deep love; and to them all I owed a heavy debt of gratitude for the pleasure they had given me. Now that they were in danger, I felt obligated to help them, if I could; just as they, without knowing it, had often helped me."

Yet the part of Fry's story that I found most unsettling was that this "debt of gratitude" turned out to be less than mutual. The mystery of why so few people today have heard of Varian Fry—despite his Americanness, his youth, his good looks, his Harvard degree, and the fact that the people he rescued were the guiding lights of Western civilization—seems linked to a peculiar lack of gratefulness among the many famous people whose lives he saved. The stories about how most of these rescued celebrities later avoided or ignored him are numerous and detailed. One of the more glaring examples is Hannah Arendt. Arendt, her mother, and her then-husband were on Fry's lists. As it happened, Arendt and her family left France after Fry's expulsion, saved by the Emergency Rescue Committee's remaining French staffers before the committee itself was shut down by the Vichy police. Yet despite spending a lifetime writing about the philosophical implications of fascism, Arendt never once acknowledged in a single public word that she owed her life to Fry's committee—nor, it appears, in a private one, even when presented with the opportunity. In a collection of letters between Arendt and her fellow public intellectual Mary McCarthy, McCarthy told Arendt about her experience meeting Varian Fry at a friend's home in 1952, describing him as a "perfect madman" and elaborating on that insult in ways both inaccurate and cruel. Arendt replied to McCarthy's letter, but said absolutely nothing in response to the ad hominem attack on the person whose organization had saved her life. Even if Arendt didn't know who Fry was (which would be difficult to imagine, given his immense renown among refugees in Marseille, where Arendt lived while waiting to emigrate to the United States), it is nonetheless remarkable that her Holocaust-related writings, many dealing directly with righteous Gentiles, never mention that she was among the rescued.

Sauvage, who deeply admires Arendt, appeared truly pained when he shared this detail with me. "In one of her letters, Arendt says that 'A writer is his life,'" he said, in the tone of a confessional. "How can you spend your whole life without acknowledging this part of your life? Not one word about Fry, or about the attempt to save a culture of which one were a part?"

It was, indeed, very odd. Because of the people Fry saved, New York became the international center of the postwar art world; because of the people Fry saved, American universities became the premier research institutions on Earth; because of the people Fry saved, Hollywood was reconfigured into global hegemony. Varian Fry, essentially, saved not only thousands of people but the culture of Europe. Weren't we supposed to have seen this movie already?

I asked Sauvage who had been involved in Fry's mission, and he ticked off the various American players who had worked with Fry in France. Every one of them had already died, and Sauvage was the only one who had filmed interviews with them. So I decided to explore the only evidence left, in the hopes of unraveling the mystery of Fry's heroic actions—as well as the darker mystery of why the culture he saved has largely forgotten him. The answers were far more disturbing than I could possibly have imagined.

———

In 1935, Varian Fry was twenty-seven years old, spending a month in Berlin as the editor of the American magazine *The*

Living Age, when he witnessed how a modern civilized country executes a pogrom. Near the end of his trip, storm troopers engineered a riot that happened to take place right outside his hotel. Hearing shouting in the streets from inside the lobby, he stepped out onto the Kurfürstendamm, which at that time was among the city's most expensive and fashionable streets.

"I found a large crowd lined up on both sides of the street, forcing each car which came by to run the gauntlet, stopping all cars in which Jewish-looking men or women were riding, and dragging out the Jews and beating them up," he reported by phone for the Associated Press. "I saw one man brutally kicked and spat upon as he lay on the sidewalk, a woman bleeding, a man whose head was covered with blood. . . . All along the Kurfürstendamm, the crowd raised the shout 'Jude' whenever anyone sighted or thought he'd sighted a Jew. The cry sent the crowd converging on the poor victim, who was asked for his identification papers. If he could not prove himself a good 'Aryan,' he was insulted, spat upon, roughly handled, and sometimes knocked down, kicked, and beaten." Since gore is now routinely depicted in the news, I found it difficult to appreciate the level of brutality Fry was describing when I first read this article in the archives of the *New York Times*, in which it ran on July 17, 1935. It turned out that Fry, or maybe his editors, had understated the violence for the benefit of delicate New York readers, who perhaps preferred a minimum of blood with their morning coffee. Several of those Fry saw being beaten had died of their wounds by the following day.

But what astonished Fry was less the raw cruelty than the

organized spectacle of it. "At times," he reported, "a chant would be raised . . . 'the best Jew is a dead Jew'—precisely like a Christian liturgy, with a leader speaking the lines first and the crowd chanting them over and over again, line for line, after he had finished. Everywhere the people were in a holiday mood; in fact, one German youth said to me, 'This is a holiday for us.' Old men and young men, boys, Storm Troopers, police, young girls of the domestic servant type, well-bred women, some even in the forties and over—all seemed to be having a good time."

The parallels with Christianity that Fry reported for the Associated Press and the *Times*—the liturgical chanting, the sense of a "holiday"—were not the only ones to come up during his stay in Berlin. Mary Jayne Gold, one of Fry's co-rescuers, wrote in her memoir of Fry telling her of another incident he witnessed in 1935 Berlin that later motivated him to return to Europe. "At a café, Varian watched a pair of storm troopers approach the table of a Jewish-looking individual," Gold recalled. "When the poor man reached nervously for his beer, with a quick thrust of his knife one of the storm troopers pinned the man's hand to the wooden table. The victim let out a cry and bent over in pain, unable to move. . . . I think the mental image of that hand nailed to the table beside the beer mug had something to do with [Fry's] decision to go." I read this story and could not help but notice what the hyperliterate Fry had surely noticed himself. He'd witnessed a crucifixion. It was, of course, too gory a tale for the *New York Times*.

Visiting the Nazis' foreign press office the day after the pogrom, Fry was granted an interview with Ernst Hanfstaengl, the Nazis' chief foreign press officer. Hanfstaengl was delighted to speak with Fry, because he and Fry had much in common. Like Fry, Hanfstaengl was an American—and like Fry, he was a Harvard alumnus. Also like Fry, he was passionate about the creative arts: Hanfstaengl was a talented pianist and composer, and in his younger days he had written several popular fight songs for the Harvard football team. In his current job, he had used his talents again, adapting his Harvard fight songs into anthems for the Hitler Youth. Hanfstaengl cheerily informed Fry that there were two groups within the Nazi Party: a "moderate" wing that wanted to expel the Jews, and a "radical" wing, led by Adolf Hitler, that wanted to murder them. Fry's report of this conversation also appeared in the *New York Times*. In 1942, in an article for the *New Republic* titled "The Massacre of the Jews," Fry admitted that "when Hanfstaengl told me, in his cultured Harvard accent, that the 'radicals' among Nazi party leaders intended to 'solve' the 'Jewish problem' by the physical extermination of the Jews, I only half believed him. I learned better in November 1938." The "cultured Harvard accent" of this Nazi leader haunted Varian Fry, but what haunted me were the Harvard fight songs. It would take me much longer to fully understand their implications.

In 1940, what American intellectuals viewed as an "emergency" finally arose. In June of that year, the Germans took over the northern half of France. With the June 14 fall of Paris, which had become a capital for refugee intellectuals, thousands who had fled Germany suddenly found themselves in what Fry would later call "the greatest man-trap in history." These refugees fled Paris (as did much of the city's population) for France's southern "unoccupied" half, where the French government then herded any German nationals they could find—most of whom were Jews or political refugees—into French-run concentration camps, ostensibly to prevent German espionage. When France surrendered to Germany on June 25, the Nazis established a puppet government based in the southern city of Vichy in exchange for complete collaboration with the Nazis. In Article 19 of the armistice terms, the Vichy government agreed to "surrender on demand all Germans named by the German Government in France"—that is, anyone who was on the Gestapo's list, a list that would quickly expand beyond German Jews and anti-Nazi activists to include just about anyone the Nazis didn't like, including many well-known "degenerate" artists. The window for escape from Europe was closing, but the refugees had nowhere to go. Few countries wanted them, even though they were some of the premier artistic, scientific, and intellectual minds of the world. As Herbert Pell, then the U.S. envoy to Portugal, put it, "There is a fire sale on brains here, and we are not taking full advantage of it."

The Emergency Rescue Committee was created in New York the day after the French surrender—at a prescheduled benefit event for the American Friends of German Freedom—to provide emergency visas for prominent refugees and to escort them out of France. Ingrid Warburg, niece of the German Jewish financier Felix Warburg and a well-known patron of the arts, took the reins in developing lists of endangered European artists as well as in raising money for the cause—mainly, as she told Sauvage decades later, from Jewish donors. Curators at the Museum of Modern Art assembled lists of artists thought to be most in danger. The Nobel Prize–winning German novelist Thomas Mann, teaching at Princeton, provided lists of similarly endangered German-language writers; the leaders of various universities compiled their own wish lists of scholars, thinkers, and scientists.

The committee proposed many people for the job of traveling to the "unoccupied zone" and distributing visas to those on the genius lists, but anyone qualified to do it—those with personal connections to the intellectuals to be rescued, or those who had been refugees themselves—would by definition also be in danger. When no one else stepped up, the committee reluctantly said yes to a thirty-two-year-old volunteer, a nobody who had no relevant experience and no qualifications for the job: Varian Fry.

But who was he?

This is a question that confused me—and it seems to have confused Fry too. One could call him a journalist, but that descriptor would be only somewhat true. He worked for several American magazines, but he never held any such job for more than a year or two, barely ever worked as a reporter, and abandoned journalism entirely before he turned forty. One could call him an intellectual, but that also doesn't quite fit, as the term is usually reserved for academics, pundits, or prolific critics, and Fry was none of these. His teaching was limited to high school, and the vast majority of the writing he would publish in his life was for school textbooks or Coca-Cola Company reports. One could call him a lover of the arts, and that is surely true, but it hardly counts as a profession or an identity—and he was not a patron of the arts, except in the sense that he saved artists' lives.

One could call him a WASP blue blood, but that isn't accurate either; while he was certainly a WASP, he had grown up in a comfortable but not terribly wealthy family, with no pedigree to speak of and no fortune to finance his whims.

Pierre Sauvage insists that rescuers "almost uniformly had a role model that influenced them, because if one doesn't have an image of how one should behave in a similar situation, one simply doesn't know what's possible." If Fry had any such person in his life, it was his grandfather, a man who worked for the "orphan trains," a Children's Aid Society operation that collected abandoned children from city streets and exported them to foster homes in the American west. Fry's second wife, Annette, who divorced him two weeks before his unexpected death in 1967, claimed that Fry was very much inspired by his grandfather's work. (She herself was clearly inspired by it; she published a children's book about the orphan trains thirty years after Fry died.) But as I picked my way through Fry's vast personal papers at Columbia University, to which Annette had donated them after his death, I found no mention of this wonderful grandfather who supposedly inspired the greatest act of Fry's life. Andy Marino, who wrote one of the two biographies of Fry published in the past twenty-five years, points to Fry's experience being bullied at the prestigious Hotchkiss boarding school as a defining moment for him; Fry's hazing there, which included being forced to traverse a room hand-over-hand hanging from a scalding steam pipe across the ceiling, supposedly made him drop out. But this life-changing instance of teenage torment isn't something I could find Fry ever mentioning either—and if being bullied in school motivates people toward heroism, there ought to be a whole lot more heroes in the world. And all of Fry's supposed marginalization did not prevent him from graduating from Harvard, where he not only earned a degree in classics but also co-founded a nationally important literary journal, *Hound and Horn*, with Lincoln Kirstein, who later co-founded the New York City Ballet and was an important member of the planning committee for Lincoln Center. The more I learned about Fry, the less I believed that he was a noble loner isolated by his principles. His personal correspondence covers multiple microfilm reels, featuring many, many people. For an outcast, he sure seemed to have had a lot of friends.

After college Fry moved to New York, where he had a job writing *Reader's Digest*–style books on current events for a publisher and think tank called the Foreign Press Association before taking the position at *The Living Age*. At twenty-three, in 1931, he married thirty-one-year-old Eileen Hughes, an editor at the *Atlantic*. Their marriage was childless and, by all accounts, challenging. It became even more difficult when Fry left for France in August of 1940, planning to bike around Provence and bird-watch while delivering visas to the people on the Emergency Rescue Committee's lists. He expected the job to be completed in four weeks. As it happened, he stayed for more than a year.

———

In our current century, it is difficult to appreciate the vast renown of many of the people on Varian Fry's lists, only a few of whom are still household names. Today, for instance, few readers outside of Germany have heard of the German Jewish novelist Lion Feuchtwanger. But by 1940 he was the most widely read German-language writer in the world and, in translation, one of the most widely read writers in the world, period—a fact especially noteworthy because nearly all of his pre-1940 novels deal with explicitly Jewish themes. The book that catapulted him to fame was *Jew Süss*, published in the United States as *Power*, a fictionalized biography of Joseph Süss Oppenheimer, a Jewish financier for Prussian royalty in the eighteenth century. Oppenheimer's sentencing on fraudulent antisemitic charges included hanging and "gibbeting," or the public display of his hanged corpse in a suspended human-size birdcage for six years. Refusing a last-minute conversion that would have averted his death sentence, Oppenheimer died *al kiddush hashem*, in sanctification of God's name, reciting the Sh'ma, Judaism's central statement of faith in one God. Oppenheimer's story had been fictionalized before and was later the subject of a Nazi film, but Feuchtwanger's 1925 version, in which Oppenheimer is a complex figure forced to choose between power and dignity, became an international bestseller. When Feuchtwanger's close friend Sinclair Lewis won the 1930 Nobel Prize for Literature, Lewis declared in his acceptance speech that Feuchtwanger should have received the prize instead.

Having never heard of Feuchtwanger before, I read *The Oppermanns*, one of the few contemporary novels he ever wrote. Set in 1932 and 1933, *The Oppermanns* is the story of four Jewish siblings in Berlin, scions to a successful furniture business founded by their grandfather in the nineteenth century.

As the Nazi influence grows, each family member's sense of self-worth is degraded or destroyed in a dramatic way—from the elder brother, who is forced to surrender the family's firm to a competitor, to the teenage son, who resorts to suicide to end his humiliation at the hands of his high school teacher, to a younger brother, who signs a petition and ends up in a concentration camp, where he is tortured into madness. Ultimately, the close-knit family is scattered across the world as they flee the country they had always considered home. The novel's events are described as taking place at the *end* of "fourteen years of antisemitic incitement" in Germany, tracing back to Germany's devastating defeat in the first World War. The book is full of references to Judaism, including quotations from the Talmud, yet Feuchtwanger's writing is conventional, engaging but not artistic. Today, the story feels familiar, even trite—until one remembers that it was first published, in German, in November of 1933.

In my own novels, I often struggle with the desire to write current events into fiction. Usually I chicken out, too nervous about branding myself politically or making statements I might later regret. If this is how a writer feels in peacetime in the freest of societies, then the courage required to write a novel like Feuchtwanger's when he did is almost unfathomable. I saw a hint of the scope of that courage in the author's note that appears on the book's first page: "After the type of this volume had already been set, a family by the name of Oppermann advised the publisher that Oppermann is a strictly Christian name and that they would, therefore, like to have it avoided that bearers of the name Oppermann be branded before the general public as belonging to a Jewish family. In view of the existing circumstances, the publisher readily understands this attitude on the part of the Oppermann family and herewith advises the readers of this novel of the facts which the Oppermann family wishes to have readily understood."

Reading the delicate wording of this "author's note" is like watching someone balance on a tightrope over a bonfire. Feuchtwanger had spent 1932 in America on a lecture tour; when Hitler came to power, his house in Berlin was confiscated and he was among the first public figures to be stripped of his German citizenship. With his wife, Marta, he went to France, and by 1940 he had landed in a French concentration camp. A photograph of him behind barbed wire moved one of his readers, Eleanor Roosevelt, to offer her support to the Emergency Rescue Committee—support she would withdraw once the State Department began to turn against the committee's work. By the time Fry arrived in France, Feuchtwanger had already escaped from the camp with the help of the American vice consul in Marseille, Hiram Bingham IV, a righteous Gentile from a long line of traveling truthseekers. (Hirams I and II introduced Christianity to Hawaii; Hiram III discovered the ruins of Machu Picchu.) Bingham had nabbed Feuchtwanger while prisoners were bathing in a river and then disguised him in women's clothes for the trip to Marseille. Fry found him and Marta hiding in Bingham's house; Feuchtwanger was so famous that he was terrified of being recognized on the street.

The internationally acclaimed and similarly bestselling Austrian Jewish novelist Franz Werfel was even more pas-

sionate in warning his readers about the totalitarian men-
ace. His best-known book, *The Forty Days of Musa Dagh*, has
rightly become an enduring classic. As an anti-Nazi novel it
is more coy than *The Oppermanns*, hiding behind a historical
story about the persecution and murder of Armenians by the
Ottoman Turks. In his own author's note, Werfel goes out of
his way to maintain the charade, declaring, "This book was
conceived in March of 1929, in the course of a stay in Damascus" that inspired him to write about "this incomprehensible destiny of the Armenian nation." It was published in 1933
and became a phenomenon around the world. Werfel lost his
citizenship when Germany annexed Austria in 1938, and, like
Feuchtwanger, he fled to France.

For all of Werfel's international fame, he became even
better known through his wife, Alma, who was notorious
among Europe's creative elite. A non-Jew with family connections high in the pre-Nazi Austrian government and a
romantic attraction to fame, Alma Schindler had already broken the heart of the Austrian painter Gustav Klimt before she
turned eighteen. From that humble beginning, she moved on
to Jewish artists, embarking on a lifetime of screwing over
brilliant Jewish men. She dumped Klimt for the prominent
Austrian Jewish composer Alexander Zemlinsky, whom she
subsequently dumped to marry the rather more prominent
Austrian Jewish composer Gustav Mahler. Mahler, besotted,
dedicated the movements of various symphonies to her; her
later control over Mahler's legacy became so intense and dis-
torting that the term "the Alma Problem" is now a concept
in musicology. Alma dumped Mahler in 1910 to marry one

of her few non-Jewish lovers, the Bauhaus architect Walter Gropius—an experience that exiled Mahler to Freud's couch and, five months later, an untimely death. To no one's surprise, Gropius failed to meet Alma's standards, and she dumped him for the artist Oskar Kokoschka (also later saved by Fry's mission)—whom she subsequently dumped because, as she told a friend, "Oskar is not a genius, and I only marry geniuses." Being no genius, Kokoschka took this even harder than Mahler. He coped by custom-ordering a nude life-size Alma mannequin, accurate down to the breasts and genitals, which he dressed in haute couture and lived with for over a year, dining with it in restaurants and sitting beside it at the opera. He then threw a party during which he and his friends smashed wine bottles over the mannequin's head until it was destroyed. Alma, meanwhile, had already moved on to genius husband number three, Franz Werfel—to whom she wasn't faithful either, but who at least earned her a place on the lists of the Emergency Rescue Committee.

It was at his first meeting with the Werfels that Varian Fry would realize, as he put it in a letter to his wife, "the shock of my own inadequacy"—that is, just how over his head he actually was in Marseille.

They met in the hotel where the Werfels were staying, as Fry mentions in his memoir, "under the name Mr. and Mrs. Gustav Mahler." When I read this line, I laughed out loud. The dark comedy routine between the famous and the star-struck had only just begun.

"Werfel looked exactly like his photographs," Fry wrote,

"large, dumpy and pallid, like a half-filled sack of flour." The half-filled sack of flour was completing a novel, *Song of Bernadette*, the sale of movie rights to which would later buy him a house in Beverly Hills. Meanwhile he and Alma were hiding from death. In a bathrobe and slippers—because really, why bother getting dressed for the person offering to save your life?—he explained to Fry the precise nature of the problem. Thanks to the Emergency Rescue Committee, the Werfels had obtained their American visas at the U.S. Consulate in Marseille. But leaving Vichy France actually required *two* visas—one for the destination country, and the other an exit visa from the French government. The French government would not issue exit visas to persons whom it had promised to "surrender on demand" to the Gestapo; in fact, applying for one was among the more efficient ways to alert the Gestapo to one's presence in Marseille. Alma knew of people who had left France without exit visas by crossing into Spain over the Pyrenees, but no one knew what happened to them once they arrived in Fascist Spain. Should the Werfels apply for exit visas and hope for the best? Or should they risk leaving the country illegally—and if so, how was this half-filled sack of flour going to climb over a mountain range?

"You must save us, Mr. Fry," Franz Werfel pleaded in English, according to Fry's memoir. "Oh, *ja*, you must save us," Alma said casually, and poured them more wine. Reading this, I could almost hear Alma's languid voice—the voice of a jaded celebrity, accustomed to using other people as means to ends. It seemed clear to me, if not immediately to Fry, that the Werfels in their pajamas had instantly recognized the awed young American as their latest hired help. Unfortunately, Fry had no idea how to help them.

After his date with the Werfels, Fry met Frank Bohn, an American who had been sent by the American Federation of Labor to rescue refugees who were labor activists. Bohn already knew the drill. He explained to Fry that he didn't need to go out looking for the people on his lists. Word would spread quickly that an American had arrived with dollars and visas, and they would come to him. Bohn also explained that the only way to get these refugees out of France was through illegal means. Fry's would be an underground operation, with the cover story that it was a humanitarian mission to provide refugees with money while they waited for legal visas—money they desperately needed, as most had had no income source for months. Fry set up shop at the Hotel Splendide and referred to his "office" by the abbreviated name "Emerescue," or in French, "Centre Américain de Secours."

The visa game had complicated rules. The U.S. State Department had authorized special emergency visas, but the American Consulate in Marseille, eager to please its allies in the Vichy government, took its time issuing them. Even refugees who were able to obtain French exit visas often found that by the time they did so, their American visas had expired. Sometimes a third "transit visa" was also required for travel through Spain and Portugal to Lisbon, from where New York sailings departed. Many stateless refugees could not even obtain the necessary papers for travel within France. Refugees with no papers at all, or whose names were well known—true of many on Fry's lists—needed false passports, which

Fry obtained from a disgruntled former Czech consul. One of the first refugees who met Fry helpfully provided him with a map of the French-Spanish border town of Cerbère, where refugees could bypass border patrols by crawling behind a cemetery wall.

Fry hid the map behind the mirror in his room at the Splendide. Within days, he looked outside his hotel room window and saw long lines of terrified refugees, all waiting for him. The hotel concierge was becoming irate. Fry decided to hire a staff. And these were the dead people whom I was able to meet in Los Angeles, courtesy of the many hours of filmed interviews conducted by Pierre Sauvage.

Fry's right-hand man was Albert Otto Hirschmann, a young German Jew who had come to France to study before the war and then enlisted in the French Army, where he invented an identity for himself as a Philadelphia-born Frenchman named Albert Hermant. With the army's defeat, he had become a specialist in all things illegal, making friends with Marseille's gangsters and becoming Fry's chief link to Marseille's underworld, supplying bogus passports and fenced cash. "My advantage," he told Sauvage, "was that I was not easily scared."

To manage the crowds at the door, Fry hired Charlie Fawcett, a wrestler from the American Volunteer Ambulance Corps. Fawcett recalled a Vichy policeman who tried to deter refugees from entering Fry's offices; Fawcett's professional wrestling hold on the man's head ended his interference. Fawcett also found ingenious ways to conceal messages between Marseille and New York. Since the Vichy police censored all telegrams and mail, important information about the mission had to be sent to New York with refugee couriers—in condoms inserted into toothpaste tubes, for example. Fawcett found a way to hide messages in a trumpet and even learned to play a few tunes that didn't require the third valve, where the messages were hidden, in case border guards got suspicious. It worked. "No one takes you seriously when you're holding a trumpet," Fawcett said with a laugh. Working hard at not being taken seriously, he was once asked by a refugee if he were a bachelor. "When I said I was, she said she had two Jewish women for me to marry, to get them out of France," he remembered. "In the end I married six of them."

Another essential employee was Miriam Davenport, a Smith College graduate from a poor family and recently orphaned who had been studying art on scholarship in Paris when the Nazis took over. "In France I suddenly realized that I had lost my future," Davenport told Sauvage years later. She came to Fry only because she had fled to Marseille with her next-door neighbor Walter Mehring, a German poet who was on Fry's lists. Noticing her knowledge of art, Fry hired her to process the steady stream of refugees who came through his door—to decide who was famous, who was talented, and who was really in danger. "I'm terrible with names," she said, "but at that time I remembered every client's name, because all these people had were their names." When asked why she took the job, Davenport recalled her Christian upbringing. "The Book of Ruth was read to me as a fairy tale," she told Sauvage. "I felt very strongly that these people were my people, and that I had to do something about it."

One of the things she did about it was to approach Mary

Jayne Gold, a young WASP heiress from Illinois whom she had met during the flight from Paris, for funding to expand Fry's A-list to include a B-list. Gold, who had already donated her private plane to the French Resistance, provided more than cash. When four refugees on Fry's list were imprisoned in the Vernet concentration camp, Fry suggested that Gold use "feminine wiles" to persuade the camp's commandant to release them. Gold, who at thirty-one had already had a dozen years' worth of sexual adventures across Europe, was game. When she arrived at the camp, she saw guards filling in graves before she met the commandant and gave him the names of the prisoners she wanted freed. "I tried to be something between sexy and ladylike," she said. It worked: He asked her out to dinner that night, and their agreement was as tacit as it was clear. But that evening she waited in the restaurant for hours; the commandant had stood her up. The following day she went back to the camp. "I asked him, 'What about our date?' He told me, 'Mademoiselle, I assure you I would have rather had dinner with you, but I had to dine with the Gestapo. Your friends will be on the train at noon, on my honor.' It killed me that he had a sense of honor."

One of the strangest things about watching Sauvage's interviews with these brave people is, as Sauvage had told me, how happy they appear. Articulate, warm, witty, and brilliant, they are impossible to dislike. Their age—most were in their eighties when Sauvage interviewed them—makes their happiness seem almost a part of them, a defining feature of who they are, as Sauvage would have it. But as I watched them recall their experiences, often laughing at their antics from their youthful days, I couldn't help but feel as though something were missing. Perhaps it was merely the filter of an interview, or the polite reserve of a generation taught to hide emotions. But as they spoke of a time that was surely full of impossible choices, the horror of imminent arrest, and the devastating reality of being forced to turn away many thousands of people and leave them to their deaths, these genuine heroes gave no signs on-screen that they had been involved in something that didn't have an entirely happy ending. Instead, they seemed to see the whole thing as a fun adventure, free from any anger or regret.

But then I began to wonder if I were the one being fooled.

"We were misfits," Miriam Davenport said of Fry's team. "We didn't fit the pattern of human behavior, of staying out of trouble and keeping your mouth shut." The biggest misfit was Fry himself. Wanting to project an air of authority, he dressed in dapper pinstriped suits each day, with a pressed handkerchief in his pocket and a flower in his lapel, shaking refugees' hands with supreme confidence and sending them off with a "See you in New York." It was a stage act of the breezy American, at a time when Fry's organization was still so under-resourced that he and his staffers, unable to afford their own office space, were holding meetings in the bathroom of his hotel room with the faucets running to avoid being overheard by Vichy spies. Fry's confident persona with his refugee clients was incongruous, both in war-torn Marseille and with the difficult personality his closest associates knew. "One of the secrets of his success," Gold told Sauvage, "was that he was an ornery cuss." As I was starting to realize, Fry had trouble getting along with anyone.

His need to micromanage the mission led Fry to escort the Werfels over the Pyrenees himself. He also took three other refugees: Heinrich Mann, a phenomenally popular anti-Nazi novelist and brother of Thomas Mann; Heinrich Mann's wife, Nelly; and Thomas Mann's son Golo, a renowned historian. The Feuchtwangers would follow shortly along the same route.

To say the refugees were unprepared for the journey would be a massive understatement. Alma Werfel arrived at the train station with seventeen pieces of luggage, all of which she insisted were essential. (Some contained manuscripts of Mahler's symphonies, though most contained her clothes.) For her stealth trek over the Pyrenees, she wore a bright-white dress, which even the least attentive border guard scanning the horizon would be hard-pressed to miss. Heinrich Mann, unaccustomed to the name on his forged documents, had forgotten to relabel his clothes; Fry, gladly playing the part of hired help, picked Mann's monogram out of his hat for him. When they reached the border, Fry and Alma's luggage took a train through the mountains while the refugees traversed them on foot—with the athletically challenged Werfel nearly being carried over the hills by Leon Ball, another American expatriate who worked for Fry. Along with the luggage, the refugees also ditched their old identification papers with Fry on the train. Their noble servant then had to torch them all in the train's bathroom. "The paper burned with an acrid, choky smoke," Fry later wrote, "and not daring to open the door, I had to get down on the floor for air." When the refugees met him at the Spanish border post many hours later, "we almost fell into one another's arms, as though we were old friends who had been separated for years and had met by accident in some strange city where none of us ever expected to be." That warm embrace, I was beginning to understand, was not quite what it seemed.

Not everyone was so lucky. Fry's most famous failure was the literary critic Walter Benjamin. Fry had provided Benjamin with his American visa, as well as a forged passport and a personal escort over the Pyrenees. Unfortunately, the day in September 1940 that Benjamin arrived in Spain was the day Spain decided to close the frontier. (Prior to that, refugees without French exit visas who slipped past French border patrols could present their Spanish transit visas to Spanish border guards a few miles down the hill without comment.) Terrified of turning back, even more terrified of sneaking further into Fascist Spain, and completely convinced that posterity would take note of the crucial manuscripts in his suitcase, Benjamin killed himself on the Spanish border. The suitcase and its manuscripts were discarded by a customs official after his body was found.

Back in Marseille, Fry had acquired an actual office—which the Vichy police soon raided. He decided that a refuge outside the city was necessary. With funding from Mary Jayne Gold's fortune, he rented a large suburban house full of gilded mirrors and taxidermed animals. It was called Villa Air-Bel.

Air-Bel became a haven for refugee Surrealists, who passed the frigid winter of 1940–1941 there with Varian Fry. Occupants included André Breton, poet and founding father of French Surrealism, who decorated the premises with jars of live praying mantises; his wife, Jacqueline Lambda Breton, a dancer and artist; their five-year-old daughter, Aube, whom the Surrealists already lauded as a brilliant painter; Wilfredo Lam, a French-Cuban painter who was Picasso's only protégé; Benjamin Péret, whom Fry described as "the French poet whose verses sometimes read as though they had been copied down from the walls of public toilets"; Victor Brauner, "the one-eyed Romanian painter whose women and cats all had one eye"; and many other writers and artists, nearly all of whom later became fixtures in the American art establishment. On Sundays, the Saint-Exupérys, Henri Matisse, André Masson, and André Malraux would stop by, staging impromptu exhibits and performances.

Under rationing, they nearly starved. As Fry wrote to his father, "If you want to get some idea of what it is like living in France now, drink Postum with saccharine and no milk, eat stale bread without jam or butter, and treat yourself on rare occasions to a slice of an old horse." But when I read this letter, I could see that for Fry, the exotic austerity was clearly part of the fun. And it was a glamorous austerity indeed. The house came with a cook, a waitress, a laundress, a maid, a gardener, and even a barber who shaved the male occupants daily—and while there wasn't much food, there was somehow plenty of wine. The artists in residence played Surrealist games: putting on exhibits by hanging their paintings from trees, making playing-card decks featuring Baudelaire and Freud, and singing long into the night. It was the high point of Fry's life.

Fry's delight in holing up with the stars of the European intellectual establishment did not go over well at home. The State Department made it clear that it would cooperate with its Vichy French "allies" in tracking Fry's illegal activities. The evil of the American government in discouraging desperate immigrants haunts many students of this period, including Sauvage. In fact, Sauvage believes that the reason Fry is so unknown is precisely because he reveals U.S. complicity in the Holocaust. "We live on two myths—that we didn't know, and that we couldn't do anything even if we did know," Sauvage said to me as soon as I sat down in his office. "This is the religion, and it isn't true. We knew plenty and could have done a lot. Varian Fry was a hero, but he was also a maverick who flew in the face of American policy. He shouldn't be allowed to acquit everyone who wasn't with him." There is a painful and undeniable truth to this. When Fry's American passport expired, the American embassy refused to renew it, placing him in mortal danger. The Emergency Rescue Committee in New York, aware that Fry had run afoul of the State Department, also began calling for his immediate return. But Fry refused to leave.

Fry's wife, Eileen, who stayed in New York and acted as his liaison to the Emergency Rescue Committee, had been begging him to return home for months. She also had her own ideas about Fry's mission. In one of her first letters, before Fry had settled himself in Marseille and disabused her of the common notion that France was full of orphans dying in the streets, she wrote, "I did not speak to you about bringing back a French child, which probably seems an absurd suggestion to you at the moment, and of course I don't want you to go around *looking* for one—I just mean that you must be sure to bring one back if you feel that *is* what you want to do with any special one." The idea of Fry bringing back a French child as a kind of souvenir for his childless wife does seem absurd, compared to rescuing Europe's greatest minds. But in fact Eileen's suggestion was not so different from what Fry was actually sent to France to do—to handpick the "special ones" among the refugees who had been deemed worth saving, and, by necessity, to leave the rest behind to nearly inevitable murder.

The inevitability of murder, of course, is the premise of all narratives of Holocaust rescue—and part of what makes me so uncomfortable with them. The assumption in such stories is that the open maw of death for Europe's Jews and dissidents was something like a natural disaster. These stories, in some sense, force us—people removed from that time by generations—to ask the wrong questions, the kind of questions that we might ask about a shipwreck or an epidemic. *Someone* has to die, this thinking goes, and the only remaining dilemma is who will get the last seat on the lifeboat or the last vaccine. But these questions fall short by assuming that the perpetrators were irrelevant. As long as we are questioning the choices that were made, shouldn't we be considering the possibility of the Holocaust not happening at all? If someone was in a position to choose whether to save person A or person B, shouldn't whole societies have been in the position to reject the notion of genocide altogether? Why didn't everyone become Denmark?

The stories of rescuers that we find most satisfying, as Sauvage points out, are those in which rescuers agonize over what they are doing, wondering whom to save; there is a kind of prurient appeal in observing that agony at a distance, which explains the popularity of fictions like *Sophie's Choice* and fictionalizations like the movie version of *Schindler's List*. Sauvage is disturbed by this notion of the anguished rescuer, because it assumes that people cannot possibly be as good as they actually are. But the unarticulated and more disturbing idea here is that rescuers like Fry do not actually call into question the premise that innocent people are doomed to be murdered. Instead, such stories simply reveal that the most righteous people available could do no more than provide, for a tiny number of people, the possibility of remaining alive. In this sense, rescuer stories are the opposite of inspirational. They are stories that make painfully clear everything that might have been.

Sauvage is not merely dismissive of the elitism charge against Fry's mission; he is positively hostile to it. "I think that devaluing rescuers who only do a little bit is a way for the person who does nothing to make himself feel less guilty," he told me. "You judge a mission by what it accomplished, not by what it didn't accomplish."

This is an odd position for Sauvage to take, given that his own parents, non-famous Jews who did not fit Fry's mission of rescuing celebrity intellectuals, were among the thousands of desperate people Fry had to turn down. In this sense, the survival of Sauvage himself, born in Le Chambon after his parents were lucky enough to find someone who said yes to saving their lives, is one of the things that Fry's committee "didn't accomplish." More confusingly, Sauvage also seems to want us to judge America by what it didn't accomplish—while arguing that the exclusive nature of Fry's mission was exactly what made him want to accomplish it.

When I challenged Sauvage on this, he defended Fry along

with his own theory of the righteous. "Altruism is a bogus concept," he said. Most rescuers, according to Sauvage, derived great joy from what they did. "Fry was meeting a deep need within himself. He was an intellectual, and a lover of the arts. He was really helping the people he loved."

It is certainly true that Fry was saving people who felt like family to him. As he put it in one of his many telegrams to Eileen refusing to leave France, COULD NO MORE ABANDON MY PEOPLE HERE THAN COULD MY OWN CHILDREN STOP HUNDREDS HAVE COME TO DEPEND ON ME FOR MONEY ADVICE COMFORT STOP LEAVING NOW WOULD BE CRIMINALLY IRRESPONSIBLE STOP. In another telegram, begging the committee for more time, he put his concerns more bluntly: PLEASE MAKE THEM REALIZE WE HAVE UNDERTAKEN IMMENSE TASK SAVING CULTURE EUROPE STOP.

"We had hardly any money," Miriam Davenport said in one of Sauvage's interviews, describing how she felt when forced to make decisions about who was worth saving. It was one of her few moments on-screen without a smile. "Our money was specifically designated for people whose art had put them in danger. We were helping the people who were in danger because of what they believed." But what, I wondered, did these rescued people believe?

In December of 1940, Vichy police turned up at Villa Air-Bel and arrested Varian Fry, along with several artists and colleagues. Fry managed to burn some address lists and forged passports in the bathroom, but an anti-Vichy doodle by André Breton was sufficient to imprison the entire group in the Marseille Harbor, on a steamship crammed with six hundred prisoners. Fry threw a note for the American consul onto the pier, wrapped in a ten-franc bill. Receiving the message, the consul sent nothing more than a tray of sandwiches. Once the group was arbitrarily freed three days later, Fry found himself tailed by eight plainclothes cops.

Fry continued sending refugees out of France, and the Vichy police began tapping his phone. They arrested several of his French colleagues, and their aggression toward the refugees worsened when Vichy passed its first anti-Jewish laws in October 1940. Police ran dragnets in Marseille, arresting every foreign Jew they could find. One of those arrested was Marc Chagall. In high dudgeon, Fry contacted the police and yelled at them for arresting one of the world's greatest living artists. "If the news of his arrest should leak out, the whole world would be shocked," Fry informed them. Half an hour later Chagall was freed. The team rejoiced. But Mary Jayne Gold spotted Fry wiping his glasses, muttering under his breath, "No, we should be able to save them all. Why just the world's greatest painter?" When I read Gold's account of this moment, I saw in it the very first evidence of a crack in Fry's façade. Before this point, he had seemed like one of the happy rescuers of Sauvage's theory and then some, a cheerful servant of the famous. But now something new was happening to Fry, or perhaps he was merely revealed for who he always was: a person blessed, or cursed, with the ability to see what no one else could see.

On August 29, 1941, the Vichy police arrested Varian Fry on behalf of the U.S. government. As the police chief had explained to Fry a few weeks earlier, "You have caused my good friend the Consul-General of the United States much annoyance. Your government and the committee you represent have both asked you to return to the United States without delay." He was warned then that he would be arrested if he did not leave the country on his own. Fry's wife had attempted to ensure her husband's safety by soliciting help from powerful people in America. The only tangible result was a letter from Eleanor Roosevelt: "I am sorry to say that there is nothing I can do for your husband. I think he will have to come home because he has done things which the government does not feel it can stand behind." Fry's entire staff escorted him to the Spanish border, along with the Vichy police. The Vichy government had thoughtfully provided him with an exit visa.

When I drove through Beverly Hills in the depths of winter, searching for the house where Franz and Alma Werfel settled, the yards were mobbed by gardeners, tending to grass planted in a desert at the foot of a tectonic cliff. The Werfels' house is on a road like many others in Beverly Hills, lined with fifty-foot palm trees that form a colonnade like a Roman cardo. The street is a kind of movie set, with each house done in a different style—a Tudor, a Spanish mission, a Bauhaus type, an Italian palazzo, all side by side, with gated driveways and intercom systems. I thought of ringing the bell at the Werfels' former home, a large colonial, until I noticed the foot-wide sign on its lawn from the Greater Alarm Company.

Down the hill in Santa Monica, it's clear that not all of Fry's famous clients shared this level of success. Heinrich Mann had crossed the Pyrenees with the Werfels, sharing their joyous escape from doom. But when I arrived at his home, a ground-level apartment in a concrete bloc painted in unpleasant pastel colors, I saw no alarm-company logos. Instead, the door to Heinrich Mann's final home was marked by a large sign that read THIS AREA CONTAINS CHEMICAL SUBSTANCES KNOWN TO THE STATE OF CALIFORNIA TO CAUSE CANCER, REPRODUCTIVE TOXICITY, BIRTH DEFECTS AND OTHER REPRODUCTIVE HARM. I left quickly, as did Mann. Dismayed by his Pacific paradise, in 1950 he accepted a position as the first president of the German Academy of the Arts in Communist East Berlin, just to escape America—though he died in Los Angeles before returning home.

At the end of my tour I drove to Villa Aurora, Lion Feuchtwanger's palatial home in Pacific Palisades, which is open to visitors. The house was a demonstration home built by the *Los Angeles Times* in 1927, modeling the most modern living in the Los Angeles suburbs. It has specially insulated walls, earthquake-protection features, and a living room that converts to a silent movie theater, with an organ for accompaniment. But when the Depression hit, no one wanted an enormous house—and when the Depression lifted, wartime gas rationing made the suburbs undesirable. The Feuchtwangers bought the Spanish-style mansion on the cheap when they arrived in 1940; today it is used as an artists' colony for German writers and filmmakers. Feuchtwanger quickly resumed his pre-exile success and reassembled his thirty-thousand-volume library, which I explored while enjoying his panoramic views of the Pacific. The house's property used to be much bigger, I was told by the young German woman who took me through the villa, but at some point most of the backyard fell off the cliff, leaving behind only a strip of grass on the edge of a precipice. My guide told me this in a bland voice, as though it were perfectly normal for the earth to fall away beneath one's feet. Feuchtwanger, formerly an outspoken Communist, apparently had no doubts about his new home or his sense of self. As I left the villa, I was handed a postcard photograph of a middle-aged Feuchtwanger smiling, posing in his California courtyard with his two pet turtles.

The triumph of the refugee artists in New York is legendary. "Not only was there a new mode of painting developing in New York by the mid-1940s," wrote the art historian Martica Sawin, "but it was emerging among those artists who had the greatest amount of contact with Surrealist émigrés." Artists like Marc Chagall redefined urban spaces with public art around the world; works by Max Ernst, Marcel Duchamp, André Masson, and others entered the permanent exhibitions of major museums. Their students had names like Robert Motherwell and Jackson Pollock.

Even the rescuers themselves went on to enjoy fruitful careers. Hirschmann, Fry's right-hand man and expert in all things illegal, became an architect of the Marshall Plan and a world-renowned economics scholar at Harvard, Columbia, Yale, and Princeton. Miriam Davenport worked for Albert Einstein's Emergency Committee on Atomic Energy, exhibited as a sculptor and painter, and eventually earned a doctorate in French literature. Charlie Fawcett, Fry's police-wrestling doorman and ace messenger, went to Hollywood and acted in more than one hundred movies, at one point starring opposite Sophia Loren.

But Varian Fry, if he had lived to be interviewed by Pierre Sauvage, would probably not have been smiling. His life after his return from France was not a story of success. He could barely keep a job or a wife. He worked for a time at the *New Republic*, but resigned when he became more anti-Communist than his fellow traveling journalists. He later worked for other publications, but could never get along with a boss well enough to avoid being fired. Leaving journalism in a principled huff, he bought a small TV production company and ran it into the ground. He took a job writing corporate literature for Coca-Cola, but was fired for insisting that Pepsi's drink dispensers were more effectively illuminated. He took a job teaching Latin at an Episcopal school, but was fired for playing his classes a recording of Tom Lehrer's "Vatican Rag." Lucy Frucht, a relief worker who had been in awe of Fry's commanding presence in Marseille, told Sauvage how disoriented she was when she ran into Fry in New York in the 1950s, making cigarette commercials. "In Marseille, he was God," she said. "And then suddenly he was nobody."

His personal life was worse. His marriage to Eileen dissolved almost immediately upon his return from France in 1941. His second marriage, to a much younger woman named Annette Riley in 1952, lasted longer and produced three children, but was difficult long before it ended. Annette, who for decades took charge of Fry's papers and legacy thanks to the fact that Fry did not amend his will in the two weeks between his divorce and his death, insisted in her interviews with Sauvage that Fry was happy in his postwar life. "Guys who came back from the war didn't talk much about it," she told Sauvage. "I do not get the feeling that he was living his life pining after the glory days. He used to say to me that he wanted to be a regular guy and learn about baseball." But Fry went through extensive psychoanalysis and even participated in Alfred Kinsey's study on human sexuality. To his children, Fry was a distant father at his best, and he was not often at his best. His daughter still refuses to speak of him. Ultimately the Fry family's attempt at a suburban postwar idyll could not endure the vagaries of Fry's restless soul. He had thrived in the knife-edge world of wartime Marseille, but what he could less easily endure was what Auschwitz survivor Gerda Weissmann-Klein once described as every concentration camp inmate's abiding fantasy, and the title of her 2004 memoir: *A Boring Evening at Home*.

Behind Fry's angst was something far more profound than the boredom of a war hero living out his days behind a picket fence. He was genuinely anguished over the fate of the thousands, even millions, whom he had been forced to leave behind—and in his anguish I saw what was missing from Sauvage's interviews with his smiling colleagues decades later. "I have tried—God knows I have tried," Fry wrote in his memoir's unpublished preface, "to get back again into the mood of American life. But it doesn't work. . . . If I can make others see it and feel it as I did, then maybe I can sleep soundly again at night." After his expulsion from France, he tried in every way he could to continue fighting. He tried to enlist in the army, but an ulcer kept him out. The Office of Strategic

Services, the CIA precursor that had run covert operations in Europe during the war, didn't want him either; his work in Marseille had involved too many contacts with possible Communists. He then tried screaming his head off about what was happening in Europe. In 1942, he wrote a cover story for the *New Republic* titled "The Massacre of the Jews," reporting with hard evidence on the murders of over 2 million Jews in Europe. "There are some things so horrible that decent men and women find them impossible to believe," he wrote. "That such things could be done by contemporary western Europeans, heirs of the humanist tradition, seems hardly possible." He pleaded for the one thing he knew would have saved the Jews of Europe: offering them asylum in the United States. His plea was roundly ignored, to the tune of 4 million more murders. He then devoted himself to writing *Surrender on Demand*, his memoir, in the hope that it would bring more attention to those he had abandoned in Europe. But the war had been won by the time it was published, and the book received little attention. Americans wanted to hear about their own heroism, not about their failures.

What was perhaps most painful for Fry after his return from France was the dissolution of his relationships with the artists and intellectuals he had saved—or, rather, the revelation that these relationships were themselves a sort of fiction.

Franz Werfel, whom Fry had personally escorted out of France, refused to return Fry's wife's phone calls on Fry's behalf while Fry was still in Marseille. Walter Mehring, the celebrated German poet who had been not only Fry's client but also a personal friend of Fry and his staff, settled in Los

Angeles and signed a lucrative contract to write screenplays for Warner Bros. Fry's committee had advanced him 30,000 francs to establish a new life in the United States. Refugees were not expected to pay back such loans, but Mehring's deadbeat status became harder to swallow when he began cruising around Pacific Palisades in a pricey new convertible while Fry's committee was still scrounging for money to save more lives. When Fry found a publisher for his memoir, his former client Lion Feuchtwanger wrote him a complimentary note: "Your narrative of the events is so impressive that the reader can't help experiencing them with you." But the narrative was apparently not impressive enough for the world-famous author to offer Fry any help getting the book reviewed or read, or even a blurb for the jacket. Their personal correspondence begins and ends with that congratulatory note. Feuchtwanger's lack of gratitude toward Fry's mission actually dated back to 1940, when he first stepped off the boat in New York. At the pier, he began giving interviews in which he thoroughly detailed his escape from France, down to the route he took over the Pyrenees. The risk this posed for Fry's committee's safety, along with that of Feuchtwanger's fellow refugees, can hardly be measured. But Feuchtwanger couldn't have cared less. Soon he would be living in a model home built by the *Los Angeles Times*, with his wife and turtles.

The refugees' ingratitude became painfully clear in 1966, the year before Fry's death. That year Fry decided to raise money for the International Rescue Commission—a philanthropic group loosely evolved from Fry's Emergency Rescue Committee— by putting together a fundraising album of original lithographs from the artists the committee had rescued. Fry thought this would be simple; after all, he had saved these people's lives.

It wasn't.

Reading Fry's papers from that year is an enraging experience. Nearly every page involves some frustrated effort to convince refugee artists to support the group that had saved their lives. There are endless telegrams and letters back and forth in 1966 and 1967 between Fry and Chagall, Fry and Chagall's agent, and Fry and Chagall's second wife, all of whom provide various excuses as to why the renowned artist was unfortunately unable to provide Fry with the time of day. Chagall, whom Fry had not merely supplied with a visa but even personally sprung out of jail when the French police were about to hand him over to the Gestapo, finally did agree to provide a lithograph—but refused to sign it, deliberately reducing its value by orders of magnitude. Fry asked André Breton, with whom he had lived at Air-Bel along with Breton's wife and young daughter, to write an introduction to the album. No amount of begging could convince Breton to do it. In a letter to a friend, Fry tried to justify the failure of so many artists to respond to him: "Artists don't answer letters, usually, if they even read them; and the telephone is no substitute for physical presence—repeated physical presence." He began searching for funding for a trip to France. After seeing Max Ernst's show at New York's Jewish Museum, Fry begged Ernst to participate, eventually yelling at him by mail, I DO NOT WANT THE ALBUM TO COME OUT WITHOUT SOMETHING FROM YOU IN IT!

Ernst would ultimately have even more reason to be grateful to Fry. As I read through Fry's maddening correspondence from 1966 and 1967, the last two years of his life, I came across an incident that I had not seen reported anywhere else. In 1966, a German newspaper published an article claiming that Ernst, who was not Jewish, had deliberately abandoned Luise Straus, his Jewish ex-wife and the mother of his son, to the Gestapo—while he trotted off to fame and fortune in America. In reality, Ernst had offered to remarry Straus for visa purposes, despite being involved with Peggy Guggenheim at the time. Fry was concerned that a fake marriage by the likes of Max Ernst would endanger the whole rescue operation, but in the end he was willing to try it. The only unwilling party was Luise Straus, who preferred taking her chances with the Nazis to remarrying Max Ernst. (She died in a Nazi prison camp.) When Fry heard of the smear against Ernst, he went to tremendous lengths to clear the artist's name. I saw how Fry had dug up copies of the receipt for Luise Straus's American visa, and then how he had solicited affidavits on Ernst's intentions from the former U.S. consuls and vice consuls, from the curators of the Museum of Modern Art in New York, and from witnesses who had been with them in Marseille. But Ernst still largely rebuffed Fry's requests for a lithograph, capitulating only when Fry arrived in France in September of that year to beg him in person. The stress of pleading with Ernst and many other artists during his trip to Europe became so severe that it landed Fry in a French hospital with his first heart attack.

Part of this ingratitude was mere celebrity vanity. As Fry once wrote to his wife, "Mrs. Guggenheim says Chagall is a shit. (Jewish ladies are *so* outspoken!) I guess she's right." But the lack of graciousness was not unique to celebrities. Pierre Sauvage, recalling his research in Le Chambon, pointed out to me that many of those rescued declined to even acknowledge their rescuers in later years. This was partly because they simply wanted to forget the greatest horror of their lives.

But there is also something inherently shameful in the rescuer-rescued relationship—the humiliation of being reduced to depending on another person for survival—and that shame expresses itself in resentment toward rescuers. "Gratitude is what makes you hate someone," Hannes Stein, a German Jewish journalist with whom I shared my bafflement about the legacy of Varian Fry, told me. Stein argued that this type of resentment was completely natural, and he offered his own country as a prime example. "Germans hate America," he went on. "They have three reasons to be grateful to America: America saved them from themselves, rebuilt their country after the war, and saved them from the Soviets. And that's exactly why Germans hate America." If we are honest, we must admit that there is a profound shame in the fact of the Holocaust from the Jewish point of view as well—and I wondered if my discomfort with rescuer stories came directly from that shame. How on earth, Fry's rescued Jews and dissidents must have wondered, could we wildly successful adults have gotten ourselves into this pathetic situation—where our lives suddenly depend on the religious commitments of a pig farmer, or the intellectual ambitions of an oddball like Varian Fry?

The shame is only highlighted by the enormous difference in the experience for the rescuers and the rescued. For those rescued, it was the worst time of their lives, when their lives had the least significance. For the rescuers, it was the best time of their lives, when their lives mattered most. Everyone Fry saved had been living a nightmare. Yet as he left France, Fry wrote to his wife, "I have had an adventure—there is no other but this good Victorian word—of which I had never dreamed."

That good Victorian word, and the literature it evokes, brought me back to the question raised for me when I'd read Fry's *New Republic* article on the massacre of the Jews, in which he marveled that such atrocities could be committed by the "heirs of the humanist tradition." It was to preserve that very tradition, of course, that Fry had gone to France—to, as he put it, SAVE CULTURE EUROPE STOP. As I peered into the chasm between rescuers and rescued, I saw that there was something equally strange about this grand goal of saving Western civilization. What, after all, did that "humanist tradition" consist of? What were its greatest achievements, its highest values? What did those rescued intellectuals actually believe? And when Fry was trying to save European culture, what was he trying to save?

———

Searching for answers, I looked to the writings of one of the Emergency Rescue Committee's biggest success stories, Hannah Arendt—and realized that I had somehow managed to reach adulthood without ever reading *Eichmann in Jerusalem: A Report on the Banality of Evil*, the book for which Arendt is best remembered. Amazed by this gaping hole in my education, I read it on my flight to Los Angeles, and then read it again on my way home. Based on Arendt's reporting for *The New Yorker* on the trial of the high-ranking Nazi, the book is most famous for its assertion of how "banal" the accused appeared, that Eichmann was not a cackling evil genius but rather a boring bureaucratic man, and that this sense of tedium was itself the Holocaust's prime novelty of horror. I knew this before opening the book; more than half a century later, this insight has become almost banal itself. But Arendt's chief argument in that book, I discovered, is actually to convince her readers that the source of Eichmann's—and by extension the Nazis'—evil was Eichmann's "inability to think." "He was genuinely incapable of uttering a single sentence that was not a cliché," she writes, and later elaborates: "He *merely*, to put the matter colloquially, *never realized what he was doing.* . . . It was sheer thoughtlessness—something by no means identical with stupidity—that predisposed him to become one of the greatest criminals of the period."

Yet Eichmann as Arendt describes him, spewing clichés at his trial, did not appear to me at all as a person with an "inability to think." He seemed rather like the opposite—that is, someone who had spent a rather astonishing amount of time thinking, absorbing ideas and translating them into action. It was just that he had been thinking about bullshit, and in the process had become buried so deep in it that extraction had become impossible. Arendt did refine this idea in her later writings, but as I read and reread this book in my airless middle seat on my cross-country flights, I found myself wondering why it was so important to her to claim that Eichmann wasn't capable of thinking. What if he were?

Arendt also claims that the premise of Eichmann's trial in Jerusalem, where he was tried not only for crimes against humanity but for "crimes against the Jewish people," was fundamentally flawed, because for the Jews, as Arendt put it, "the catastrophe that had befallen them under Hitler . . . appeared not as the most recent of crimes, the unprecedented crime of genocide, but on the contrary, as the oldest crime they knew and remembered." Arendt calls this "the misunderstanding at the root of all the failures and shortcomings of the Jerusalem trial." But I couldn't help wondering if the Petliura pogroms in Ukraine in 1919–20, in which more than fifty thousand Jews were murdered in an explicit attempt at genocide, hadn't looked incredibly unprecedented; or if the vast totalitarian brainwashing of the Inquisition, as it used creative rhetoric to convince people to turn their Jewish neighbors in to be burned at the stake, hadn't seemed impressively novel at the time; or if the populist innovation of the Roman Empire, turning the torture and murder of rabbis into public stadium-filling spectacles, didn't strike a philosopher or two as "the most recent of crimes." I thought of more examples like this—two or three per century, just off the top of my head—but it soon became tedious, and I bored even myself. "Evil" may or may not be banal, but killing Jews sure is.

I finished Arendt's book wishing I had liked it—and worried that my failure to appreciate her perspective was a reflection of my own "inability to think." I went to my local library and read a collection of her essays titled *Responsibility and Judgment*, which was on the shelf next to Ayn Rand's *The Virtue of Selfishness*. In her 1971 essay "Some Questions of Moral

Philosophy," Arendt reflects on the disappearance of morality in Nazi Germany: "All this collapsed almost overnight . . . as though morality suddenly stood revealed in the original meaning of the word, as a set of mores, customs and manners, which could be exchanged for another set. . . . Did we finally awake from a dream?"

As I returned the book to its place beside *The Virtue of Selfishness*, I recalled that the American Yiddish poet Yankev Glatshteyn had thought the opposite: For those who had been awake enough, there had never been any dream at all. Born in Poland in 1896, Glatshteyn was a secular man, and American enough to have enrolled in law school at New York University. In his searing April 1938 poem "A gute nakht, velt" (Good night, world) he wrote:

> Good night, wide world
> Big, stinking world.
> Not you, but I slam the gate.
> . . . .
> Go to hell with your dirty cultures, world.
> . . . .
> Flabby democracy, with your cold
> compresses of sympathy.
> Goodnight, electric impudent world.
> Back to my kerosene, tallowed shadows,
> . . . .
> To my pages inscribed with the divine name, my biblical
>   books,
> . . . .

To judgment, to deep meaning, to duty, to right.

World, I step with joy toward the quiet ghetto light.

Good night. I'll give you a parting gift of

    all my liberators.

Take your Jesusmarxes, choke on their courage.

Croak over a drop of our baptized blood.

. . . .

From Wagner's idol-music to wordless melody, to humming.

I kiss you, cankered Jewish life.

It weeps in me, the joy of coming home.

This poem has haunted me since I first read it as a twenty-year-old student at Harvard—a place, I slowly came to understand, that could teach me many things, including how to think, but that could not teach me goodness. Not because it taught the opposite, but because moral education is simply not what secular Western education or secular Western culture is for.

Varian Fry, my fellow alumnus, had noticed this too. When he'd interviewed the Nazi press official Ernst Hanfstaengl in 1935 in Berlin, he was alarmed to discover that Hanfstaengl saw him as a fellow Harvard man. An American with German parents, the clearly bright Hanfstaengl graduated from Harvard College twenty-seven years before Fry and later earned a doctoral degree. Like many Harvard-educated children of immigrants, Hanfstaengl decided to return to his parents' native country in its time of trouble in order to improve it, to do the most good he could with his education. He soon became a personal friend of Hitler's and rose to the level of chief foreign officer at Joseph Goebbels's Ministry of Propaganda. The American had done well for himself, and for his parents' country.

In 1934, a year into his job as head of Hitler's foreign press department, the Harvard Alumni Association appointed Hanfstaengl vice marshal at Harvard's commencement for his twenty-fifth class reunion—in recognition of his achievements as a high-ranking official overseas. Though Hanfstaengl declined the honor after much controversy, he did attend his reunion, trailed by dozens of reporters and a security squad of local and state police. The Nazi regime's attitudes were no secret by 1934; 1,500 protesters met him at the dock in New York. But at Harvard, the sanctity of free intellectual inquiry prevailed, along with hallowed respect for diversity of opinion. The *Harvard Crimson* urged the university to give him an honorary degree. When Hanfstaengl withdrew from the vice-marshalship, he made up for it by donating $1,000 to the university as a "Dr. Hanfstaengl Scholarship," for students to spend a year in the new Germany. In his letter to Harvard president James Conant accompanying his donation, Hanfstaengl wrote, "It is my profound conviction that my years at Harvard have since given me incalculable advantages, not the least of which consist in a knowledge of America and the world and in the spirit of discipline and fair play."

According to the Fry biographer Andy Marino, Hanfstaengl opened his 1935 interview with Varian Fry by asking after their shared alma mater, and then bragging to Fry that the Führer himself enjoyed hearing him play "Three Cheers for Harvard" on the piano. Who wouldn't give three cheers for Harvard, after all——a place where few people, no matter how enormous their other faults, could be accused of Arendt's morally damning "inability to think"? Then Hanfstaengl told Fry that he and the other Nazi moderates were hoping to do the humane thing and export the Jews to Madagascar, while Hitler's faction of more radical Nazis planned to murder them all. When Fry concluded the interview, Hanfstaengl presumably offered his best wishes to the best minds in America, asking Fry to give his regards to their fellow Harvard men.

———

One problem with Pierre Sauvage's theory of the righteous—that they are happy individuals with a deep sense of who they are—is that it doesn't describe Varian Fry at all. With the powerful exception of his time in France, there don't appear to be many periods in Fry's life when he was "rooted in a clear sense of identity"——or even simply happy.

When I challenged Sauvage on this in his office in Los Angeles, he was defensive, even irritated. "I really believe that when you read Fry during the war years, he does embody that sense of self-knowledge and happiness," Sauvage insisted. "I think it's absurd to suggest that Varian Fry in Marseille wasn't the very embodiment of self-knowledge, of confidence, of knowing he was where he should be."

It is certainly true that Fry was supremely happy and confident in Marseille. But there is something tautological about claiming that the traits that foster righteousness can be expressed only in a situation involving righteous conduct, and they can be painfully absent during the remainder of one's life. How relevant can such traits be, then, if we really want to learn from the righteous—to cultivate their traits in ourselves or to value these traits in others, as Sauvage is so eager for us to do? What really makes a person become a Varian Fry?

Not long after my return from Los Angeles, I spoke with Varian Fry's son Jim. Jim Fry is an evolutionary biologist and the proprietor of his father's papers. When I mentioned Sauvage's idea of rescuers being motivated by a deep sense of who they are, Jim Fry laughed. "When I think of my father," he said, "I don't think of someone who was secure in his sense of self."

Jim Fry was only nine years old when his father died, but I asked him for his impressions of his father, both from his own memories and from those of others who had known him. "He was a difficult person," he began, but clarified: "My father was mentally ill, and struggled with that his whole life. If he were alive today, he'd probably be diagnosed with high-functioning bipolar disorder."

Jim Fry is convinced that his father's mental illness was not something that he developed in middle age, but that he had struggled with it for most of his life, since long before his work in Marseille. He pointed to his father's departure from Hotch-kiss, which biographers have made into a bold stand against hazing, and his yearlong suspension from Harvard, which, as his father told it, was solely because of a prank he pulled by putting a For Sale sign on the president's house. "Something about these stories doesn't add up," Jim Fry told me. "He was always alienating people, having weird outbursts."

In one of Sauvage's interviews, Miriam Davenport recalled some of Fry's wackier antics—like receiving visitors in his boxer shorts—and said that she felt Fry was influenced by the screwball comedies of the 1930s. Yet the more I read about Fry, the more apparent it seemed that no such influence was necessary. The memoirs and biography of Lincoln Kirstein, the Harvard friend (and future New York City Ballet founder) with whom Fry created the *Hound and Horn* literary journal, paint a picture of someone whose personal troubles long preceded the war. In his diaries, Kirstein recalled a party at Harvard during which Fry "went nuts," ripping a telephone out of the wall and throwing it out a window. (This seemed innocuous to me at first, until I realized that we were talking about a 1920s institutional wall-mounted telephone—the kind made of cast iron and wood.) The episode was far from unique; the frequency of such incidents led to Fry's suspension from school.

Kirstein's diaries also make it abundantly clear that Fry was gay—something that nearly everyone in Fry's life seems to have known, though the pieties of the era prevented nearly anyone from admitting it. In Marseille, it was obvious enough that even the Vichy police, trying everything to entrap Fry, sent both girls and boys to seduce him. I was raised in an era with its own silly pieties, so just as Annette Fry felt obligated to insert a notice into the papers at Columbia insisting that her husband wasn't "foppish," I feel obligated to make one of the two currently pious claims about an important figure's homosexuality: either that it was irrelevant to his heroism and that considering it relevant makes one a bigot, or that it was the fundamental influence in creating his sense of empathy for others. Neither of these is really true. Instead, Fry's sexuality seems like yet another aspect of his personality—like his intense intelligence, and also like his mental illness—that made it impossible for him to lead a conventional life. And if one believes Pierre Sauvage's claim that the righteous are people who are rooted in an unshakable sense of their own identity, then the mental double life led by the twice-married Fry, and the torment he put his wives and children through in its service, would hardly seem to be a recipe for righteousness.

Fry was sounding less and less like Sauvage's man jumping onto the subway tracks, the one so anchored in the world that he considers risking his life to save someone else to be no big deal. When I asked Jim Fry what motivated his father to take on the rescue mission, his answer did not conjure up anyone resembling the humble peasants of Le Chambon. "I'm sure some of it was a desire to be important and to hang out with famous people," he said. "There were genuine humanitarian reasons, too, but there was a synergy between those reasons and the desire to hang out with famous people and to be recognized for it. My mother has said that he would never admit this, but the year he spent in Marseille was the best year of his life. I can imagine that he was on a manic high the whole time."

Yet something about the manic side of Fry's personality crossed the boundary from troubled into visionary. "As my mother would say, he didn't suffer fools gladly," Jim Fry said. He mentioned his father's resignation from the *New Republic* as an example of his father's "theatrical breaches with people." Fry had quit the magazine in a rage over his fellow editors' refusal to condemn Stalin during the war. "In hindsight," Jim Fry mentioned, almost as an afterthought, "he was right."

That rightness gave me pause. It was the same rightness that Fry exhibited in Berlin in 1935, when the rest of the world was looking for ways to compromise with Hitler; the same rightness that took him to Marseille; the same rightness that wouldn't let him leave. Perhaps Fry actually was, as Mary McCarthy wrote, "a perfect madman." To perceive the blinding irrational vastness of absolute evil, one almost needs to be mad.

When I asked Jim Fry for his own version of what makes a righteous person like his father take the risks that he took to save others, his answer was equally unsteadying. "Maybe you need to be a little unhinged to do something foolhardy like that," he said. He added with a laugh, "If Prozac had existed in the 1920s, we wouldn't be having this conversation."

Then, in a quieter tone, he told me: "What I think is most interesting is how someone so troubled could have done something so valuable and important." He read me his favorite quote about his father, from a review of Andy Marino's biography by Christopher Caldwell: "Fry was impossible to work with, mentally troubled, locked in himself," Caldwell wrote. "But let us not forget that he was a prophet, too, and put himself in harm's way to prevent the future he saw unrolling before him. Not the ideal person, maybe. But certainly the kind that every generation has always had too few of."

Fry may have been a "misfit," as Miriam Davenport put it—a trivial word often used for artists and the other thinkers Fry saved. But artists and thinkers at least know that Western culture honors their kind of work. Indeed, Western culture views them as the guardians of civilization, as the Emergency Rescue Committee did. There is no such cultural assumption, among "heirs of the humanist tradition," about righteousness. It is considered unremarkable, banal. It is not considered what it actually is: prophecy.

Varian Fry's oddness was not that of a Marcel Duchamp. It was that of an Ezekiel. The real reason that no one today has heard of Varian Fry is because the gift he had is not one that we value.

It is easy to forget that there are other values a culture might maintain, other people whom one could consider guardians of civilization instead of artists and intellectuals—and that a large proportion of the people who were actually murdered in the Holocaust adhered to one of these alternatives. Fry tried to save the culture of Europe, and for that he should be remembered and praised. But no one tried to save the culture of Hasidism, for example, with its devotion to ordinary, everyday holiness—or Misnagdism, the opposing religious movement within traditional Eastern European Judaism, whose energy in the years before the war was channeled into the rigorous study of *musar*, or ethics. Entire academies devoted to the Musar Movement were destroyed, their books burned out of the world, their teachers and leaders and scholars murdered—all the things that everyone feared would happen to the vaunted culture of Europe. No rescue committee was convened on behalf of the many people who devoted their lives and careers to what Pierre Sauvage laments that no one pursues—the actual study of righteousness. For them, there were no Varian Frys.

I returned from Los Angeles in the depths of a frigid winter night, coming home to the warm, dim nightlights of my children's rooms. My five-year-old had papered my refrigerator with incomprehensible drawings to honor my return. I thought of Aube Breton, the five-year-old budding Surrealist with whom Fry once lived at Air-Bel, who now sells her father's papers at auctions for enormous sums. My own five-year-old's work surely could never compare. Yet on the long, dark flight home from Los Angeles, as I took part once more in the obscenely false game of wondering who should have been saved, I could not help wishing that instead of an emergency rescue committee saving Europe's greatest artists, that there had been an emergency rescue committee saving Europe's greatest prophets—that perhaps what should have been saved was not more of the culture of Europe, but more people like Varian Fry.

———

On the morning of September 13, 1967, Varian Fry failed to appear at his new job at a high school in Connecticut, where he had recently been hired to teach Greek and Latin—his greatest love, the foundation of European civilization. When he didn't answer his telephone at home, his former wife Annette, who had divorced him two weeks prior, received a phone call. She immediately contacted the Connecticut State Police, who sent an officer to knock on his door.

The policeman found Fry sitting in bed with his glasses on his lap, his sheets and blankets covered with typewritten pages from his autobiography. He was revising it for an abridged school edition, which was later published under the title *Assignment: Rescue*. He had suffered another heart attack, this one fatal. He was fifty-nine years old.

The police officer noted the contents of the manuscript scattered around the bed, detailing Fry's valiant attempt to save the culture of Europe.

"It appeared," the policeman reported, "to be a work of fiction."

Chapter 9

DEAD JEWS
OF THE
DESERT

ON A NARROW STREET IN DAMASCUS, SYRIA, THE OLDEST city in the world, I pull open a heavy iron door in a cinderblock wall and enter an ancient synagogue. Behind the door, just past a tiled courtyard shaded by a large tree, I step inside and am stunned by what I see.

I'm standing in a jewel box. The small room is illuminated by dozens of elaborate beaded chandeliers; its walls are covered with thick red velvet drapes, its stone floor with richly patterned carpets. In front of me is a large flat stone topped with a golden menorah: Here, an inscription informs me, the Hebrew prophet Elijah anointed his successor Elisha, as described in the biblical Book of Kings. (Legend dates this synagogue to biblical times, though historical sources confirm its age at a mere five hundred years.) For a place that drew Jewish pilgrims for centuries, it is remarkably well preserved—and startlingly intimate. There are no "pews" here; instead, there are low cushioned couches facing one another, as though this were a sacred living room. A raised marble platform at the room's center has a draped table for public Torah readings; on the room's far end is an ornate wooden cabinet filled with ancient Torah scrolls, their parchments concealed inside magnificent silver cases. On the walls are framed Hebrew inscriptions, featuring the same prayers my son is currently mastering for his bar mitzvah in New Jersey. I read the familiar ancient words and feel my breath leave me with the jerking motion of a dream, tripping on a missing step as I fall through a hole in time.

I should mention here that I've never been to Damascus. Also, this synagogue no longer exists.

I'm on the website of a virtual museum called Diarna, a Judeo-Arabic word meaning "our homes." The flagship project of the nonprofit group Digital Heritage Mapping, Diarna is a vast online resource that combines traditional and high-tech photography, satellite imaging, digital mapping, 3-D modeling, archival materials and oral histories to allow anyone to virtually "visit" Jewish historical sites throughout the Middle East and North Africa, though the project has now expanded to include sites around the globe.

As I write this, with the coronavirus pandemic trapping millions of people in quarantine, virtual tours have become de rigueur for hundreds of international tourist attractions. Diarna might seem like just one more online field trip, a fun

way to hop on a screen and explore. But Diarna is something entirely different from a digital playground: It is an
outrageously difficult and utterly thankless effort in pre-
serving places that apathy and malevolence have almost
erased from the world. The places Diarna documents aren't
merely threatened by political instability, economic hardship,
authoritarianism, and intolerance. In many cases, Diarna's
virtual records are all that stands between these centuries-
old treasures and total oblivion. That synagogue I "visited,"
the Eliyahu Hanavi-Jobar Synagogue in Damascus, was documented by one of Diarna's photographers in 2010. In 2016, Syria's civil war transformed the priceless five-hundred-year-old
site to rubble—photos of which you can also find on Diarna.
The implications of this project are enormous, not only for
threatened Middle Eastern minorities, but for everyone. It
has the power to change the very nature of how we understand the past.

Diarna was created in 2008 by Jason Guberman-Pfeffer,
who was then a recent college graduate active in Middle Eastern human rights circles, and Fran Malino, a Wellesley professor (now emerita) studying North African Jewish history. That
year, a mutual acquaintance of theirs traveled to Morocco to
explore her family's Moroccan-Jewish roots, and she found
that many of the places she visited—synagogues, schools, and
cemeteries—were startlingly decayed. And the elderly people
who remembered the places best were dying off. At that point,
the senior scholar Malino and the young activist Guberman-
Pfeffer put their heads together and realized that by combining their archival skills, their contacts in the region, and newly

available technologies, they could create, in the virtual world, a way to preserve these places forever.

"It morphed almost immediately into this huge project," remembers Malino, who is now Diarna's head of Digital Heritage Mapping. Malino began by recruiting among her own students, but was soon startled by how many young people—including American photographers and budding scholars, and also locals in the region—signed on. "In very short order with a very small budget, we had a number of people working for us so we could set up a website and accumulate a lot of information and photos."

Over a decade later, with Guberman-Pfeffer as its director, Diarna has run more than sixty field expeditions, sending photographers and researchers to collect information and visual evidence of the remains of Jewish communities in dozens of countries, and the organization has now documented nearly three thousand sites throughout the Middle East and North Africa, as well as elsewhere in the world. On Diarna's online interactive map, anyone can zoom in and explore them all. Some of these locations include little more than a town's name and basic information about its Jewish history, with research still in progress. But many include beautiful photography showing physical sites from many angles, bibliographies of historical resources, and oral histories from former Jewish residents describing lives lived in these places. Other sites are being elaborately documented in ways unimaginable even just a few years ago. Today, Diarna's photographers and researchers are using tools like a portable no-parallax camera that creates a fully immersive 360-degree view of a building's interior, drone photography for bird's-eye views of ancient ruins, and design software that can turn traditional photography into vivid 3-D models.

Social media has also made it newly possible, even easy, to collect amateur photos and videos of places otherwise inaccessible, and to locate people who once lived in these Jewish communities. Diarna's interactive map often includes links to amateur videos and photos when no others exist, giving people a window on sites that are otherwise invisible. And as former Jewish residents of these places age beyond memory's reach, Diarna's researchers are conducting as many in-person interviews with such people as they can, creating a large backlog in editing and translating these oral histories to make them accessible to the public. "We're in a race against time to put these places on the map," Guberman-Pfeffer says, "and to preserve these stories before they're forever lost."

———————

I've been thinking about time and loss since I was six years old, when it first dawned on me that people who die do not ever return—and that this was also true for each day I'd ever lived. My obsession with this unforgiving fact turned me into a writer, chasing the possibility of capturing those disappearing days. These efforts inevitably fail, though I stupidly keep trying. When I first learned about Diarna, I was a bit alarmed to discover an entire team of people who not only shared my obsession, but who were entirely undeterred by the relentlessness of time and mortality—as if a crowd of chipper, sane people had barged into my private psych ward. The bright, almost surreal hope that drives Diarna is the idea that, with the latest technology, those lost times and places really can be rescued, at least virtually, from oblivion.

Jews have lived throughout the Middle East and North Africa for thousands of years, often in communities that long pre-dated the Islamic conquest. But during the mid-twentieth century's tumultuous power shifts in the region between colonial and postcolonial control, political instability and anti-semitic violence intensified to create a vast exodus, driving nearly a million Jews to emigrate to Israel and elsewhere, leaving entire countries all but devoid of Jews—and leaving behind synagogues, schools, and cemeteries that served these communities for generations. The circumstances of this mass migration varied. In some places, like Morocco, the Jewish community's flight was largely voluntary, driven partly by sporadic antisemitic violence but mostly by poverty and fear of regime change. At the other extreme are countries like Iraq, where Jews were stripped of their citizenship and had their assets seized, and where, in the capital city of Baghdad, a 1941 pogrom left nearly two hundred Jews murdered and hundreds of Jewish-owned homes and businesses looted or destroyed.

Today, people and governments in the region have varying attitudes toward the Jewish communities that once called them home. Morocco publicly honors its Jewish history; there, the government has supported Jewish site maintenance, and Diarna cooperates with a local nonprofit called Mimouna (named for a Moroccan-Jewish post-Passover celebration), a group devoted to documenting Jewish life. In other places, there is public denigration or even denial of a Jewish past. In

Saudi Arabia, for instance, decades of pan-Arabist and Islamist propaganda have left the public ignorant that Jews even lived in the kingdom at all after the Islamic Conquest, despite recent official efforts to recognize the kingdom's remarkable Jewish historical sites, including the ruins of entire Jewish cities.

Diarna is officially apolitical, refusing to draw conclusions about any of this—which for a novelist like me is maddening. I want the past to be a story, to mean something, especially a past as dramatically severed from the present as this one is. So do lots of other people, it turns out, from Zionists to Islamic fundamentalists. Guberman-Pfeffer politely declines to engage. "It's not our job to give a reason why this particular village doesn't have Jews anymore," he tells me. "We just present the sites." Malino, as a historian, is even more rigorous in defending Diarna's neutral approach. "In my mind the goal is to make available to all of us, whether they're in ruins or not, the richness of those sites, and to preserve the wherewithal of accessing that information for the next generation. We are not taking a political position, not trying to make a statement. Absolutely not."

Every Diarna researcher I talked to stood firm on this point. But the choice to present these Jewish sites is itself a statement, one that underscores an undeniable reality: "The Middle East is becoming more homogenous."

I hear this from Diarna's chief research coordinator, Eddie Ashkenazie, himself a descendant of Syrian Jews, who emphasizes the project's value for people in the region today. "We're pointing out that the store next to your grandfather's in the market was once owned by the Cohen family," he says.

"Whether they got along or it was fraught with tension is going to vary depending on the time and place, but it testifies to a society that had other voices in it, that had minorities in it, that was heterogeneous. Today you have whole societies that are only Libyan Muslims, or only Shi'ite Arabs. But they used to be incredibly diverse. All Diarna is trying to do is say that Jews once lived here."

Diarna is currently focused on documenting Jewish communities in rural areas of North Africa and the Middle East, where establishing this simple historical reality is a radical act. "We are rewriting the history books," Ashkenazie claims, and then corrects himself: "Or, not rewriting; we're just writing this history, period. Because no one else has yet."

Ashkenazie walks me through an elaborate presentation that spells out exactly how Diarna does its current work. He tells me about the Libyan town of Msellata, where a former Jewish resident, interviewed by one of Diarna's researchers, mentioned that the synagogue was once located "near the police station." On-screen, Ashkenazie shows me how he used the mapping tool Wikimapia to find the town's police station and calculate a walking-distance radius around it. Next came diligence plus luck: While scouring Libyan social media, he came across an archival photo that a current Msellata resident happened to post on Facebook, which clearly showed the synagogue across the street from a mosque. Ashkenazie then identified the still-standing mosque from satellite images, thereby confirming the synagogue's former location. "What you don't see are the hours of interviews before we got to the guy who mentioned the police station,"

Ashkenazie says. "It's the work of ants. It's very tedious, but it works."

As I listen to the awe in Ashkenazie's voice, I find myself wondering why anyone would do this "work of ants." Is there really value to documenting this level of detail? At what point does one simply accept that what's past is past? My own great-grandparents, Jewish immigrants from Eastern Europe at the turn of the last century, wanted at all costs to forget the "old country"; this was true for many Middle Eastern Jewish refugees as well, especially those with memories of societies that had turned on them. Ashkenazie admits that many of Diarna's interviewees—mostly elderly Israelis—have to be convinced to sit down with his researchers, baffled as to why anyone would care. The idea that some earnest archivist would call them up and ask them to identify the street corner where their synagogue once stood would have seemed absurd to my ancestors too.

But my cynicism about this "work of ants" quickly slides into a horrifying hopelessness—which is exactly why Diarna's work is so emotionally wrenching to observe. The frightening reality beneath Diarna's efforts is the same one that haunted me as a child. The disappearance of these communities is just an acute (and sometimes violent) version of what eventually happens to every community, everywhere. All of us will die; all of our memories will be lost. Today it's a synagogue in Tunisia that's crumbling; eventually the sun will explode. Why even try?

These questions haunt me as I sift through Diarna's site, along with some unedited video interviews that Ashkenazie shares with me to give me a sense of the enormous scope of oral history material Diarna is working to collect, translate, and post: a man describing Yom Kippur in rural Yemen, a woman detailing the Tomb of Ezra in Iraq, someone else recalling the Hebrew textbooks he studied in a Cairo school. The speakers in these videos are deeply foreign to me, elderly people with Arabic accents describing daily lives I can barely imagine. Yet they often mention things I recognize: a holiday, a biblical figure, a prayer, a song. It occurs to me that Jewish tradition, like every tradition, is designed to protect against oblivion, capturing ancient experiences in ritual and story and passing them between generations—and that Diarna is simply a higher-tech version of what everyone's ancestors once did, passing along memories around a fire, technology expanding that warm, bright circle around the fire to the world at large. I zoom in, listen, as if leaning toward the warmth, the light. And then, as I click idly from one Diarna file to another, an invisible curtain rises.

In an oral history video not yet translated or posted online, an elderly Israeli man speaks in Arabic-accented Hebrew about his hometown of Yefren in Libya. Up the hill from his family's branch-ceilinged stone house, he says, was the tiny town's eight-hundred-year-old synagogue and adjoining ritual bath. As he sits with a Diarna researcher at his kitchen table in Israel, he scribbles maps and floorplans, describing the synagogue with its interior arches, its columns, its holy ark for Torah scrolls. Listening to this man's rambling voice is like hearing someone recount the elaborate details of a dream—which is why it is utterly unnerving to click on the town of

Yefren on Diarna's interactive map and find a recent YouTube clip by a British traveler who enters the actual physical ruins of that very synagogue. In the video, the building is a crumbling wreck, but its design is exactly as the Israeli man remembered it. I follow the on-screen tourist in astonishment as he wanders aimlessly through the once-sacred space; I recognize, as if from my own memories, the arches, the columns, the alcove for the Torah scrolls, the water line still visible in the remains of the ritual bath. The "work of ants" actually, magically, works. The effect is humbling, shocking—like seeing a beloved dead relative in a dream. The past is alive, trembling within the present.

The problem is that Diarna's hardworking ants are often digging on a live volcano. This is a region where ISIS and other groups are hell-bent on wiping out minorities, where political upheaval has generated the greatest human migration stream since the end of World War II, and where the deliberate destruction of priceless cultural artifacts sometimes happens because it's Wednesday.

Mapping sites in this environment can require enormous courage—not only because of instability and war, but because the hatred that prompted the Jews' flight has surreally outlived their departure. Libya is one of many places where Jews were violently rejected by their society: Tripoli was more than 25 percent Jewish before World War II, but in 1945 more than a hundred Jews in the city were murdered and hundreds more wounded in massive pogroms, prompting the community's flight. Later, the dictator Muammar Gaddafi expelled all remaining Jews and confiscated their assets. In

2011, after Gaddafi's ouster, a single Libyan Jew who returned and attempted to remove trash from the wreckage of the city's Dar Bishi Synagogue was hounded out of the country by angry mobs waving signs reading NO JEWS IN LIBYA; apparently one was too many.

Earlier that year, a journalist in Tripoli offered to provide Diarna with photos of the once-grand Dar Bishi—a feat requiring Marvel-superhero tactics. "She slipped her minders and broke into the synagogue, which was strewn with garbage, and took pictures of it all," Guberman-Pfeffer told me of the reporter. "Gaddafi's men captured her and confiscated her camera—but the camera was the decoy, and she had pictures on her cell phone." From her photos, Diarna built a 3-D model of the synagogue; the reporter still refuses to be named for fear of repercussions. Other Diarna researchers have resorted to similar subterfuges or narrow escapes. One Kurdish journalist who helped document Iraqi Jewish sites had to flee a poison gas attack.

Even those well beyond war zones often feel on edge. As I spoke with Diarna's devoted researchers—a mix of professionals, student interns, and volunteers—I was alarmed by how many of them warily asked me to let them review any quotes, knowing how haters might pounce on a poorly worded thought. One photographer, who cheerfully told me how he'd gotten access to various Diarna sites by "smiling my way in," suddenly lost his spunk at the end of our conversation as he insisted that I not use his name. If people knew he was Jewish, he confided, he might lose the entrée he needed for his work.

"There's a lot of blood, sweat, and tears to get these images out to the public," says Chrystie Sherman, a photographer who has done multiple expeditions for Diarna—and who took the pictures of the destroyed synagogue in Damascus. Sherman was documenting Tunisian sites for Diarna in 2010 when she decided on her own to go to Syria, despite rumblings of danger. "I was terrified," she remembers. "I left all of my portrait equipment with a friend in Tunis, and just took my Nikon and went to Damascus and prayed to God that I would be OK." Following a lead from a Syrian woman in Brooklyn, she went to the country's last remaining Jewish-owned business, an antique shop in Damascus. The owner took her with other family members to the synagogue, which was no longer used for worship—and where his elderly father, remembering praying there years earlier, sat in his family's old seats and broke down in tears. At another synagogue, Sherman was followed by government agents. "They asked why I was there, and I just told them I was a Buddhist doing a project on different religions. I didn't tell them I was Jewish. You have to think on your feet."

Sherman's photographs for Diarna are incandescent, interiors glowing with color and light. Even her pictures from rural Tunisia, of abandoned synagogues in states of utter ruin, radiate with a kind of warmth, a human witness holding the viewer's hand. "It's hard to describe this feeling, which I have over and over again," she says of her work for Diarna. "You are seeing centuries of Jewish history that have unfolded, and now everything—well, the world has just changed so dramatically and a lot of things are coming to an end." Her words remind me of a Talmudic passage brought to life by the Hebrew poet Chaim Nachman Bialik in an epic work called "Dead of the Desert." The passage and poem describe desert wanderers who discover the mythically preserved bodies of the generation of Israelites—"ageless generation, a people awesome in power, ancient of days"—who died in the desert before reaching the Promised Land, silent witnesses to a forgotten past.

"I was only in Syria for five days, and I was so excited to return with my portrait equipment," Sherman told me. "But then the Arab Spring began, and I couldn't go back."

————————

You can't go back. No one ever can. But it's still worth trying.

Because of Diarna, I see my own American landscape differently. I pass by the tiny eighteenth-century cemetery near my home with its Revolutionary War graves, and I think of the histories that might lie unseen alongside the ones we enshrine, wondering whether there might be a Native American burial ground under the local Walgreens, whether I am treading on someone else's ancient sacred space. I know that I must be. We are always walking on the dead.

Yet something more than time's ravages keeps me returning to Diarna. As I explored the photographs and oral histories Diarna has collected, I found myself reeling from yet another antisemitic shooting attack in my own country, this one at a kosher market twenty minutes from my home—its proximity prompting me to hide this information from my children, avoiding mentioning our murdered neighbors. Within days of those murders, my social media feed was already full of pictures from a different attack, at a Los Angeles synagogue where someone—whether hatefully motivated or simply unstable—trashed the sanctuary, dumping Torah scrolls and prayer books on the floor. The pictures reminded me of Sherman's jarring Diarna photos of a ruined synagogue in Tunisia, its floor strewn with holy texts abandoned in piles of dust. Our public spaces today, online and off, are often full of open derision and disrespect for others, of self-serving falsehoods about both past and present, of neighbors turning on neighbors. It's tough these days not to sense an encroaching darkness. Peering through the windows of Diarna's many ruins, I'm looking for more light.

"It's hard to recognize other viewpoints if you're in a bubble where everyone thinks like you," Diarna's research coordinator Ashkenazie tells me. He's talking about homogenized societies in the Middle East, but he could be talking about anywhere, about all of us. "By raising this Jewish history, we're puncturing these bubbles, and saying that in your bubble at one time not long ago, there once were others with you," he says. "It's not so crazy to welcome others."

It's not so crazy. I look through the images of our homes, all of our homes, the windows on my screen wide open.

Chapter 10

BLOCKBUSTER
DEAD
JEWS

THE WEEK I BOUGHT MY ADVANCE TIMED-ENTRY TICKETS for *Auschwitz: Not Long Ago, Not Far Away*, the massive blockbuster exhibition at the Museum of Jewish Heritage in downtown Manhattan, there was a swastika drawn on a desk in my children's public middle school. It was not a big deal. The school did everything right: It informed parents; teachers talked to kids; they held an already scheduled assembly with a Holocaust survivor. Within the next few months, the public middle school in the adjacent town had six swastikas. That school also did everything right. Six swastikas were also not a big deal.

Auschwitz: Not Long Ago, Not Far Away is a big deal. It is such a big deal that the Museum of Jewish Heritage had to alter its floor plan to accommodate it, making room for such items as a reconstructed barracks. Outside the museum's front door, there is a cattle car parked on the sidewalk; online, you can watch video footage showing how it was placed there by a crane. The exhibition received massive news coverage, including segments on network TV. When I arrived before the museum opened, the line for ticketholders was already snaking out the door. In front of the cattle car, a jogger was talking loudly on a cell phone about pet sitters.

When I was fifteen years old, I went to the Auschwitz-Birkenau site museum in Poland. I was there with March of the Living, a program that brings thousands of Jewish teenagers from around the world to these sites of destruction. It is the sort of trip that clever people can easily critique. But I was fifteen, and deeply invested in Jewish life, and I found it profoundly moving. Being in these places with thousands of Jewish teenagers felt like a thundering announcement of the Holocaust's failure to eradicate children like me.

That was in the 1990s, when Holocaust museums and exhibitions were opening all over the United States, including the monumental United States Holocaust Memorial Museum in Washington. Going to those new exhibitions then was predictably wrenching, but there was also something hopeful about them. Sponsored almost entirely by Jewish philanthropists and nonprofit groups, these museums were imbued with a kind of optimism, a bedrock assumption that they were, for lack of a better word, effective. The idea was that people would come to these museums and learn what the world had done to the Jews, where hatred can lead. They would then *stop hating Jews.*

It wasn't a ridiculous idea, but it seems to have been proven wrong. A generation later, antisemitism is once again the next big thing, and it is hard to go to these museums today without feeling that something profound has shifted.

In this newest Auschwitz exhibition, something has. *Auschwitz: Not Long Ago, Not Far Away* originated not from Jews trying to underwrite a better future, but from a corporation called Musealia, a for-profit Spanish company whose business is blockbuster museum shows. Musealia's best-known show is the internationally successful *Human Bodies: The Exhibition*, which consisted of cross-sectioned, color-fully dyed cadavers (sourced, it was later revealed, from the Chinese government) that aimed to teach visitors about anatomy and science. Its other wildly popular show is about the *Titanic*. This is, of course, not a disaster-porn company but rather an educational company—and who could argue against education?

Perhaps the earlier Holocaust museums built by the Jewish community were unsuccessful simply because of their limited reach; despite the 2 million annual visitors to the United States Holocaust Memorial Museum, two-thirds of American Millennials in one recent poll were unable to identify what Auschwitz was. Six hundred thousand people saw Musealia's Auschwitz exhibition during its six months in Madrid before it arrived in New York. Those six hundred thousand people have all now heard of Auschwitz. There is apparently a need for more education, despite the efforts of a generation of nonprofit museum educators. As Musealia has demonstrated, there is also public demand.

And the Musealia people clearly know what they are doing. The Auschwitz exhibition was produced in cooperation with numerous museums, most prominently the Auschwitz site museum in Poland, and was carefully curated by diligent historians who are world-renowned experts in this horrific field. It shows.

The Auschwitz exhibition is everything an Auschwitz exhibition should be. It is thorough, professional, tasteful, engaging, comprehensive, clear. It displays more than seven hundred original artifacts from the Auschwitz site museum and collections around the world. It corrects every annoying minor flaw in every other Holocaust exhibition I have ever seen. It does absolutely everything right. And it made me never want to go to another one of these exhibitions ever again.

The exhibition checks all the boxes. There are wall texts and artifacts explaining what Judaism is. Half a room describes premodern antisemitism. There are sections on persecuted Roma, homosexuals, the disabled; the exhibition also carefully notes that 90 percent of those murdered in killing centers like Auschwitz were Jews. There are home movies of Jews before the war, including both religious and secular people. There are video testimonies from survivors.

The exhibition is dependable. There is a room about the First World War's devastation, and another on the rise of Nazism. The audio guide says thoughtful things about bystanders and complicity. There are cartoons and children's picture books showing Jews with hooked noses and bags of money, images familiar today to anyone who has been Jewish on Twitter. There are photos of signs reading KAUFT NICHT BEI JUDEN (Don't buy from Jews), a sentiment familiar today to anyone who has been Jewish on a college campus with a boycott-Israel campaign. There is a section about the refusal of the world to take in Jewish refugees. Somewhere there is a Torah scroll.

The exhibition is relentless. After an hour and a half, I marveled that I was barely past Kristallnacht. *What the hell is taking so long?* I found myself thinking, alarmed by how annoyed I was. *Can't they invade Poland already? Kill us all and get it over with!* It took another hour's worth of audio guide before I made it to the Auschwitz selection ramp, where bewildered Jews were unloaded from cattle cars and separated into those who would die immediately and those who would die in a few more weeks.

Somehow after I got through the gas chambers, there was still, impossibly, another hour left. (*How can there still be an hour left? Isn't everyone dead?*) Slave labor, medical experiments, the processing of stolen goods, acts of resistance, and finally liberation—all of it was covered in what came to feel like a forced march (which, yes, was covered too). It was in the gas-chamber room, where I was introduced to a steel mesh column that, as the wall text explained, was used to drop Zyklon B pesticide pellets into the gas chamber, killing hundreds of naked people within fifteen minutes, that I began to wonder what the purpose of all this is.

I don't mean the purpose of killing millions of people with pesticide pellets in a steel mesh column in a gas chamber. That part, the supposedly mysterious part, is abundantly clear: People will do absolutely anything to blame their problems on others. No, what I'm wondering about is the purpose of my knowing all of these obscene facts, in such granular detail.

I already know the official answer, of course: Everyone must learn the depths to which humanity can sink. Those who do not study history are bound to repeat it. I attended public middle school; I have been taught these things. But as I read the endless wall texts describing the specific quantities of poison used to murder 90 percent of Europe's Jewish children, something else occurred to me. Perhaps presenting all these facts has the opposite effect from what we think. Perhaps we are giving people ideas.

I don't mean giving people ideas about how to murder Jews. There is no shortage of ideas like that, going back to Pharaoh's decree in the Book of Exodus about drowning Hebrew baby boys in the Nile. I mean, rather, that perhaps we are giving people ideas about our standards. Yes, everyone must learn about the Holocaust so as not to repeat it. But this has come to mean that anything short of the Holocaust is, well, not the Holocaust. The bar is rather high.

———

Shooting people in a synagogue in San Diego or Pittsburgh isn't "systemic"; it's an act of a "lone wolf." And it's not the Holocaust. The same is true for arson attacks against two different Boston-area synagogues, followed by similar simultaneous attacks on Jewish institutions in Chicago a few days later, along with physical assaults on religious Jews on the streets of New York—all of which happened within a week of my visit to the Auschwitz show.

Lobbing missiles at sleeping children in Israel's Kiryat Gat, where my husband's cousins spent the week of my museum visit dragging their kids to bomb shelters, isn't an attempt to bring "Death to the Jews," no matter how frequently the people lobbing the missiles broadcast those very words; the wily Jews there figured out how to prevent their children from dying in large piles, so it is clearly no big deal.

Doxxing Jewish journalists is definitely not the Holocaust. Harassing Jewish college students is also not the Holocaust. Trolling Jews on social media is not the Holocaust either, even when it involves photoshopping them into gas chambers. (Give the trolls credit: They have definitely heard of Auschwitz.) Even hounding ancient Jewish communities out of entire countries and seizing all their assets—which happened in a dozen Muslim nations whose Jewish communities predated the Islamic conquest, countries that are now all almost entirely Judenrein—is emphatically not the Holocaust. It is quite amazing how many things are not the Holocaust.

The day of my visit to the museum, the rabbi of my synagogue attended a meeting arranged by police for local clergy, including him and seven Christian ministers and priests. The topic of the meeting was security. Even before the Pittsburgh massacre, membership dues at my synagogue included security fees. But apparently these local churches do not charge their congregants security fees, or avail themselves of government funds for this purpose. The rabbi later told me how he sat in stunned silence as church officials discussed whether to put a lock on a church door. "A *lock on the door*," the rabbi said to me afterward, stupefied.

He didn't have to say what I already knew from the emails the synagogue routinely sends: that they've increased the rent-a-cops' hours, that they've done active-shooter training with the nursery school staff, that further initiatives are in place that "cannot be made public." "A *lock on the door*," he repeated, astounded. "They just have no idea."

He is young, this rabbi—younger than me. He was realizing the same thing I realized at the Auschwitz exhibition, about the specificity of our experience. I feel the need to apologize here, to acknowledge that yes, this rabbi and I both know that many non-Jewish houses of worship in other places also require rent-a-cops, to announce that yes, we both know that other groups have been persecuted too—and this degrading need to recite these middle-school-obvious facts is itself an illustration of the problem, which is that dead Jews are only worth discussing if they are part of something bigger, something more. Some other people might go to Holocaust museums to feel sad, and then to feel proud of themselves for feeling sad. They will have learned something officially important, discovered a fancy metaphor for the limits of Western civilization. The problem is that for us, dead Jews aren't a metaphor, but rather actual people that we do not want our children to become.

The Auschwitz exhibition labors mightily to personalize, to humanize, and these are exactly the moments when its cracks show. Some of the artifacts have stories attached to them, like the inscribed tin engagement ring a woman hid under her tongue. But most of the personal items—a baby carriage, a child's shoe, eyeglasses, a onesie—are completely divorced from the people who owned them. The audio guide humbly speculates about who these people might have been: "She might have been a housewife or a factory worker or a musician . . ." The idea isn't subtle: this woman *could be you.* But to make her you, we have to deny that she was actually herself. These musings turn people into metaphors, and it slowly becomes clear to me that this is the goal. Despite doing absolutely everything right, this exhibition is not that different from *Human Bodies,* full of anonymous dead people pressed into service to *teach us something.*

At the end of the show, on-screen survivors talk in a loop about how people need to love one another. While listening to this, it occurs to me that I have never read survivor literature in Yiddish—the language spoken by 80 percent of victims—suggesting this idea. In Yiddish, speaking only to other Jews, survivors talk about their murdered families, about their destroyed centuries-old communities, about Jewish national independence, about Jewish history, about self-defense, and on rare occasions, about vengeance. Love rarely comes up; why would it? But it comes up here, in this for-profit exhibition. Here it is the ultimate message, the final solution.

That the Holocaust drives home the importance of love is an idea, like the idea that Holocaust education prevents antisemitism, that seems entirely unobjectionable. It is entirely objectionable. The Holocaust didn't happen because of a lack of love. It happened because entire societies abdicated responsibility for their own problems, and instead blamed them on the people who represented—have always represented, since they first introduced the idea of commandedness to the world—the thing they were most afraid of: responsibility.

Then as now, Jews were cast in the role of civilization's nagging mothers, loathed in life, and loved only once they are safely dead. In the years since I walked through Auschwitz at fifteen, I have become a nagging mother. And I find myself furious, being lectured by this exhibition about love—as if the murder of millions of people was actually a morality play, a bumper sticker, a metaphor. I do not want my children to be someone else's metaphor. (Of course, they already are.)

My husband's grandfather once owned a bus company in Poland. Like my husband and some of our children, he was a person who was good at fixing broken things. He would watch professional mechanics repairing his buses, and then never rehired them: he only needed to observe them once, and then he forever knew what to do.

Years after his death, my mother-in-law came across a photograph of her father with people she didn't recognize: a woman and two little girls, about seven and nine years old. Her mother, also a survivor, reluctantly told her that these were her father's original wife and children. When the Nazis came to her father's town, they seized his bus company, and executed his wife and daughters in front of him. Then they kept him alive to repair the buses. They had heard that he was good at fixing broken things.

The Auschwitz exhibition does everything right, and fixes nothing. I walked out of the museum, past the texting joggers by the cattle car, and I felt utterly broken. There is a swastika on a desk in my children's public middle school, and it is no big deal. There is no one alive who can fix me.

Chapter 11

COMMUTING

WITH

SHYLOCK

THIS PLAY TAKES PLACE IN THREE SETTINGS: IN RENAIS-sance Venice, in Elizabethan England, and in twenty-first-century New Jersey, in a minivan that smells like gummy bears. I'll begin with the most important setting, the one with the gummy bears, because that is the only one that includes a person I love.

The person I love is my ten-year-old son, though he is not always an easy person to love. Insistent, demanding, obsessive, morbid, and often too smart to be pleasant, he needs to get his way, and for a few recent months I was trapped in the car with him during a daily commute, forty minutes in each direction. We survived by listening to podcasts, which I downloaded by the gigabyte. Fortunately for me, he was fascinated by

Radiolab. Fortunately, that is, until the hosts' banter during an episode about organ ownership devolved into a tangent about a certain famous play, involving a certain famous character who insists and demands, obsessively and morbidly, on receiving default payment on a loan in the form of a pound of flesh.

My son was riveted. "We *need*," he informed me, "to download that play."

I felt slightly ill—both at the prospect of *The Merchant of Venice*, which I had not read in twenty-five years, and at the prospect of yet another showdown with my son. "It's Shakespeare," I tried to deflect. "The language is really hard. I don't think you'll understand it. Besides," I gulped, "*The Merchant of Venice* is—"

"Wait, it's *Shakespeare*?" He had heard of Shakespeare; he was aware that Shakespeare was Important. "If it's Shakespeare, then we totally have to do it! You can always pause it and explain stuff."

I tried to deflect again. "If we're going to read Shakespeare, there are better plays. *Macbeth*, *Hamlet*. This one is—"

"*This one* is about a pound of flesh! I want a pound of flesh!"

He would have his pound of flesh. What happened next is shameful, and the shame is my own.

Late that evening after my children went to sleep, I tooled around online and found a well-reviewed BBC Radio production of *The Merchant of Venice*. Before downloading, I hesitated.

I remembered the visceral feeling of physical nausea I had while reading that play as a student long ago. And I remembered believing the various teachers and professors who assured me and my peers that the play wasn't antisemitic, of course, just a product of its time—and the proof was that it was way better than Christopher Marlowe's *The Jew of Malta*, where the title character expresses his fondness for poisoning wells. After all, didn't Shakespeare make Shylock into a fully realized human, what with his oft-cited eyes, hands, organs, and dimensions? He had eyes, hands, organs, dimensions! If you pricked him, did he not bleed? What greater reassurance did anyone require? In school I accepted all this. Who was I, a teenage girl, to say that Shakespeare was wrong? Unlike teenage girls, who are apparently often mistaken about things like art and sexual assault, Shakespeare was not the sort of person who was wrong.

As I hesitated with my phone in my hand, I had a sudden brainwave. A quarter century had passed since I had read any criticism or scholarship about the play. I had aged, and so had the world. Perhaps Shakespeare was now wrong!

I quickly discovered that Shakespeare was still not wrong. In seconds, I located the enormous volume of scholarly and popular articles published about *The Merchant of Venice* in the years since I first read it. This corpus explained once more, now with an added frisson of wokeness, why the play wasn't antisemitic—not if you really understood it, as vulgar and whiny people often failed to do. It was really a critique of capitalism. It was really a commentary on the Other. It was really a tragedy buried in a comedy. It was really a satire of antisemitism. (Because it's a comedy!) The Christian characters in it were *just as bad*; therefore it was really condemning Christianity. In fact it was a recognition of our common humanity. Shakespeare was the greatest writer who ever lived, and he was simply incapable of creating a character who wasn't fully human.

There were a few outliers in this discussion, like the Yale professor Harold Bloom and the British trial lawyer Anthony Julius. But those who thought the play was irredeemably anti-semitic were, the consensus went, vulgar and whiny—and, completely coincidentally, they were also Jewish, which some-how magically invalidated their opinions on this subject. (On the other hand, Jewish scholars who praised the play for its "nuance" were fondly and repeatedly hyperlinked; such pieces, especially in British publications, often advertised the writer's Jewishness with titles like "A Jewish Reading of . . .") A 2016 piece objecting to the play in the *Washington Post* had exclu-sively negative comments trailing beneath it, most of which said righteous things about "political correctness," "censor-ship," and "historical context." "What's next," one groused, "getting rid of the Bible?" One comment attacked the arti-cle's author, a lawyer with a Jewish surname: "Another power-hungry, mischievous attorney with an axe to grind!" No moderator removed this one-line comment. Like the play, it wasn't antisemitic at all.

Staring into my phone, I sank into my own insecurity, which took the form of a belief that centuries of Shakespear-ean scholars, and Shakespeare himself, must surely know more than I do. It was a familiar feeling from being a teenage girl, except now it was worse, because now I was feeling it not for myself, but for my son. Wasn't it impressive, after all, that a ten-year-old *wanted* to listen to a play by Shakespeare? How could I, as his parent devoted to educating him, not to mention as an English-language writer myself, shut down my child's earnest desire to share with me his first experience of Shakespeare, the epitome of Western civilization? *Wasn't* Shakespeare the epitome of Western civilization, especially for an English-language writer like me? Wasn't that the whole reason that a ten-year-old wanting to read his work was impressive in the first place?

To my eternal shame, I clicked "Download Now." The next morning in the minivan, we began.

The next scene in this play takes place in Elizabethan England, where a man named William Shakespeare, for supposedly obscure reasons, decided to write a five-act verse comedy— yes, a comedy—whose events unfold largely due to a blood-thirsty Jew.

The reasons are supposedly obscure because Jews were expelled from England in the thirteenth century and were only invited to return in the seventeenth—and thus Shakespeare, the Wikipedia-level thinking goes, is unlikely to ever have met one. But Shakespeare frequently wrote about places he'd never been and types of people he'd never met; Shakespeare's plays appear to be inspired not by personal experiences but by earlier works. This is clearly the case for *The Merchant of Venice*, whose characters and plot are lifted whole-sale from a single source, a fourteenth-century Italian story collection called *Il Pecorone* ("The Simpleton") by Giovanni Fiorentino. It's all there in every tiny detail: the young man who falls in love with a rich lady in distant Belmont, the young man's merchant friend who fronts the money for his lovelorn pal's journey, the avaricious Jewish moneylender from whom the merchant gets a loan, the Jew's demand for a pound of flesh in case of default, the trial at which the lady from Belmont disguises herself as a male lawyer, the decree by the "lawyer" that the Jew may take his pound of flesh provided he draw no blood, and even the hijinks involving the woman's ring. The only major difference is that Shakespeare's lady is far nobler than Fiorentino's: in *Merchant*, Portia unhappily fulfills her father's requirements of her suitors, while in *Il Pecorone*, the lady enjoys drugging her suitors and robbing them blind. By removing this detail, Shakespeare removed the suggestion that malicious schemers come from all walks of life. In *Merchant*, there is no such confusion.

It also seems unlikely that Shakespeare was unaware of actual Jews in England, given that one of the biggest news stories in the years immediately preceding the play's composition was the public trial and execution at the Tower of London of a converted Portuguese Jew named Dr. Roderigo Lopez, chief physician to Queen Elizabeth I, who was accused of being paid by the Spanish monarchy to poison the queen. Dr. Lopez, one of the most respected physicians of the sixteenth century, had indiscreetly revealed that he once treated the Earl of Essex for venereal disease. The earl took revenge by framing Dr. Lopez for treason and arranging for his torture; while on the rack, Dr. Lopez "confessed"—though "like a Jew," as the court record states, he denied all charges at trial, while the attorney for the Crown referred to him matter-of-factly as "a perjuring murdering traitor and Jewish doctor."

His execution on Tower Hill in 1594 was accompanied by chants of "Hang the Jew!" from the raucous crowd. But as befitted traitors, Dr. Lopez was not in fact hanged to death, but was rather hanged until partially strangled, and removed from the gallows while still alive. He was then castrated and disemboweled, and his genitalia and intestines were burned before his eyes. After that he was beheaded and pulled by horses into four pieces; these segments of his drawn-and-quartered corpse, along with his severed head, were publicly displayed in separate locations until they decomposed.

It would be vulgar and whiny to overlook the nuances of this situation. After all, the historical record gives Queen Elizabeth a cookie for dawdling on signing Dr. Lopez's death warrant; her doubts about his guilt even led her to mercifully allow his family to keep his property, not unlike the equally merciful Duke of Venice in Shakespeare's play. And it is of course entirely unclear whether this trial and public humiliation of an allegedly greed-driven Jew attempting to murder an upstanding Christian, rapturously reported in the press with myriad antisemitic embellishments, had anything at all to do with Shakespeare's play about the trial and public humiliation of a greed-driven Jew attempting to murder an upstanding Christian—which Shakespeare composed shortly after Dr. Lopez decomposed. Most likely these things were completely unrelated.

———

Our next scene takes place in Renaissance-era Venice, where a fictional man named Antonio needs cash, and borrows some from another fictional man, a character in this Venetian verse play who has a non-Italian name and often speaks in prose. Shylock is certainly a fully human character, what with his speaking in prose and having a non-Italian name, and a familiar human character at that. He is insistent, demanding, obsessive, morbid, too smart to be pleasant, and needs to get his way—or as he puts it, over and over, "I will have my bond."

In reality, Shylock having his bond was never up to him. Whether a person like him would even be permitted to breathe in Renaissance Venice was subject to a charter that had to be renewed every few years; non-renewal was a real possibility, as the Jews discovered during two separate expulsions. In 1516, with Venetian treasuries starved from an expensive war, city authorities looking for new sources of revenue invited Jews—restricted to a few professions including moneylending and running pawnshops—to live in the neighborhood that gave the world the word "ghetto."

The nature of this urban prison, where overcrowding led to "skyscrapers" in which people were stuffed into homes with ceilings under full height, was apparently not well understood by British playwrights. In Shakespeare's play, Shylock's daughter Jessica escapes out her window by night to elope with her Christian lover waiting just below. In the real Venice, Jessica would have had a harder time eloping that evening, considering that the ghetto gates were locked at night and manned by four Christian guards (paid for, as required by Venetian decree, by the Jews themselves), and the ghetto was located on an island whose surrounding canals were guarded all night long by four roving patrol boats to prevent escape (the boat patrols likewise paid for by the Jews). It's possible, of course, that this story is set a bit later, when the ghetto expanded onto the mainland. Accessing the rest of Venice from the mainland part of the ghetto would have required traversing a heavily guarded tunnel under a building—low enough, according to historian Carl Nightingale's history of segregated cities, to force the Jews to bow before the Christian city as they approached it. Outside the ghetto, where Jews were forced to wear yellow badges and hats, one popular public spectacle was the burning of the Talmud, all copies of which were confiscated and torched by 1553.

Such precautions were necessary because of the enormous threat. The consequences of allowing Jews to live in one's city were vividly illustrated in 1475 in the Italian town of Trent, about 130 miles from Venice, where the entire Jewish population was subjected to elaborate tortures until they "confessed" to murdering a Christian toddler named Simon in order to eat his flesh and blood. For everyone's safety, Jews were burned alive and their assets confiscated, while Simon was canonized as a saint.

This incident occurred just in time for word to spread via high-tech new media, which could disseminate lies at an unprecedented pace. Many of Italy's very first printed books were about Jews butchering and eating this child, with lurid illustrations that today would qualify as deep fakes. These cartoons and their accompanying texts inspired the planting of children's corpses in Jewish homes and the subsequent public executions of Jews in at least five other Italian towns within ten years. Venice's government, eager to maintain its Jewish tax base, officially discouraged the blood libel, but the population was more inspired by the renowned Venetian poet laureate Raffaele Zovenzoni—whose hymn describing the sainted child's murder, which begged authorities to protect the people from bloodthirsty Jews, went viral. Within a few generations, Venice's brilliant idea of imprisoning Jews in ghettos went viral too.

Our final scene unfolds in twenty-first century New Jersey, in the minivan fragrant with gummy bears, where I decided to prepare my son for what he was about to hear.

"There's something you need to know about this play," I told him as we pulled out of the driveway. "The guy who wants the pound of flesh is Jewish. And the way this play shows this guy is . . . well, it's what we call antisemitic."

"Ooh, vocab word!" my son yelped. *Vocab word*, I thought, and tried to cheer myself up. *See, we're learning here! Learning is good!* "What's that?"

"It means people who hate Jews."

At ten, my son was familiar with this concept. "Like in the Passover story, or the Purim story, or the Hanukkah story? Or like that guy who shot people in Pittsburgh?"

"Yeah. Like that." I grimaced, thinking of the shootings, beatings, and stabbings since then, the ones I made a point of not telling him about. "Only in this play, it's not about killing people. It's about making Jews into a mean cartoon of a bad guy." It occurred to me, as I gripped the steering wheel, that these two things were in fact not at all separate, that the cartoon brand of hatred was actually the prerequisite for the killing-people brand of hatred, from Simon of Trent to Pepe the Frog. I chose not to explain this. Later it occurred to me that my failure to explain this, my perpetuation of the lie that the trolling was distinct from the danger, was part of the problem. My son did not notice.

"So that's why there's the weird stuff about a pound of flesh?" he asked.

"Exactly," I said. Then my loyalty to the glories of Western civilization kicked in, and I felt the need to add: "This play does cartoony stuff with this character, but Shakespeare also makes the guy more real. He isn't just a greedy bad guy; it's more complicated than that. He's human too."

My son was getting antsy. He would have his bond. "Play it," he demanded. I did.

The BBC production was vivid and engrossing. To my surprise I only needed to pause the playback occasionally to explain sixteenth-century puns, along with the concept of "usury," which baffled my son: "Don't people need to be able to borrow money to buy houses and stuff? Why is that bad?" I had no answer for this. But with the entrance of Shylock, there was far more to explain.

Shylock enters with the words "Three thousand ducats," and then says of the noble Antonio, "I hate him for he is a Christian; / But more for that in low simplicity / He lends out money gratis and brings down / The rate of usuance here with us in Venice. / If I can catch him once upon the hip, / I will feed fat the ancient grudge I bear him."

"Pause it," my son demanded from the backseat. "What does that mean?"

"Um, which part?" I asked, stalling. There was no good part.

"The part about why he hates the guy. He hates him for being Christian? That's dumb. But what's his other reason?" he asked, clearly hoping the other reason was a better one.

It wasn't a better one. "Because Shylock wants more money," I said.

"I thought you said he's not just a greedy bad guy."

"He's not," I claimed. "It gets better."

It did not get better. Soon I was forced to explain Shylock's self-serving interpretation of the biblical Jacob (a story my son knew in a sacred context), Antonio's fondness for spitting on Shylock, Shylock's deal for the pound of flesh, and numerous other verses that were painful to elucidate for one's Jewish ten-year-old son, such as "The Hebrew will turn Christian; he grows kind."

I drove in a daze, stunned by the sheer awfulness of it, dreading every verse my son would ask me to explain. The quality of mercy is not strain'd, the play's heroine famously says, but I strained to recall all the merciful interpretations of the play's richness, its "nuance." It was damned hard to hear the nuance while parsing lines like "Certainly the Jew is the very devil incarnal," or "My master's a very Jew; give him a present, give him a halter," or explaining what Shylock meant when he planned to "go in hate, to feed upon / The prodigal Christian." About an hour in—after Shylock's daughter escaped her evil father ("Our house is hell"), but before Shylock declared that he'd rather his daughter be "dead at my foot . . . and the ducats in her coffin"—we made it to Shylock's famous monologue, the part that makes it all OK.

I hit Pause, knowing I needed to build this up. "This speech changes what you think of Shylock," I told my son. "It makes him more human."

My son put on his game face. "OK, let's hear it."

The actor began the brief soliloquy that every English-speaking Jew is apparently meant to take as a compliment: "I am a Jew. Hath not a Jew eyes? Hath not a Jew hands, organs, dimensions, senses, affections, passions? . . . If you prick us, do we not bleed? If you tickle us, do we not laugh? If you poison us, do we not die? And if you wrong us, shall we not revenge?"

"Wait, *that's* the part where he's more human?"

I hit Pause again. "Sure," I told my son, game-facing him back in the rearview. "He's reminding us how he's like everyone else. He's a normal person with normal feelings."

My son laughed. "You seriously fell for that?"

I swallowed, sickened. "What do you mean?"

"Shylock's just saying he wants revenge! Like, 'Oh, yeah? If I'm a regular human, then I get to be *eee-vil* like a regular human!' This is the evil monologue thing that every supervillain does! 'I've had a rough life, and if you were me you would do the same thing, so that's why I'm going to KILL BATMAN, mu-hahaha!' He's just manipulating the other guy even more!"

"No, he's—" I fumbled, remembering the monologue's final words: *The villainy you teach me, I will execute; / and it shall go hard but I will better the instruction.* For the first time I heard the unspoken phrase that followed it: *Mu-hahaha.* I'd been trolled, betrayed, like Shylock at court.

"You're not supposed to *fall for* the evil supervillain monologue! What idiot would fall for that?"

What idiot would? I would. I did. I stared at the road in shame. My son is insistent, demanding, obsessive, morbid, too smart to be pleasant. It had not previously occurred to me that those traits are also his greatest strengths, the sources of his integrity.

"That was pathetic," he muttered. It was unclear whether he was referring to the play, or to his mother. "Are there other parts where he actually acts normal? Or is that it?"

Was that it? I reviewed the other moments scholars cite to prove Shylock's "humanity." There were two lines of Shylock treasuring his dead wife's ring, unlike the play's Christian men who give their wives' rings away. But unlike the other men, Shylock never gets his ring back—because his daughter steals it, and becomes a Christian, and inherits what remains of his estate at the play's triumphant end. Then there was the trial scene, where modern actors often make Shylock seem tragic rather than horrific. But that was performance, not text. Finally, scholars point to the many times Shylock *explains* why he is so revolting: Christians treat him poorly, so he returns the favor. But for this to satisfy, one must accept that Jews are revolting to begin with, and that their repulsiveness simply needs to be explained. None of it worked. And then I saw just how deep the gaslighting went: *I* felt obligated to make it work, to contort this revolting material into something that excused it.

I have a doctorate in literature. I am aware that Shakespeare's plays contain many layers and mean many things. But the degrading hideousness of this character is obvious even to a ten-year-old, and no matter how many more layers the play contains, that is unambiguously one of them. Why, I wondered, should I feel obligated to excuse this blindingly obvious fact, like some abused wife explaining why her darling husband beat her up? Why did I need to participate in this perverse historical mind trick of justifying my own people's humiliation—a humiliation that was never just a cartoon but that cost so many of my ancestors their dignity and even their lives?

"Nope," I told my son. "That's it."

"That's *it*?! That totally sucked."

"It gets worse," I said. How could I have exposed my child to this? "Let's turn it off."

But my son is insistent, demanding, obsessive, morbid. When you prick him, he does not bleed. "I need to hear how it ends."

The trial scene was agonizing. We listened together as Shylock went to court to extract his pound of flesh; as the heroine, chirping about the quality of mercy, forbade him to spill the Christian's blood as he so desperately desired; as the court confiscated his property, along with his soul through forced conversion; as the play's most cherished characters used his own words to taunt and demean him, relishing their vanquishing of the bloodthirsty Jew. My son stopped asking me to explain. Twenty minutes of congratulatory hijinks followed Shylock's final exit, as the cast reveled in their victory and his seized assets. At last it was over.

The minivan fell silent. Then my son announced, "I never want to hear that again."

"You will definitely hear that again," I said.

It's true. "Censorship" is beside the point, the insane extremes of "cancel culture" extravagantly irrelevant, because this double helix of hate and art is built into our world. My son will read this play in school. Or he will hear about a new performance; it's one of the most performed plays Shakespeare ever wrote. He will encounter headlines and jokes using the phrase "pound of flesh." He couldn't even make it through a season of *Radiolab* without it. There is a terrible bond at work here, tying us inexorably to a long history of ugly caricatures and spilled blood. And there is also a much subtler and more insidious bond, tying us to the need to justify and accept it. But unlike me, my son insists on integrity, demands it. He is not afraid to be unpleasant; he knows evil when he hears it. He is ready for this bond.

I told him, "At least now you know."

"Yes," he said, and smiled. "Next can we download *Dracula?*"

Chapter 12

DEAD AMERICAN JEWS,
PART THREE:
Turning the Page

THE THIRD TIME THERE WAS A SHOOTING ATTACK AGAINST American Jews, the *New York Times* did not call me to ask for a quick op-ed, and neither did anyone else. I presume this was because when something happens three times, it is no longer news. Perhaps these news outlets realized just how unnewsworthy this story actually was. People murdering Jews, as a three-thousand-year-old global phenomenon, is pretty much the opposite of news. When no one called me, I felt profoundly relieved, because the things I wanted to say about it were no longer things that I could actually say.

The third shooting attack, and the dozen or so other physical attacks on American Jews that followed in rapid succession after it—some barely reported—were what privately changed me, perhaps because that third shooting happened at a kosher grocery store about twenty minutes from my house.

Unlike after the Pittsburgh and San Diego attacks, information on the Jersey City attack was slow to accumulate. The two assailants first killed a livery driver (it was later discovered that they had Googled his Jewish-sounding surname), then progressed to killing a police officer who had noticed their stolen U-Haul, and then proceeded to attack the grocery store, resulting in a protracted gun battle in which the grocery's owner, a customer, and a store worker were killed, along with the two assailants, who were killed by police after an exchange of fire that lasted well over an hour. The scene in the city was dramatic: Entire neighborhoods swarmed with state troopers and the National Guard, and children in nearby schools were held in lockdowns until late at night.

The event was initially reported as a kind of perp chase gone horribly wrong, during which criminals outrunning cops ducked into a random store for cover. But antisemitic screeds found in the attackers' vehicle and in their social media postings told a different story, as did the tactical gear they wore, the massive stash of ammunition and firearms they brought along, and security camera footage showing them driving slowly down the street, checking addresses before parking and entering the market with guns blazing. Then there were the enormous quantities of explosives in their stolen truck, which an FBI agent later said had the capacity to kill people within a range of five hundred yards. Their real targets, authorities surmised, were likely the fifty Jewish children in the private elementary school at the same address, directly above the store—all of whom huddled in closets for the entire gun battle, listening to their neighbors being murdered below.

The delayed clarity on what exactly happened in Jersey City muted some of the public empathy that instantly followed the previous attacks. So did the identities of the attackers, both of whom were Black, and their targets, who were Hasidic Jews—who, it has progressively become clear, many otherwise enlightened Americans view as absolutely fair game for bigotry.

This was obvious from reporting within hours of the attack, which gave surprising emphasis to the murdered Jews as "gentrifying" a "minority" neighborhood. This was remarkable, given that the tiny Hasidic community in question, highly visible members of the world's most consistently persecuted minority, in fact came to Jersey City *fleeing* gentrification, after being priced out of long-established Hasidic communities in Brooklyn. More tellingly, as the journalist Armin Rosen has pointed out, the apparently murderous rage against gentrification has yet to result in anyone using automatic weapons to blow away white hipsters at the newest Blue Bottle Coffee franchise. What was most remarkable about this angle, however, was how it was presented in media reports as providing "context."

The "context" supplied by news outlets after this attack was breathtaking in its cruelty. As the Associated Press explained in a news report about the Jersey City murders that was picked up by NBC and many other outlets, "The slayings happened in a neighborhood where Hasidic families had recently been relocating, amid pushback from some local officials who complained about representatives of the community going door to door, offering to buy homes at Brooklyn prices." (Like many homeowners, I too have been approached by real estate agents asking me if I wanted to sell my home. I recall saying no, though I suppose murdering these people would also have made them go away.) New Jersey's state newspaper, the *Star-Ledger*, helpfully pointed out that "the attack that killed two Orthodox Jews, an Ecuadorian immigrant and a Jersey City police detective has highlighted racial tension that had been simmering ever since ultra-Orthodox Jews began moving to a lower-income community"—even though the assailants never lived in Jersey City, and apparently chose their target simply through internet searches for Jewish institutions in the New York area. The *Washington Post* began its analysis of the murders by announcing that Jersey City "is grappling with whether the attack reflects underlying ethnic tensions locally and fears that it could spark new ones"—even though the rest of the article described in detail how "longtime black residents and ultra-Orthodox implants alike say that they haven't experienced significant ethnic tensions here." Nonetheless, readers were informed, "the influx of Hasidic residents comes as many of the longtime black residents feel increasingly squeezed." This was all about gentrification, the public learned. The assailants, who wore socially acceptable clothing, were expressing an understandable communal sentiment. The newly dead Jews, on the other hand, were members of the unharassed majority, despite being the country's top hate-crime target according to the FBI. They were also rich, despite experiencing the same poverty rates as the rest of New York and New Jersey. On top of that, they wore unfashionable hats. So it kind of made sense that people wanted to murder their children with high-impact explosives.

I was not able to find any similar "context" in media reports after the 2015 massacre at a Black church in Charleston, South Carolina, or the 2016 massacre at an LGBTQ nightclub in Orlando, Florida, or the 2019 massacre at a Walmart in El Paso, Texas frequented by Latino shoppers— all hate-crime attacks that unambiguously targeted minority groups. In each of those cases, as was true in Jersey City, media coverage included sympathetic pieces about the victims, along with investigative pieces about the perpetrators, the latter focused on how perpetrators were drawn into violent irrational hatred. But in reviewing media reports from the aftermath of these events, I found no coverage of how straight people in Orlando other than the perpetrator—in other words, reasonable, non-murderous, relatable "normal" neighbors—were understandably upset about gay couples setting up shop in the neighborhood and disrupting their "way of life," or about how white people with deep family roots in Charleston felt understandably wistful about the Black community's "takeover" of certain previously white neighborhoods, or about how non-Latinos in El Paso felt "squeezed" by ongoing "tensions" with Latinos who had pushed for more bilingualism in schools.

No one covered this "context," because doing that would be bonkers. It would be hateful victim-blaming, the equivalent of analyzing the flattering selfies of a rape victim in lurid detail in order to provide "context" for a sexual assault.

That doesn't mean that intergroup tensions (or the problems with flattering selfies) aren't ever worth examining. It simply means that presenting such analysis as a hot take after a massacre is not merely disgusting and inhuman, but also a form of the very same hatred that caused the massacre—because the sole motivation for providing such "context" in that moment is to inform the public that those people got what was coming to them. People who think of themselves as educated and ethical don't do this, because it is both factually untrue and morally wrong. But if we're talking about Hasidic Jews, it is quite literally a different story, and there is one very simple reason why.

The mental gymnastics required to get the Jersey City attack out of my head were challenging, especially when the Jewish community in the New York area was treated in the two weeks following this massacre to more than a dozen other assaults of varying degrees, most of them coming during the festival of Hanukkah. These included Jews being slapped, punched, kicked, and beaten on the streets by people who made their motives clear by shouting antisemitic insults, and many other variants on this theme that received much less attention. (One that shook me personally was when a young white man broke into my students' dormitory at Yeshiva University at four a.m. and started a fire—using matches from the dorm lobby's Hanukkah candle-lighting.) All this was merely an intensified version of physical assaults on Hasidic Jews in New York that had been happening regularly for over a year—incidents that ranged from run-of-the-mill acts of knocking elderly people to the ground to the rather more advanced tactic of clobbering someone over the head with a large paving stone, causing a fractured skull.

This new normal culminated in a particularly horrifying attack, when a man entered a crowded Hanukkah party at a Hasidic rabbi's house in Monsey, New York, wielding a four-foot machete, and stabbed or slashed five people, all of whom were hospitalized; one victim, who fell into a coma, died several months later from his wounds. Stabbing Jews was apparently in vogue in Monsey, as this was actually the second antisemitic knifing in town in just over a month. The previous attack's victim was beaten and stabbed while walking to morning prayers, winding up in critical condition with head injuries. Media coverage of these attacks also sometimes featured "context" (read: gaslighting), mentioning heated school-board or zoning battles between Hasidic and non-Hasidic residents—even after the perpetrator was identified as a resident of a town forty minutes away. One widely syndicated *Associated Press* article situated the previous week's bloodbath by informing millions of readers that "The expansion of Hasidic communities in New York's Hudson Valley, the Catskills and northern New Jersey has led to predictable sparring over new housing development and local political control. It has also led to flare-ups of rhetoric seen by some as antisemitic." In other words, the cause of bloodthirsty antisemitic violence is . . . Jews, living in a place! Sometimes, Jews who live in places even *buy land on which to live*. To be fair, there were many countries and centuries in which this Jews-owning-land monkey business was illegal, though twenty-first-century Hudson Valley, the Catskills, and northern New Jersey are sadly not among those enlightened locales. Predictably, this leads to sparring, and flare-ups. Who *wouldn't* express frustration with municipal politics by hacking people with a machete?

After the first attack in Pittsburgh, I was devastated. After the second attack in San Diego, I was angry. But after the third attack near my home and the season of horror that followed, I simply gave up.

There was no way I could write about any of this for the *New York Times*, or any other mainstream news outlet. I could not stomach all the "to be sures" and other verbal garbage I would have to shovel in order to express something acceptable to a non-Jewish audience in a thousand words or less. I could no longer handle the degrading exercise of calmly explaining to the public why it was not OK to partially amputate someone's arm with a four-foot-long blade at a holiday party, even if one had legitimate grievances with that person's town council votes. Nor could I announce, as every non-Jewish media outlet would expect, that these people whose hairstyles one dislikes are "canaries in the coal mine," people whose fractured skulls we all ought to care about because they *serve as a warning*—because when Jews get murdered or maimed, it might be an ominous sign that *actual* people, people who wear athleisure, might *later* get attacked! I was done with this sort of thing, which amounted to politely persuading people of one's right to exist.

The thought of writing about any of this for Jewish media outlets was sickening for a different reason. It was demoralizing to confront the American Jewish community's ongoing and escalating panic, the completely justified intergenerational PTSD freak-out voiced constantly from every point on the political spectrum, the repetitive anxiety attacks expressed on social media, the nonstop discussion about whether this was like Berlin in 1935. This facile comparison was of course ridiculous on its face, as well as insulting to the overwhelming majority of Americans who responded to these attacks in exactly the opposite fashion from the mass state-sponsored violence of Nazi Germany. If anything, this felt more like Paris in 2005—a place where there was no shortage of legal protections and official goodwill, but where one wouldn't be crazy to occasionally hide a yarmulke under a baseball hat. Yet the thought of explaining this was exhausting too, and also beside the point. Was I really going to expend energy delineating why this wasn't like the Third Reich, but perhaps resembled, say, second-century Egypt or tenth-century Spain? To what end? To reassure everyone that "only" a few Jews were actually maimed or dead, so everything was cool? Nitpicking over sloppy historical analogies was a convenient distraction. The fact was that a communal memory of multiple millennia had been activated, and it was deep and real.

Of all the tedious and self-serving explanations for why this scourge was apparently reemerging in American life (*Guns! Trump! Trolls! Twitter!*), the most convincing was actually the most boring, and also the most disturbing: The last few generations of American non-Jews had been chagrined by the enormity of the Holocaust—which had been perpetrated by America's enemy, and which was grotesque enough to make antisemitism socially unacceptable, even shameful. Now that people who remembered the shock of those events were dying off, the public shame associated with expressing antisemitism was dying too. In other words, hating Jews was normal. And historically speaking, the decades in which my parents and I had grown up simply hadn't been normal. Now, normal was coming back.

A week after that horrific Hanukkah, the Jewish community organized a "No Fear, No Hate" march in New York City, which twenty-five thousand people attended—though almost no one from the particular Hasidic communities which had been attacked, whose adherents generally don't go in for that sort of thing. I didn't go in for it either, though for somewhat different reasons. It interfered with Hebrew school carpooling, for one thing. And while I knew the march was intended as an act of pride and defiance and that those who attended found it empowering and inspiring, its mere existence felt profoundly depressing to me, almost like an admission of defeat. I watched the photos and videos pouring in from the march with a kind of uncomfortable schadenfreude: happy that so many people had attended, and even happier that I was obligated to drive seven children home from Hebrew school instead.

But another massive Jewish gathering near my home a few days earlier had also caught my attention, one whose attendance, ninety thousand people at MetLife Stadium in New Jersey's Meadowlands, dwarfed that of the march in New York. And unlike the march, it was attended by many of the people who had been directly targeted during those horrific weeks. This event, which was mirrored in parallel events around the world, was the "Siyum HaShas," or "Conclusion of the Talmud," a ceremony celebrating the completion of communally studying the Babylonian Talmud in a program called Daf Yomi, or "A Page a Day."

Begun by a Polish rabbi in 1923 as a way to democratize Talmud study, the premise of Daf Yomi is to study one "page" of the Talmud each day—really two sides of one large physical page, which with the necessary commentaries is more like fifteen pages of dense material—thereby completing the Talmud's 2,711 "pages" in a very reasonable seven and a half years. To those unfamiliar with Jewish text study, this probably sounds like a big commitment. But when one considers that Talmud study was traditionally a full-time affair, it was quite radical to suggest that one could actually complete the entire Talmud while still, say, holding down a job. Today, Daf Yomi, the "world's largest book club," consists of hundreds of thousands of people around the world who spend those seven and a half years quite literally on the same page. When they finish, most of them start right over again.

I wouldn't have fit in much at MetLife Stadium's Daf Yomi ceremony, which consisted almost entirely of men wearing black hats. Since the fifth century when the Talmud was compiled (its sources span the previous six centuries), women have rarely studied it. Its study was also long deemphasized in Judaism's more liberal modern movements, including those in which I was raised. But in recent years, both of those things have been changing. This most recent Daf Yomi cycle concluded with a large-scale women's ceremony in Jerusalem, attended by thousands of women—many of whom had not only studied the text for the previous seven and a half years, but had also taught it, sometimes for large audiences in person and online. My brilliant friend Ilana Kurshan, an accomplished literary agent and translator in Jerusalem, published an award-winning memoir, *If All the Seas Were Ink*, describing how the daily routine of Talmud study carried her through challenging years in her own life—and her book inspired many less-traditional people who had previously assumed that Talmud study wasn't for them. In the years since the last cycle began, Daf Yomi resources had also blossomed online, including a fantastic variety of podcasts, Facebook discussion groups, Instagram stories, Twitter accounts, and more, much of it geared toward people with no previous background in Talmud study. Daf Yomi was going viral.

I had toyed with doing Daf Yomi at the start of the last cycle, but I had just given birth to my fourth baby, and the thought of slogging through fat volumes or clunky websites with mediocre translations was unappealing at best. But this time around, there was a game-changing free app called Sefaria (Hebrew for "library"), which contained nearly the entire canon of traditional Jewish texts in their original Hebrew and Aramaic, along with more accessible English translations. I'd downloaded Sefaria years earlier—it was created by an acquaintance—and I often relied on it for biblical and other references. On that frigid and depressing Sunday in January, toggling on my phone between the vast crowds of anti-hate marchers and the vast crowds of Talmud enthusiasts, I realized with utter wonder that I already had the entire Talmud in the palm of my hand.

I suddenly knew what I wanted to do. Along with hundreds of thousands of people around the world, I opened to the very first page, and began.

––––––––––

Something magical happened when I switched over from looking online at news reports about antisemitic attacks to joining online Daf Yomi discussion groups and looking up Daf Yomi resources. The algorithms all caught on instantly, and suddenly I saw almost nothing online that wasn't related to discussions of the Talmud's opening pages—which contain a rambling, digressive, and almost bottomless conversation about when, where, and how to recite the Sh'ma, Judaism's central statement of faith in the singularity of God.

I had studied parts of the Talmud before, and in the past I'd found its structure extraordinarily annoying. The Talmud isn't written like a normal book, or even like a normal "sacred" text. It's not a story, or a manual of religious practice, or a compilation of wisdom, or a book of philosophy, or a commentary on the Bible, or even a compendium of laws. Instead, it's more like a ridiculously long social media thread, complete with pedantic back-and-forths, hashtagged references, nonstop links and memes, and limitless subthreads, often with almost no discernible arc or goal. Or to use a more timeless metaphor, it's like walking into a room full of people engaged in a heated conversation—people who are constantly interrupting one another and shunting the conversation onto different tracks, and who don't care at all if you know what they're talking about, and who therefore never bother to explain why any of this matters. For a novelist like me who spent twenty years creating artistically designed stories, engineered to draw readers in and take them to a destination, this rambling discussion passing itself off as a book (recorded, incidentally, in a nearly indecipherable shorthand style) always struck me as exasperating in the extreme. But after the dark weeks I had just sleepwalked through and all the inexpressible anxiety that had accompanied them, I walked into this irritating conversation and experienced a strange and unexpected feeling: an undeniable sense of welcome and relief. It was like coming out of a cold, dark night into a warm and lighted room. Six centuries of sages seemed to move over, still talking, and make space for me at the study-hall table strewn with open books. I sat down, exhausted, and listened.

––––––––––

From when, the Talmud's first page begins, does one recite the evening Sh'ma? From when the priests in the Temple consumed the daily sacrifices. From when the remains of those sacrifices finished burning on the altar. From dusk, which is different from sunset. Up until midnight. Actually, up until dawn. How do you know when it's dawn? When you can distinguish between a white and a blue thread, the threads sewn into the corner of a garment to remind one of God's commandments––but they'll get back to that later. Really it's up to the end of the first night watch. Or the second watch. How long is a "watch," and how many are there in the night? A biblical verse suggests there are four. Another verse suggests there are three. During the first night watch, donkeys bray; during the second night watch, dogs bark; during the third, babies wake to nurse and wives whisper with their husbands. Maybe this is beautiful imagery, or maybe it corresponds to constellations moving across the night sky. This whole conversation is about how to tell time without clocks—or, to put it another way, how to find one's place in the world while the world is in motion, how to hold fast to that constant point of stillness as all else changes. It's a skill, a science, an art. King David woke at midnight to praise God, according to the Psalms. How did he wake up? He had an Aeolian harp hanging above his bed, and the midnight wind would wake him. Would the wind always come at midnight? Or was this wind more like a writer's inspiration, moving the poet-king to rise in the dark and write his psalms of praise?

It had been a number of years since I had regularly recited the evening Sh'ma. But after spending half an hour each day following complex arguments on this point, I found myself returning to it, on my own and with my family, chanting the words as we drifted to sleep. There was a comfort here, a refuge as we recited the words. We were on a watch, awake in the dark. But someone was also watching us.

There was also something comforting in the endlessness of the rabbis' conversations. The obsessive-compulsive thought patterns of these people felt familiar to me, a personality tic that I knew well from myself and from many of my relatives and friends—one that had always frustrated me, both in myself and in others. I'd perceived it as a fault to be corrected. But now I saw clearly what it was expressing: grief, fear, and resilience.

Until the year 70 CE, Judaism had been centered at the ancient Temple in Jerusalem, where worship was mediated through priests offering sacrifices. This was a visceral, physical process involving livestock and grain and wine and incense and fire and smoke. There was nothing metaphorical or intellectual about it. Even the location itself was mandated by God. After the Romans destroyed this temple and exiled the people, there was no particular reason for this religion, or even simply this people, to survive in any form. But on the eve of this temple's destruction, one sage, Rabbi Yochanan ben Zakkai, had himself smuggled out of the besieged city of Jerusalem in a coffin, after which he convinced the Roman general Vespasian to allow him to open an academy for Torah scholars in a small town far from Jerusalem. Both Rabbi Yochanan ben Zakkai and Judaism faked their own deaths in order to survive this cataclysm. The small cadre of scholars in that small town reinvented this religion by turning it into a virtual-reality system, replacing temple rituals with equally ritualized blessings and prayers, study of Torah, and elaborately regulated interpersonal ethics. The sages frantically arguing about when and how to recite which prayers are survivors and descendants of survivors, remnants of a destroyed world. They are anxious about remembering every last detail of that lost connection to God, like mourners obsessing over the tiniest memories of a beloved they have lost. One might expect that this memory would eventually fade, that people would "move on." Instead the opposite happens. Once the process of memory becomes important, the details do not fade but rather accrue—because the memory itself becomes a living thing, enriched by every subsequent generation that brings new meaning to it.

As I followed the discussions that had previously annoyed

me, I realized something stunning: Many of the sages arguing with one another on each page didn't live in the same generation, or even in the same century. Nor were they, for the most part, quoting written texts of what those in the previous generation or century had said. Instead, they relied on designated people who served as mental court reporters, tasked with sitting in study halls where these discussions took place and mentally recording entire conversations between sages. These records were then passed down almost entirely orally, written down only generations later.

This would simply be a fascinating historical fact, except that as I turned the Talmud's pages, I discovered that it wasn't—because these people's elaborate communal memory overlapped with mine. As I followed along while the rabbis debated how prayers should be said, I frequently bumped up against actual prayers and practices—blessings of gratitude for different foods, words to recite before a journey, how and when to bow and rise—that I myself had learned when I was young, that I too knew by heart. Of course, I only knew these things because of these people on this page, and all the people after them, who had made the conscious decision to pass these things down to me. None of these sages needed to say what was obvious from the mere existence of this process. Destruction and humiliation didn't matter. Only memory and integrity did. Was the hour I was living through right now different from the hour they were living through then? Did it matter? From what hour does one recite the evening Sh'ma?

God prays, the sages say. There is a lot on God's mind. According to the Psalms, God is furious every day. How long does God remain angry? the sages ask. For one fifty-eight thousand eight hundred and eighty-eighth of an hour. In God's prayer, God says: *May it be my will that my compassion will overcome my anger.* I wondered: Was I furious every day too? (I was, then.) Could I try to be furious only for one fifty-eight thousand eight hundred and eighty-eighth of an hour? Could my compassion overcome my anger? Was my life a mere fifty-eight thousand eight hundred and eighty-eighth of an hour? From what hour does one recite the evening Sh'ma?

I was surprised by how little I was bothered by the things I didn't agree with. Perhaps you will be shocked to hear, for instance, that a fifth-century text says unenlightened things about women. I could hardly have cared less. I was more surprised by how enchanted I was by what almost no one would agree with, or at least no mature adult.

The Torah wasn't given to the ministering angels, one sage points out, but to people with bodies. Bodies that fart. There is an entirely unreasonable amount of material about bathrooms, farting, peeing, and pooping, which is off-putting if you are not a twelve-year-old boy. Luckily I live with a twelve-year-old boy, along with a ten-year-old boy and a seven-year-old boy—and boy, were they entertained. This opened my eyes to how much of the text was perhaps intentionally funny. One extended discussion, for instance, dealt with whether one could recite the Sh'ma while in the presence of a pile of poop. Walk at least four steps away from it, one sage advised. But what if you still smell it? another asked. Well, another chimed in, perhaps it depended whether the poop was wet or dry. But how can you determine *how* dry the poop is? somebody asked——and then promptly supplied a story of how one revered teacher once sent his devoted pupil to check the specific crustiness of a piece of poop. My sons' amusement at this made me realize what should have been obvious: The discussion might have been serious (especially before indoor plumbing), but there might well be a point where it crossed over to deadpan humor. Or bedpan humor. Either way, then as now, the world really is full of shit. We can pretend the shit is not there, or we can think through how to live with it without making ourselves sick.

Similar trolling powers arose in a discussion of invisible "demons" that, various sages claimed, surrounded each person by the thousands, causing all kinds of pains and illnesses. (Two major twentieth-century sages suggested a new translation for the ambiguous word "demons": germs.) Was there a way to see these "demons"? the sages wondered. Yes, one rabbi announced. One simply had to take the placenta of a firstborn black cat, who was herself the firstborn daughter of another firstborn black cat, grind up and burn that cat placenta, and then rub it into one's eyes——"and then one will see them." Another sage followed these instructions and saw the demons, which probably looked like ground-up burnt cat placenta. I could not stop laughing. On the other hand, demons really are everywhere. They are invisible, and there are thousands of them, on every screen in every pocket, spreading lies and causing pain. God is furious every day. Who isn't?

The sages mourn. One rabbi, father of ten dead sons, carries around a bone of his tenth dead son and shows it to every mourner he meets, sharing his pain and also his compassion.

Another, asked to serenade a bride and groom, stands and sings, *You are going to die*. Others smash glasses at weddings, a practice continued at my own. Ours is a broken world. Rebuilding is hard, daily, constant, endless, the marriage that follows the wedding, which is not a happy ending but an imperfect beginning. From when can we recite the evening Sh'ma?

The sages obsess over how to disagree without humiliating others. One announces that it is preferable to enter a fiery furnace than to embarrass another person in public. When the president of the rabbinic high court publicly insults a sage who disagrees with him, the court impeaches him and throws him out of office. Seeing the respect given to the sage he insulted, the former president visits him at home, where he discovers the sage is a humble blacksmith. He begs forgiveness, repeating his apologies even when the man at first ignores him. When the man accepts his apology, the court restores him to his position, though on the condition that he share the post with the sage appointed in his stead——a compromise reached with the help of the blacksmith sage, on behalf of his former enemy. I studied these passages as the news blared with a presidential impeachment trial, our public life a sickening spectacle of corruption and insult with no interest in reconciliation or even integrity. There are ways to rebuild a broken world, and they require humility and empathy, a constant awareness that no one is better than anyone else. That constant awareness requires practice, vigilance, being up at all the watches.

The sages know the world is broken. They hold the broken pieces tight. An old scholar who has forgotten the Torah he studied is compared to the broken pieces of the Ten Commandments, the stone tablets that Moses smashed in frustration when the people turned to idol worship. These broken tablets were put into the Ark of the Covenant along with the new tablets that replaced them, the shattered pieces also part of the contract with God. The old scholar who has forgotten what he knew is still honored, carried by the people on their journey. Sometimes I felt like that old scholar, my memory of what I once valued fading, diminished, broken. I turned the page, and these long-dead scholars carried me along.

The comforting thing about Talmud study, and Daf Yomi in particular, is that you are never alone with it. Online, instead of people yelling at one another about why they were right, Daf Yomi learners gather to ask one another what this sentence means, whether their interpretation works, what the deeper meaning is. I am stunned by these strangers, by their sincerity and candor—qualities one rarely encounters today, online or off. To my amazement, many are non-Jews in the process of converting to Judaism, voluntarily joining this journey even in the darkness.

To my even greater amazement, one of my fellow Daf Yomi learners is my mother—the world's least pedantic person, who signed on and has shown no signs of quitting. A grandmother of fourteen, she also now recites the Sh'ma, because, as she says, "I'm up at all the watches." A mere mother of four, I'm up for many of them too. When we finish, we will be seven years older in our respective generations, and also, God willing, seven years wiser—even if we forget what we have learned, even if we are broken, even if our forgotten wisdom rattles around inside our minds like shards of broken tablets.

I still follow today's old, old news. But now I also turn away from it, toward the old, the ancient. I am forever haunted, as all living people always are, our minds the dwelling-places for the fears and hopes of those who came before us. I turn the page and return, carried by fellow readers living and dead, all turning the pages with me.

Acknowledgments

MORE THAN MOST BOOKS, THIS ONE OWES ITS LIFE TO generous editors who urged me to follow paths I might never have traveled on my own. As always, I am grateful to my agent, Gary Morris, who assured me that this project was worth pursuing, and to my Norton editor, Alane Salierno Mason, whose generosity and insight never fail to impress me.

But this time I also have many more editors to thank—those who urged me to explore topics I never considered, and also those who trusted me to follow my own questionable instincts. For their faith in the stories in this book which they were the first to publish, and also for their own brilliant ideas and suggestions, I am grateful to Alana Newhouse, David Samuels, and Matthew Fishbane of *Tablet*, Abe Socher of *Jewish Review of Books*, Jennie Rothenberg Gritz of *Smithsonian*, Rachel Dry of the *New York Times*, Marla Braverman of *Azure*, Stuart Halpern in his role as editor of the anthology *Esther in America* (Maggid Books, 2021), and Yoni Applebaum of the *Atlantic*. I am likewise indebted to the many thoughtful people who gave generously of their time to share with me their own brave and innovative approaches to understanding the Jewish past, as well as their insightful reflections on their own experiences—including Dan Ben-Canaan, Pierre Sauvage, Ala Zuskin Perelman, Jason Guberman-Pfeffer, Fran Malino, Irene Clurman, Bonnie Galat, Jean Ispa, Alex Nahumson, Jim Fry, Eddie Ashkenazie, and Chrystie Sherman, along with many others whose words did not reach these pages. And I am also grateful to those who lent me their expertise without knowing it, including the authors listed in the bibliography.

This book bears the imprint of my parents, Susan and Matthew Horn, and especially my mother, who raised me to be curious about the Jewish past and conscious of how we make use of it in the present. It also bears the influence of my husband, Brendan Schulman, who has tolerated hearing me talk about dead Jews for over twenty years, and who has always urged me to be braver than I otherwise would be.

This book is dedicated to our children, Maya, Ari, Eli, and Ronen. It is my fervent hope that they will never feel the need to read it.

Works Consulted

Alexander, Sidney. *Marc Chagall: An Intimate Biography.* New York: Paragon House, 1978.

Arendt, Hannah. *Eichmann in Jerusalem: A Report on the Banality of Evil.* New York: Viking, 1963.

Arendt, Hannah, and Mary McCarthy. *Between Friends: The Correspondence of Hannah Arendt and Mary McCarthy, 1949–1975.* Edited by Carol Brightman. New York: Harvest Books, 1996.

"As Jewish Enclaves Spring Up Around NYC, So Does Intolerance." Associated Press, January 2, 2020.

Ault, Alicia. "Did Ellis Island Officials Really Change the Names of Immigrants?" *Smithsonian,* December 2016.

Bale, Rachael, and Jani Hall. "What You Need to Know About Tiger Farms." *National Geographic,* February 2017.

Bar-Itzhak, Haya. *Jewish Poland—Legends of Origin: Ethnopoetics and Legendary Chronicles.* Detroit: Wayne State University Press, 2001.

Ben-Canaan, Dan. "The Jewish Experience in China and Harbin: The Chinese Perception of the Other." Lecture delivered at the Hong Kong Jewish Historical Society (text provided to the author). May 29, 2016.

———. *Jewish Footprints in Harbin.* Harbin: China Education Press, 2018.

———. *The Kaspe File: A Case Study of Harbin as an Intersection of Cultural and Ethnic Communities in Conflict, 1932–1945.* Harbin: Heilongjiang University People's Publishing House, 2008.

Bialik, Chaim Nachman. *Shirot Bialik: A New and Annotated Translation of Chaim Nachman Bialik's Epic Poems*. Edited and translated by Steven L. Jacobs. Columbus, Ohio: Alpha Publishing Company, 1987.

Bloom, Harold. *Shakespeare and the Invention of the Human*. New York: Riverhead, 1998.

Bowd, Stephen, and J. Donald Cullington. *"On Everyone's Lips": Humanists, Jews, and the Tale of Simon of Trent*. Phoenix: Arizona Center for Medieval and Renaissance Studies Press, 2012.

Bresler, Boris. "Harbin's Jewish Community 1898–1958: Politics, Prosperity and Adversity." *Jews of China, Volume 1: Historical and Comparative Perspectives*. Edited by Jonathan Goldstein. London: Routledge, 1998.

Caldwell, Christopher. "Hero and Oddball." Review of *A Quiet American: The Secret War of Varian Fry* by Andy Marino. *Policy Review*, February 1, 2000.

Chagall, Marc. *My Life*. 1925. Translated by Elisabeth Abbott. New York: Orion Press, 1960.

Cohen, Gerson D. "The Story of the Four Captives." *Proceedings of the American Academy for Jewish Research* 29 (1960–61).

Dautch, Aviva. "A Jewish Reading of *The Merchant of Venice*." *British Library*, March 15, 2016.

Diarna: The Geo-Museum of North African and Middle Eastern Jewish Life. Forthcoming oral-history interviews screened for the author. Publicly available archive at www.diarna.org.

Duberman, Martin. *The Worlds of Lincoln Kirstein*. New York: Knopf, 2007.

Fallon, Scott. "Tensions Within a Changing Community Are Heightened in the Wake of the Jersey City Tragedy." *The Star-Ledger* (New Jersey), December 20, 2019.

Fermaglich, Kirsten. *A Rosenberg by Any Other Name: A History of Jewish Name Changing in America*. New York: New York University Press, 2018.

Feuchtwanger, Lion. *The Oppermanns*. New York: Viking, 1934.

Frank, Anne. *Diary of Anne Frank: The Revised Critical Edition*. Edited by David Barnouw and Gerald Van Der Stroum. Netherlands Institute for War Documentation, translated by Arnold J. Pomerans et al. New York: Doubleday, 2003.

Frank, Steve. *"The Merchant of Venice* Perpetuates Vile Stereotypes of Jews." *The Washington Post*, July 28, 2016.

Fry, Varian. *Surrender on Demand*. New York: Random House, 1945.

———. Varian Fry Papers, 1940–1967. 20 boxes. Columbia University Rare Book and Manuscript Library, New York.

Gitelman, Zvi. *Jewish Nationality and Soviet Politics: The Jewish Sections of the CPSU, 1917–1930*. Princeton: Princeton University Press, 1972.

Glatshteyn, Yankev. "Good Night, World." 1938. *American Yiddish Poetry: A Bilingual Anthology*. Edited and translated by Barbara Harshav and Benjamin Harshav. Berkeley: University of California Press, 1986.

Gold, Mary Jayne. *Crossroads: Marseille, 1940*. New York: Doubleday, 1980.

Gradowski, Zalman. "The Czech Transport: Chronicle of the Auschwitz Sonderkommando." 1944. Translated by Robert Wolf. In *The Literature of Destruction: Jewish Responses to Catastrophe*, edited by David G. Roskies. Philadelphia: Jewish Publication Society, 1989.

Graetz, Heinrich. *History of the Jews, Volume 1*. Translated by Bella Lowy. Philadelphia: Jewish Publication Society, 1891.

Ibn Daud, Abraham. *The Book of Tradition: Sefer Ha-Qabbalah*. Translated by Gerson D. Cohen. Philadelphia: Jewish Publication Society, 1967.

Isenberg, Sheila. *A Hero of Our Own: The Story of Varian Fry*. New York: Random House, 2001.

Julius, Anthony. *Trials of the Diaspora*. Oxford: Oxford University Press, 2010.

Kermode, Frank. *The Sense of an Ending*. New York: Oxford University Press, 1966.

Kirstein, Lincoln. *Mosaic: Memoirs*. New York: Farrar, Straus & Giroux, 1994.

Kurshan, Ilana. *If All the Seas Were Ink*. New York: St. Martin's Press, 2017.

Lowery, Wesley, Kevin Armstrong, and Deanna Paul. "Jersey City Grapples with Ramifications of Shooting at Kosher Market." *The Washington Post*, December 12, 2019.

Marino, Andy. *A Quiet American: The Secret War of Varian Fry*. New York: St. Martin's Press, 1999.

Mendele Moykher Seforim. *The Mare*. Translated by Joachim Neugroschel. In *Great Tales of Jewish Fantasy and the Occult*, edited by Joachim Neugroschel. Woodstock, NY: Overlook Press, 1987.

Miller, Henry B. "Russian Development of Manchuria." *National Geographic*, March 1904.

Nahman of Bratslav. *Nahman of Bratslav: The Tales*. Translated by Arnold J. Band. New York: Paulist Press, 1978.

Nightingale, Carl. *Segregation: A Global History of Divided Cities*. Chicago: University of Chicago Press, 2012.

Ozick, Cynthia. "Who Owns Anne Frank?" *The New Yorker*, October 6, 1997.

Palmieri-Billig, Lisa. "Libyans Protest Jew Attempting to Reopen Synagogue." *Jerusalem Post*, October 7, 2011.

Perelman, Ala Zuskin. *The Travels of Benjamin Zuskin*. Translated by Sharon Blass. Syracuse: Syracuse University Press, 2015.

Po-chia Hsia, R. *Trent 1475: Stories of a Ritual Murder Trial*. New Haven: Yale University Press, 1996.

Rosen, Armin. "Everybody Knows: As the Leading Targets of Hate Crimes, Jews are Routinely Being Targeted in the Streets of New York City." *Tablet*, July 15, 2019.

Rosenfarb, Chava. *The Tree of Life: A Trilogy of Life in the Łódź Ghetto*. Translated by Chava Rosenfarb and Goldie Morgentaler. Madison: University of Wisconsin Press, 1985.

Rubenstein, Joshua, and Vladimir P. Naumov, eds. *Stalin's Secret Pogrom: The Postwar Inquisition of the Jewish Anti-Fascist Committee*. Translated by Laura Esther Wolfson. New Haven: Yale University Press, 2001.

Sauvage, Pierre. *And Crown Thy Good: Varian Fry and the Refugee Crisis, Marseille 1940–41*. Forthcoming film; excerpts and additional interviews screened for the author. Varian Fry Institute, Los Angeles. www.varianfry.org.

Sawin, Martica. *Surrealism in Exile and the Beginning of the New York School*. Cambridge, MA: MIT Press, 1995.

Seidman, Naomi. "Elie Wiesel and the Scandal of Jewish Rage." *Jewish Social Studies* 3, No. 1 (Autumn 1996).

Shickman-Bowman, Zvia. "The Construction of the China Eastern Railway and the Origin of the Harbin Jewish Community." *Jews of China, Volume 1: Historical and Comparative Perspectives*. Edited by Jonathan Goldstein. London: Routledge, 1998.

Sholem Aleichem. *Tevye the Dairyman and the Railroad Stories.* Translated by Hillel Halkin. New York: Schocken, 1987.

Simpich, Frederick. "Manchuria, Promised Land of Asia." *National Geographic*, October 1929.

Skidelsky, Robert. "A Chinese Homecoming." *Prospect* (UK), January 22, 2006.

Song, Lihon. "Some Observations on Chinese Jewish Studies." *Contemporary Jewry* 29, No. 3 (December 2009).

Su, Ling. "Harbin Jews: The Truth." *Southern Metropolis Life Weekly* (China), April 2007.

The Travels of Benjamin of Tudela. Animated film (non-credited), 1978. Oster Visual Documentation Center, Museum of the Jewish People (Beit Hatfutsot), Tel Aviv.

U.S. Millennial Holocaust Knowledge and Awareness Survey. Survey commissioned by Claims Conference (Conference on Jewish Material Claims Against Germany) and conducted by Schoen Consulting, www .claimscon.org, 2020.

Weisz, George, and Donatella Lippi. "Roderigo Lopez: Physician-in-Chief to Queen Elizabeth I of England." *Rambam Medical Journal* 8, No. 3 (July 2017).

White, Nic. "Anne Frank House Banned Orthodox Jewish Employee from Wearing His Skullcap at Work." *Daily Mail* (UK), April 15, 2018.

Yehoshua, A. B. *Mr. Mani*, 1989. Translated by Hillel Halkin. New York: Doubleday, 1992.

PEOPLE LOVE
DEAD JEWS

Dara Horn

PEOPLE LOVE DEAD JEWS

Dara Horn

DISCUSSION QUESTIONS

1. What was your first reaction to the title? Having read the book, what do you take as its meaning? Why did Dara Horn choose it?

2. James Carroll says, "Because antisemitism is a Christian problem more than a Jewish one, Christian readers need this book." Broadening Carroll's statement to include non-Jewish readers in general, do you agree or disagree with this statement? Why?

3. Horn writes that she kept antisemitic experiences of her youth filed away in a mental sock drawer. Do you connect with this way of dealing with difficult situations? Why do you think Horn found that the mental sock drawer wasn't as effective for her children?

4. Horn posits that perhaps people wouldn't be as interested in seeing Anne Frank's house if she had survived the Holocaust. Do you agree? Why?

5. Were you aware of the myth of changed surnames at Ellis Island? Why do you think so many people hang on to it even though it has been debunked?

6. Horn writes that "the popular obsession with dead Jews, even in its most apparently benign and civic-minded forms, is a profound affront to human dignity" (p. xxiii). How did our society become fixated on dead Jews? Is there a way to channel this obsession into something healthier? What would that be?

7. Do you think there are other cultures or minority groups for which the dead are more interesting than the living in the popular imagination, or is this particular to the Jews?

8. Consider Horn's question, "If the purpose of literature is to 'uplift' us, is it even possible to write fiction that is honest about the most horrifying aspects of the Jewish past" (p. 73)? Is it?

9. After the shooting at the Tree of Life synagogue, Horn's children responded by simply stating it happened "because some people hate Jews" (p. 49). What do you think this reaction says about the newest generations to experience antisemitism, and how has it differed from that of previous generations? Does it reflect how antisemitism itself has changed?

10. Horn believes that Jews represent the idea of freedom. In what ways is freedom reflected through the experience of the Jews?

SELECTED NORTON BOOKS WITH
READING GROUP GUIDES AVAILABLE

For a complete list of Norton's works with reading group guides, please go to wwnorton.com/reading-guides.

Diana Abu-Jaber	*Life Without a Recipe*
Diane Ackerman	*The Zookeeper's Wife*
Michelle Adelman	*Piece of Mind*
Molly Antopol	*The UnAmericans*
Andrea Barrett	*Archangel*
Rowan Hisayo Buchanan	*Harmless Like You*
Ada Calhoun	*Wedding Toasts I'll Never Give*
Bonnie Jo Campbell	*Mothers, Tell Your Daughters*
	Once Upon a River
Lan Samantha Chang	*Inheritance*
Ann Cherian	*A Good Indian Wife*
Evgenia Citkowitz	*The Shades*
Amanda Coe	*The Love She Left Behind*
Michael Cox	*The Meaning of Night*
Jeremy Dauber	*Jewish Comedy*
Jared Diamond	*Guns, Germs, and Steel*
Caitlin Doughty	*From Here to Eternity*
Andre Dubus III	*House of Sand and Fog*
	Townie: A Memoir
Anne Enright	*The Forgotten Waltz*
	The Green Road
Amanda Filipacchi	*The Unfortunate Importance of Beauty*
Beth Ann Fennelly	*Heating & Cooling*
Betty Friedan	*The Feminine Mystique*
Maureen Gibbon	*Paris Red*
Stephen Greenblatt	*The Swerve*
Lawrence Hill	*The Illegal*
	Someone Knows My Name
Ann Hood	*The Book That Matters Most*
	The Obituary Writer
Dara Horn	*A Guide for the Perplexed*
Blair Hurley	*The Devoted*

DON'T MISS OTHER TITLES BY AWARD-WINNING AUTHOR

Dara Horn

DARAHORN.COM

A *NEW YORK TIMES BOOK REVIEW* NOTABLE BOOK

"Shimmers with Horn's signature blend of tragedy and spirituality."
————Ron Charles, *Washington Post*

"[An] intense, multilayered story. . . . Horn's writing comes from a place of deep knowledge."
————Jami Attenberg, *New York Times Book Review*

A *New York Times Book Review* Editors' Choice

"Rare and memorable."   —*Wall Street Journal*

"Nothing short of amazing."   —*Entertainment Weekly*

W. W. NORTON & COMPANY
Independent Publishers Since 1923